SOLAR AIR CONDITIONING AND REFRIGERATION

RENEWABLE ENERGY SERIES

Editor-in-Chief: A A M Sayigh

Pergamon Titles of Related Interest

BANHIDI
Radiant Heating Systems: Design and Applications

BEI
Modern Power Station Practice, 3rd edition

GRANQVIST
Materials Science for Solar Energy Conversion Systems

HARRISON
Geothermal Heating

HORIGOME
Clean and Safe Energy Forever

McVEIGH
Sun Power, 2nd edition

SAITO
Heat Pumps

SAYIGH
Energy Conservation in Buildings

SAYIGH
Energy and the Environment: into the 1990s, *5-vol set*

STECCO & MORAN
A Future for Energy

TREBLE
Generating Electricity from the Sun

Pergamon Related Journals (*free specimen copy gladly sent on request*)

Energy

Energy Conservation and Management

Geothermics

Heat Recovery Systems and CHP

International Journal of Heat and Mass Transfer

International Journal of Hydrogen Energy

Progress in Energy and Combustion Science

Renewable Energy

Solar Energy

SOLAR AIR CONDITIONING AND REFRIGERATION

Edited by

A. A. M. SAYIGH

Engineering Department
Reading University, Reading, UK

and

J. C. McVEIGH

School of Engineering
Glasgow Polytechnic, Glasgow, UK

PERGAMON PRESS

OXFORD · NEW YORK · SEOUL · TOKYO

U.K. Pergamon Press Ltd, Headington Hill Hall, Oxford OX3 0BW, England

U.S.A. Pergamon Press, Inc, 660 White Plains Road, Tarrytown, New York 10591, USA

KOREA Pergamon Press Korea, KPO Box 315, Seoul 110-603, Korea

JAPAN Pergamon Press Japan, Tsunashima Building Annex, 3-20-12 Yushima, Bunkyo-ku, Tokyo 113, Japan

First edition 1992

Library of Congress Cataloging-in-Publication Data

Solar air conditioning and refrigeration / edited by A. A. M. Sayigh and J. C. McVeigh.
p. cm. -- (Renewable energy series)
1. Solar air conditioning. 2. Refrigeration and refrigerating machinery. I. Sayigh, A. A. M. II. McVeigh, J. C. III. Series.
TH7687.9.S63 1992 697.9′3--dc20 92-8023

British Library Cataloguing in Publication Data

A catalogue record for this book is available from the British Library

ISBN 0 08 040750 1

Printed in Great Britain by B.P.C.C. Wheatons Ltd, Exeter

Contents

Foreword

Solar cooling is a major area of technology associated with high solar energy regions of the world. It goes without saying that the higher the solar flux, the greater is the need for solar cooling. Most developing countries fall within latitudes of high solar intensity and hence require cooling as well as airconditioning to preserve food and medical vaccines and to provide reasonable comfort levels at work in order to increase productivity.

During the last ten years, several developments have taken place which have advanced the subject of cooling. In this series, we have attempted to describe these developments around the world and in different application areas. The book consists of eleven chapters. Chapters One and Two outline the various means of cooling available and describe their cycles. Chapter Three explains in detail the work of Professor Chinnappa who, during the last twenty years, has made great strides in the area of solar absorption system machines. The work was carried out partly in India and partly in Australia. Chapters Four and Five go on to deal with desiccant cooling and have been written by Professors Robinson and Lof from the United States of America. Chapters Six, Seven and Eight describe in detail the concept and the importance of radiative cooling in France and in Thailand and were written by Professor Berger & Cubizoloes, Berger & Schneider and Boom & Lony. Professor Exell's experience in ice-making for rural applications is dealt with in Chapter Nine, while Chapter Ten, written by Professor Adamson from Sweden, gives an insight into passive cooling in buildings. Finally, Mr McNelis's work and experience in photovoltaic refrigeration is the theme of the eleventh chapter.

The book has many references and is illustrated by numerous photographs and diagrams. It will serve as an informative and technical book to all those who wish to increase their knowledge about solar cooling, in particular researchers in the field, funding agents and policy makers.

A A M Sayigh
University of Reading

CHAPTER 1

SOLAR ENERGY COOLING IN BUILDING

A.A.M. SAYIGH

Department of Engineering, University
of Reading, Reading, U.K.

ABSTRACT

The paper outlines the need for cooling, especially in the devel oping countries, and the electricity consumption which accompanies it. The paper discusses the importance of passive cooling, and the various means to achieve it, then concentrates on the Middle-East and reviews ASHRAE - cooling comfort zone. Also, the paper describes the energy saving in using approprite passive design. Active cooling is fully discussed.

1 - Vapour Compression System : a - The use of photovoltaic b - The use of power engine - Rankine engine

2 - Evaporative Cooling : a - The use of desert cooler b - Evaporative pond

3 - The Use of Absorption/Adsorption Machines : a - Absorption cooling b - Desiccant cooling

The effect of a cooling tower - air or water - on the use of absorption machines is then covered. A full discussion regarding collector type - chiller combination is also made and several types of collectors are listed.

Several selection recommendations are mentioned, and the final recommendation is that solar absorption systems are more preferable than other cooling systems. Also depending on the region and climate, the preferable heating device is single glazing selectively prepared flat plate collectors. Evacuated collectors are good if the cost is low and they are trouble free.

KEYWORDS

Cooling in Buildings; photovoltaic; power engine; Rankine engine; Evaporative pond; Absorption cooling; Desiccant cooling.

INTRODUCTION

Most, if not all, developing countries lie in the hotter climatic zones of the world, and without cooling facilities both productivity and comfort of the populations are hindered. Paradoxically the factor which is the cause of the heat of the region could also be the factor which could largely contribute to its alleviation, i.e. solar radiation. Thus solar air conditioning is the most fitting application of solar energy since it matches the amount of solar radiation in a given location. Where there is intensive solar radiation, the demand for solar air conditioning is greater. A human body is best able to maintain a body surface temperature of 33 °C when functioning within an environmental temperature range of 18 °C - 27 °C and a relative humidity range of 25 % to 60 %. These parameters are commonly known as the comfort zone. It is the objective of all builders to provide structures for living and working where this level is effectively maintained. At present this is largely achieved by using vast quantities of electricity often for periods of up to nine months per year. This is both wasteful and costly - for example the installed electrical power during the year 1984 by 6 of the Gulf States was 22569 Mega Watts, 50 % of which was used for cooling during the summer months.

Passive solar design is a term used since 1970 to designate bioclimatic design based on achieving comfort in a building through means such as evaporative cooling, the thermal mass of the building, ventilation and shading techniques. However, it is obvious that in zones of harsh heat passive design alone will not be enough to achieve the comfort zone. An active solar energy system must be used to complement it. This is a viable proposition because the level of solar radiation in such an area reaches over 7 $KWh/m^2/day$.

In a psychometric chart, Fig. 1, air conditioning processes can be divided into (i) sensibly cooling - achieved by passing air over chilled liquid coil reducing the temperature while keeping the moisture at a constant level; (ii) cooling and dehumidification - air is cooled sensibly until it reaches point A (the dewpoint) further cooling dehumidifies the air; (iii) cooling and humidification - this is an evaporative cooling process similar to the one which exists in the traditional desert coolers, air is blown through a spray of water and cooling is achieved by sensible heat reduction at the increase of the latent heat, the process takes place at constant wet-bulb temperatures. Other processes deal with heating, humidification and dehumidification.

Buildings and Cooling Concept

There seems to be little point to developing effective solar cooling if it is utilized in non-passive buildings. Thus building should be designed with the following objectives in mind: to resist solar heat gain by applying shading techniques; to lose heat from the building interior by ventilation; to reduce heat conduction flow into the building by using insulation and thermal mass; and to encourage radiative cooling as well as evaporative cooling through the use of ponds and plants. Figure 2 shows the bioclimatic chart which an architect must utilize to achieve maximum passive concept.

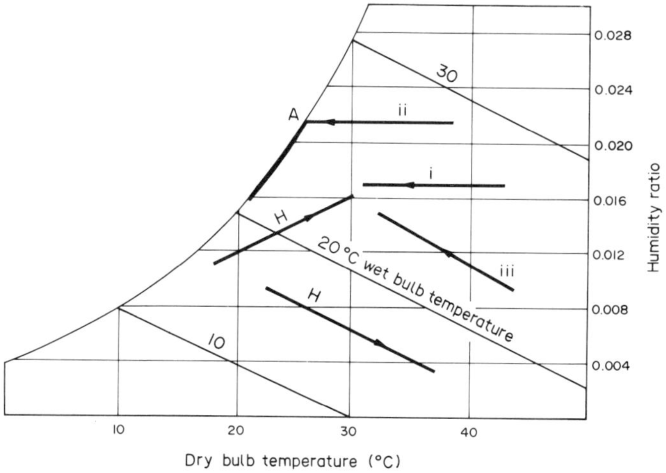

Fig. 1. Air Conditioning Processes.

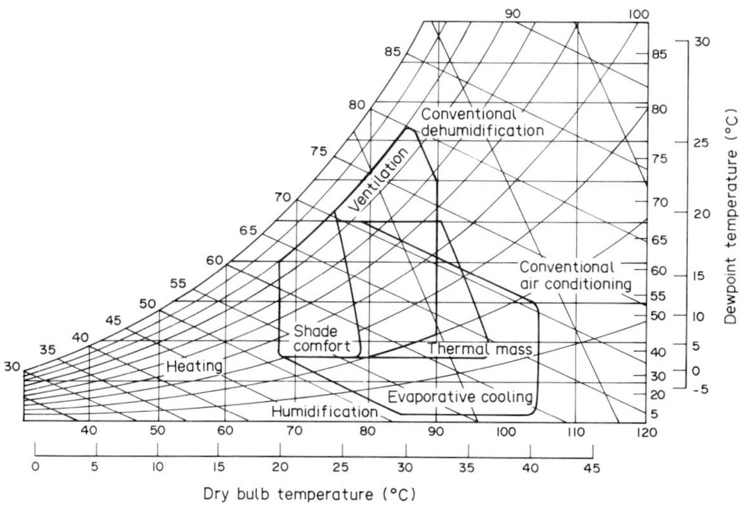

Fig. 2. Bioclimatic Chart.

Before the availability of electricity people of the hot regions used to achieve cooling by the tower concept and the earth shelter building. An honest attempt to revive these techniques would save a lot of energy and money. (1, 2, 3 and 4).

Figures 3 and 4 show the bioclimatic chart for Jedda-Saudia Arabia and Kuwait.

The inside air now returns through the saturator pads, E where it is cooled below the temperature of the living space by the evaporation of water from the pads, and as it passes through the return side of the heat exchanger wheel, D, the air receives the heat that was given up by the in-coming air-stream on the intake side. The air then is futher heated by the solar heat exchanger/gas burner, C. As the hot air passes through the return side of the drying wheel, B, it carries off the water that the wheel had previously absorbed from the incoming air.

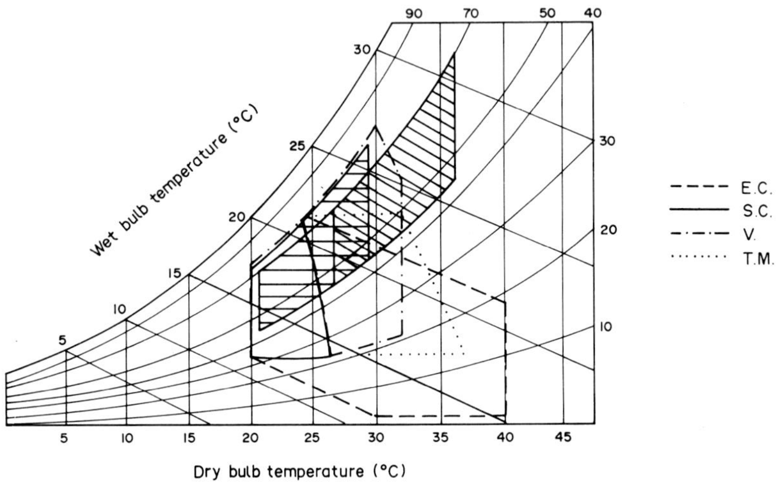

Fig. 3. Bioclimatic Chart for Jeddah.

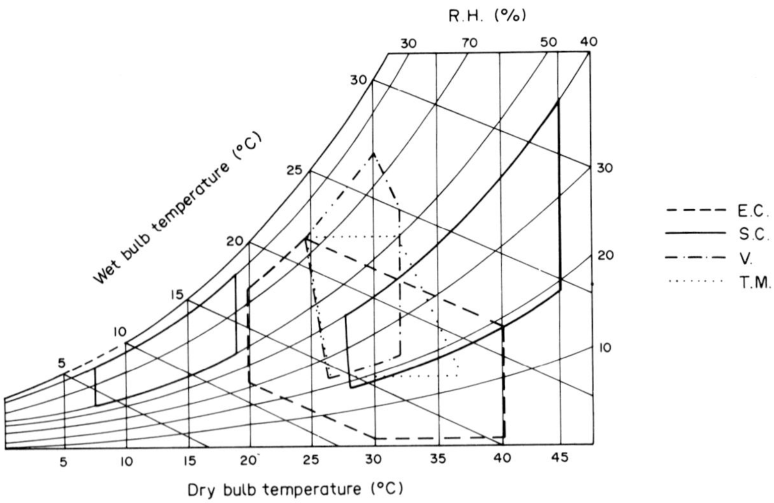

Fig. 4. Bioclimatic Chart for Kuwait.

1 - Vapour Compression System
 a - The use of photovoltaic
 b - The use of power engine such as Rankine engine

2 - Evaporative cooling
 a - The use of desert cooler
 b - Evaporative pond

3 - The use of absorption/adsotrp-tion machines
 a - Absorption cooling
 b - Desiccant cooling

4 - Passive cooling

Active Cooling Systems

Figure 5 shows the various cooling systems which lead to space cooling. Some of these systems will be discussed in detail whilst others will be briefly mentioned in order to reduce the size of this paper. Cooling can be achieved either by solar energy or by the utilization of environment. Figure 6 shows a sketch of the various paths leading to space cooling. Comfort zone is shown in Figure 7. Using solar active system for cooling requires a knowledge of the collector-engine efficiency. This is well demonstrated in Figure 8.

1 Vapour compression system
 a The use of photovoltaic
 b The use of power engine such as Rankine engine

2 Evaporative cooling
 a The use of desert cooler
 b Evaporative pond

3 The use of absorption/adsorption machines
 a Absorption cooling
 b Desiccant cooling

4 Passive cooling

Fig. 5. Cooling Systems

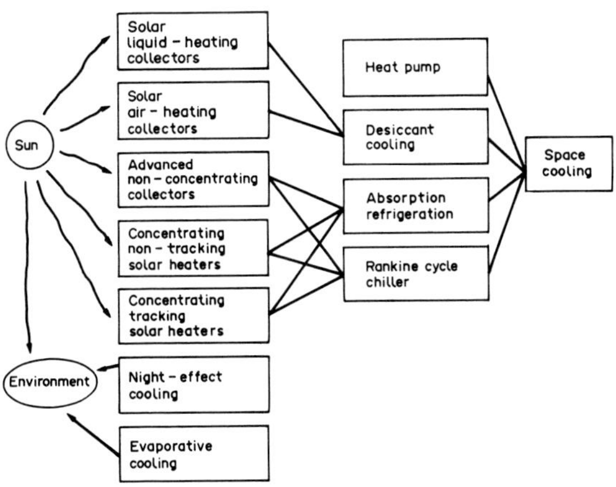

Fig. 6. Various paths leading to Space Cooling.

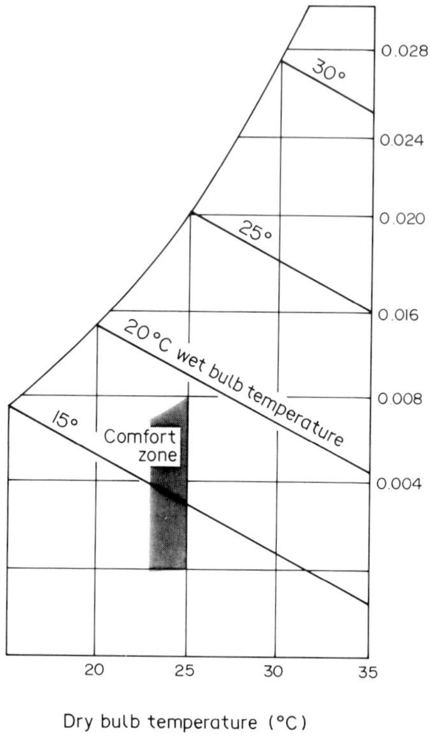

Fig. 7. Comfort Zone.

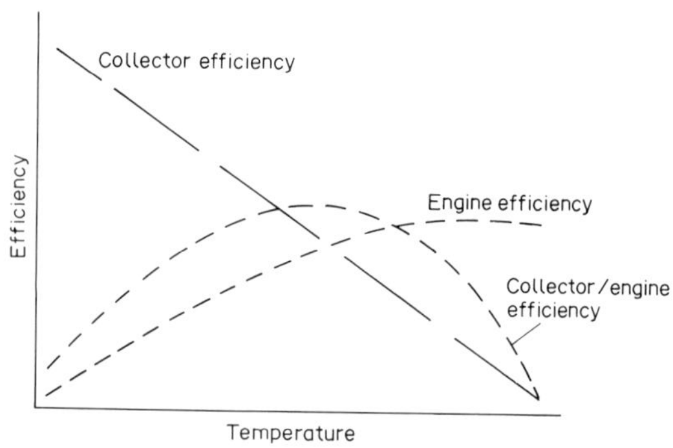

Fig. 8. Collector/engine efficiency.

MUNTERS ENVIRONMENTAL CONTROL SYSTEM - (MEC)

This is known to many as the "Lizenzia" machine which was develop by a Swedish engineer, Carl Munters, (5). The main parts of a solar-MEC are: a drying wheel; a heat exchanger wheel; an evaporative chamber, and a regeneration chamber, see Fig. 9.

The MEC system is often referred to as a desiccant system. It is important to realize that the major element of the system is the drying wheel and therefore a brief description of it is useful. Its functions and characteristics are:

* It should be capable of drying the incoming air to a very low level of moisture content, about 0.0014 kg of water per kg of dry air.

* The flowing air pressure drop across the wheel at design velocities must not be excessive so that a reduction in the parasitic power is achieved.

* The wheel must have maximum area per unit volume of desiccant.

* Assuming that the drying process takes place adiabatically, which means that the desiccant is heated as the drying is carried out, the drying capability of the desiccant at high temperatures must be good so that deep drying continues to be possible.

* The heat capacity of the desiccant system should be low and the regeneration of the desiccant should be at as low a temperature as possible.

* The system must be low in cost, have long life and be stable to gases found in air, such as CO_2, NO_x, and SO_2, (6).

Factors 4 and 5 are contradictory to each other and in practice a compromise is made. Several wheel designs are available in the market with various desiccant materials such as silicagel, ethylene glycol, special form of sodium aluminium silicate (Linde 4A), and activated alumina. Figure 10 shows water absorption isobars for various desiccant materials. There are three operational modes which can be performed by a MEC-system. These modes are:

Cooling: This is illustrated in Figure 11. How moist outdoor air is drawn into the intake section through the slowly rotating drying wheel B. The wheel heats the air and removes the moisture from it. As the hot, dry air passes through the slowly rotating heat exchanger wheel, D, the heat in the air is absorbed. The air, now cool and dry, passes through the saturator pads, E where it is further cooled by the evaporation of water. Water-saturated air at approximately 13 °C and 65 % RH is delivered to the house ducts. Inside the living space, the air warms up to the thermostat selected temperature 24 to 27 °C and its RH drops to 50 %.

The operation in this mode is shown in Figure 9 and the psychrometric chart in Figure 12. The process starts by outside air at condition E being blown through the drying wheel. This is represented (in Figure 12) by the solid line as step (1). As a result of this adiabatic drying step, the air leaves the wheel at an average condition, B, which is now dry but considerably hotter. The sensible heat gained by the air as a result of the drying step is removed by the sensible heat exchange wheel, shown in both figures as the change in states B and C and step (2). This results in a dry, cool air ar condition C. The air is then passed through a humidifier, step (3), exists as condition D, and is saturated to a temperature of about 12.8 °C.

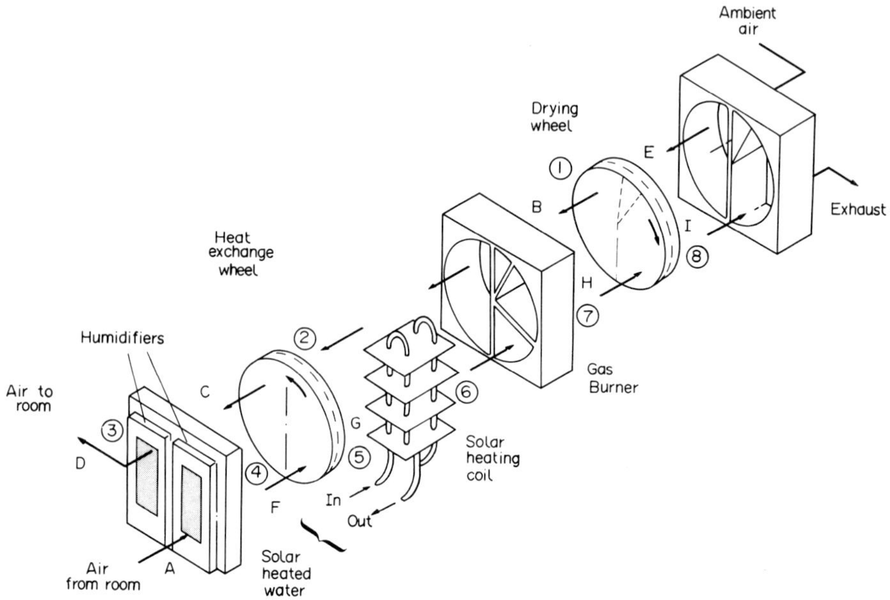

Fig. 9. Desiccant solar air conditioning.

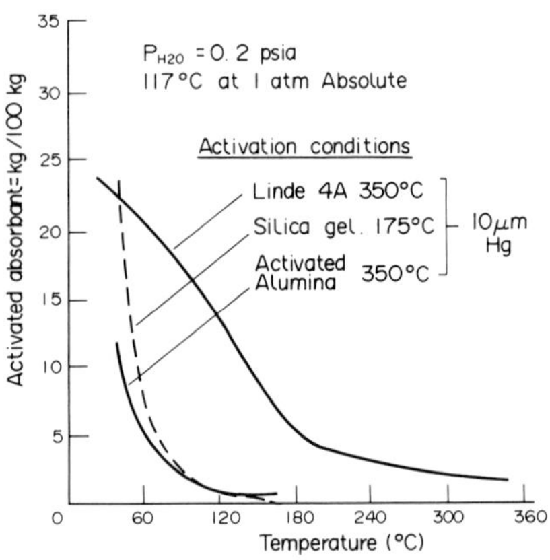

Fig. 10. Isobars for various desiccant materials.

To keep the system operating, the drying wheel, which now contains the water removed to step (1), must be regenerated by heating. The regeneration energy can be introduced in stages. This can best be seen by following the path indicated by the broken line

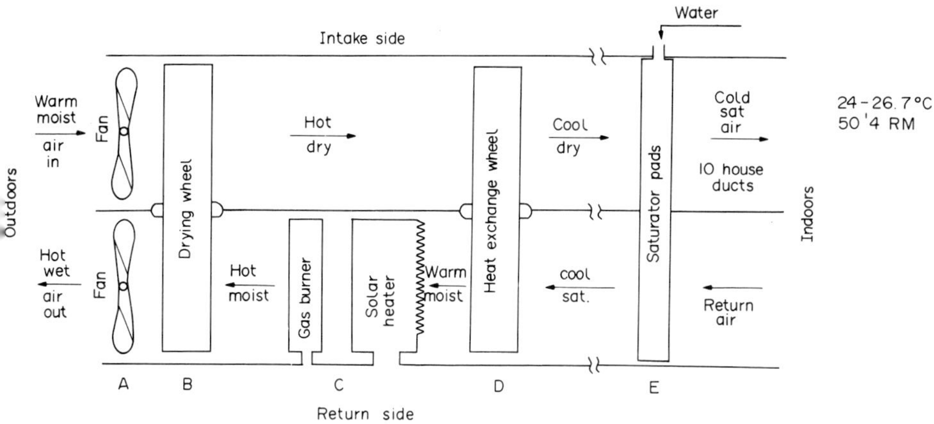

Fig. 11. Cooling action of MEC - system.

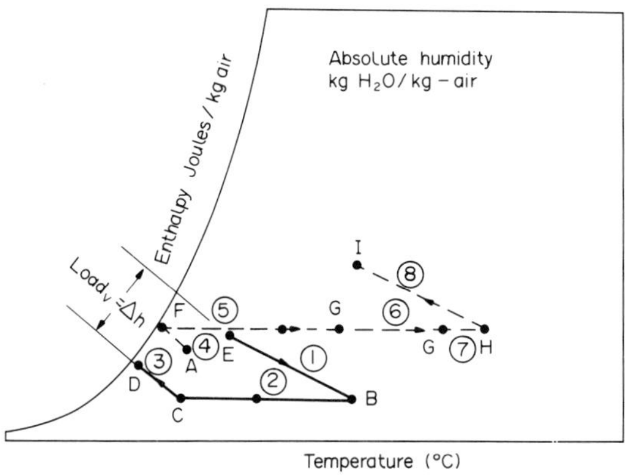

Fig. 12. Psychrometric chart.

in Figure 12 and relating this to the corresponding points in Figure 9. Room air at state A passes through a humidifier, step (4), where it is saturated and cooled to state F. This cooled, wet air becomes the heat sink for removal of the sensible energy stored previously in the heat exchange wheel from the process air in step (2). This heat removal step (5) shows the regeneration air being preheated from F to G. For equal mass flows between the process and regeneration streams, steps (2) and (5) will be equal in magnitude and opposite in direction. Additional energy, if available in the form of solar or waste heat, is now added to this preheated air stream via the heating coil between stages G and G', step (6). It may well be that the energy available for transfer is inadequate to heat the regeneration stream to a condition that will drive off the amount of water that was picked up in step (1), an auxiliary gas burner is then used. This will finish the task from G', step (7'), to the average condition just necessary to complete the regeneration, H. The final step is the rejection of regeneration stream to the outdoors. For equal mass flows, this step (8) is equal in magnitude but opposite in direction and parallel to step (1).

Heat and Humidifying: This is illustrated in Figure 13. The drying wheel B now is rotating 10 rpm faster than for cooling, while the heat exchanger wheel D is stopped completely. Cold, dry outdoor air is drawn through the intake inside of the rotating drying wheel, where it is warmed by heat transfer. Humidity is adjusted to the desired level as the air passes through the saturator pads and is delivered to the house ducts. Return air from the living space passes by the solar heat exchanger/gas burner, where it is heated. The hot air stream passes through the return side of the drying wheel, B (now acting as a heat exchanger), which picks up the heat and any combustion moisture for transfer to the incoming airstream. (One of the combustion products of gas is pure water). The return air, now cool, and any combustion products (less moisture) are discharged outside.

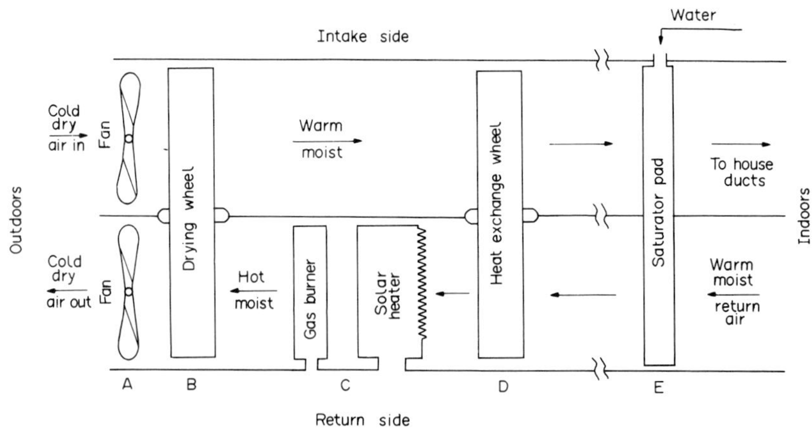

Fig. 13. MEC in a heating humidification mode.

Dehumidifying: This is illustrated in Figure 14. In this cycle the system will operate exactly the same as it does when cooling, except that water to the saturator pads is shut off. Air having a moderate temperature but high humidity is drawn through the intake side of the slowly rotating, B. Here, the moisture is removed and the air is heated. It then passes through the intake side of the slowly rotating heat exchanger wheel, D which removes the heat in the air. Moderate temperature dry air then is delivered to the house ducts. Return air from the living space passes through the return side of the heat-exchanger wheel, D, where it receives the heat previously given up to the wheel by the incoming air. The warmed air then is heated further by the solar heat exchanger/gas burner. When the hot air passes through the return side of the drying wheel, it heats the wheel and drives off the moisture that the wheel had previously absorbed from the incoming air, (7).

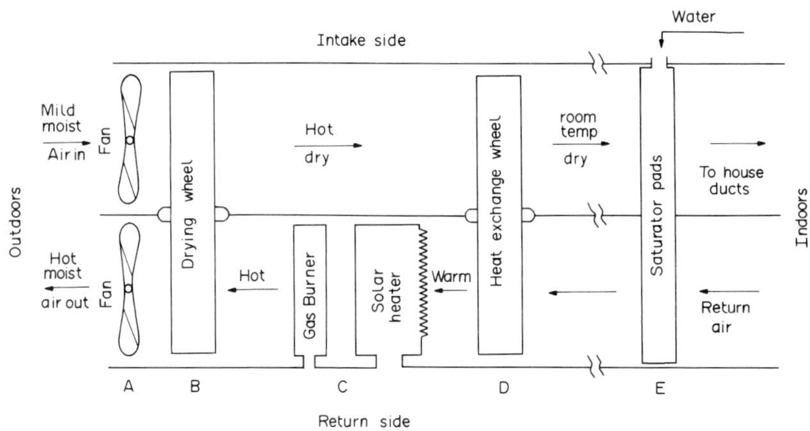

Fig. 14. MEC in dehumidification mode.

RANKINE CYCLE AND VAPOUR COMPRESSION REFRIGERATION CYCLE

This system is fully illustrated in Figure 15. It is composed by a Rankine power cycle and a compression refrigeration vapour cycle. The Rankine cycle uses R 113 pumped from a condenser (common type) by an electrically driven centrifugal feed pump through a regenerator to the boiler, (8). Vapour, with a temperature of 93 °C, from the boiler is admitted to a radial inflow turbine which provides power through a speed reducer to a conventional R 12 piston air conditioning compressor, (Standard Unit) and/or to motor/generator, (9). The low pressure vapour then passes through the regenerator which is an efficient heat exchanger and returns to the condenser to complete the cycle.

The air conditioning cycle uses R 12 supplied from an evaporator at 7.5 °C to a standard reciprocating compressor. Vapour with 53 °C and 125 psi leaves the compressor and enters the condenser, where vapour to liquid phase change takes place. When the high pressure liquid expands it leaves the condenser through a thermal expansion valve which provides the evaporator with a mixture of 7.5 °C liquid and vapour. Forced air then from the room is circulated over the evaporator at 27 °C and supplies the heat necessary to vaporize the R 12 which supplies the required tonage (1 ton/day = 12000 Btu/hr = 3.52 kW) of air condition, (10).

In order to understand the Rankine cycle/vapour compression system which is in fact a power cycle/air conditioning cycle, a brief description of each part will be carried out in the following paragraphs.

Rankine cycle: The part of solar energy here is to evaporate the working fluid, normally R-113, consistant with the collectors' temperature, while another conventional source of energy is used to add extra heat to the working fluid to achieve a higher expansion inlet temperature. Figure 16 shows the operational principles. In order to reduce the collector size and obtain much higher cycle efficiency, a vapour-to-vapour regenerator is used for the first part of the superheat, and final superheat being added by a fossil fuel, and a vapour-to-liquid regenerator is used to raise the temperature of the condensed fluid on its way to receiving heat from the solar collector.

CHAPTER 2

PRINCIPLES OF ABSORPTION SYSTEMS MACHINES

J.C.V. Chinnappa

1. Introduction

The vapour absorption system (VAS) has been the system most commonly used in solar cooling applications. In this system heat is supplied to a refrigerant-absorbent mixture in the generator (Fig. 1). The refrigerant expelled from the mixture is condensed, and then evaporated to produce a cooling effect. The cycle is completed by the re-absorption of the refrigerant in the absorber returning the refrigerant-absorbent mixture to its original concentration.

The VAS may be operated in two ways - continuously (Fig. 1(a)) and intermittently (Fig. 1(b)). In the continuous cycle the refrigerant-absorbent solution is pumped continuously from the absorber to the generator, and returns via a flow constrictor to the absorber. Meanwhile the refrigerant is continuously condensed in the condenser, flows through the expansion valve to the evaporator where it boils, producing a continuous cooling effect.

In the intermittent cycle the refrigerator in its simplest form consists of two vessels (Fig. 1(b)). One of these, the generator-absorber (or G-A) contains the refrigerant-absorbent mixture. The other vessel, the condenser-evaporator (or C-E) is initially empty. When the G-A is heated the refrigerant is boiled off from the refrigerant absorbent mixture, and condenses in the C-E. When the generation phase is over there is a quantity of refrigerant in the C-E. During the next phase, the absorption phase, the G-A is cooled and the system pressure drops. The refrigerant in the C-E now boils producing a cooling effect. This phase ends when the GE is empty, the refrigerant having been reabsorbed in the G-A. The generation phase can now be re-commenced. Cooling is therefore intermittent occurring only during the absorption phase.

The rest of this chapter reviews the considerable amount of research and development work that has been done on both forms of the VAS.

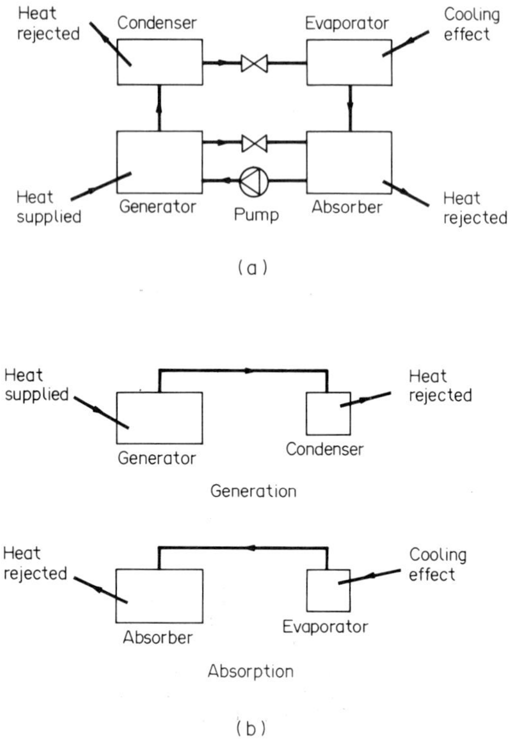

Fig. 1. Principle of vapour absorption cycle.
(a) continuous (b) intermittent.

2. The Carnot Coefficient of Performance

The Second Law of Thermodynamics states that heat cannot flow from a lower temperature to a higher temperature without work being done, and so prescribes the mode of operation, and determines the performance limits of all refrigerators.

In the vapour absorption refrigerator energy is supplied as heat, so the optimum reversible refrigerator operates on a reversible cooling cycle (R_{rev}) supplied with work from a reversible engine cycle (E_{rev}) as shown in Figure 2.

The coefficient of performance (CoP) is the ratio of the cooling effect (Q_e) to the heat supplied (Q_s):

$$CoP = \frac{Q_e}{Q_s}$$

If the reversible engine and cooling cycles are respectively Carnot and reversed Carnot cycles then

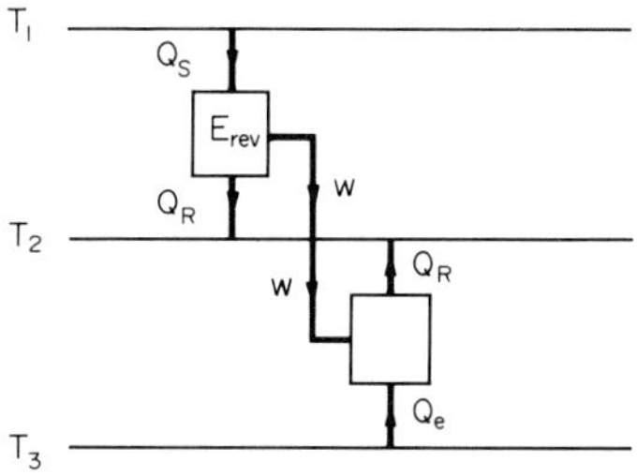

Fig. 2. Principle of reversible refrigerator. Q_S, heat supplied; Q_R, heat rejected; Q_e, cooling effect; w, work. $T_1 > T_2 > T_3$.

$$\text{Carnot CoP} = \frac{T_3}{T_1} \cdot \frac{(T_1 - T_2)}{(T_2 - T_3)}$$

Figure 3 shows the variation in CoP with increasing heat supply temperatures. The values of the CoP of actual systems are very much less, and are determined by refrigerant-absorbent properties and system design.

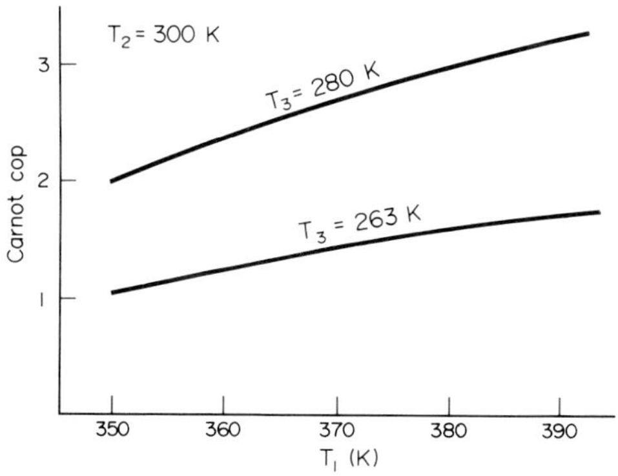

Fig. 3. Coefficient of Performance (CoP) of Carnot refrigerator.

3. Refrigerant - absorbent combinations

A full discussion of the properties of refrigerants, absorbents, and refrigerant-absorbent pairs is found in ref. 1. The desirable properties for the refrigerant are high heat of vapourisation, low specific heat capacity, and good thermal stability. For the absorbent chemical stability, high boiling point and low heat capacity are advantageous. For refrigerant-absorbent pairs the solutions should exhibit negative deviations from Raoult's Law, and a low specific heat capacity.

For use in the continuous cycle, a low viscosity for the absorbent will ensure low pump power consumption, and the most widely used refrigerant-absorbent pairs have been water-lithium bromide H_2O-LiBr) and ammonia-water (NH_3-H_2O).

Other pairs have also been proposed and their properties determined, e.g. methanol-lithium bromide [2], monomethylamine-ethylene glycol [3]. Data for the following pairs have been presented in computer compatible form [4] - dichlorofluoromethane-dimethyl formamide (R21-DMF), chlorodifluoromethane-dimethyl formamide (R22-DMF), chlorodifluoromethane-dimethyl ether of tetraethylene glycol (R22-DMTEG).

4. The single-stage continuous vapour absorption cycle

4.1. The water-lithium bromide cycle

The lithium bromide chiller employs LiBr solution as the absorbent water as the refrigerant and consists essentially of the components shown in Fig. 4.

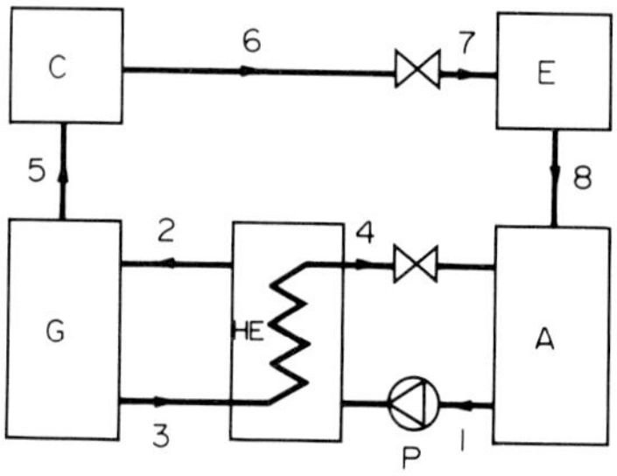

Fig. 4. The water-lithium bromide single stage cycle flow circuit. G, generator; C, condenser; E, evaporator; A, absorber; HE, solution heat exchanger; P, pump.

In solar operated chillers the generator is supplied with hot water from solar collectors. Cooling water is circulated through the absorber and the condenser (usually in that order) to absorb the heat rejected in those components. The water to be chilled is circulated through the evaporator.

Data for the H_2O-LiBr pair is presented in Figs. 5 and 6.

The basic operating cycle is shown (not to scale) in Fig. 7, the numbers indicating the corresponding state points on the flow circuit diagram (Fig. 4). The solution heat exchanger transfers heat from the hot strong absorbent leaving the generator at 3 to cooler weak absorbent leaving the absorber at 1.

Mass and energy balances may now be applied to the cycle.

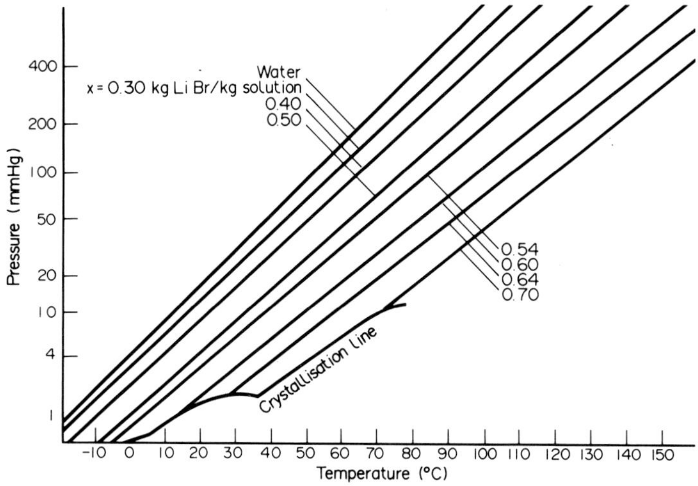

Fig. 5. Pressure - temperature - concentration data for water lithium-bromide solutions.

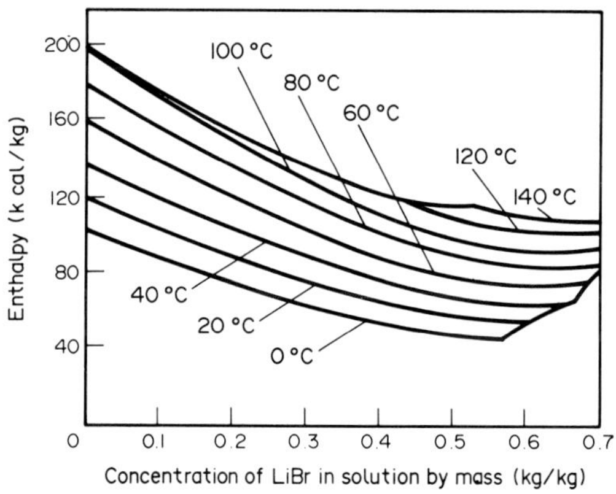

Fig. 6. Enthalpy-concentration diagram for water-lithium bromide solutions.

Mass balance on the generator:

Total flow, $\dot{m}_2 = \dot{m}_5 + \dot{m}_3$

LiBr, $x_2\dot{m}_2 = x_5\dot{m}_5 + x_3\dot{m}_3$

But $x_5 = 0$: Let $\dot{m}_1 = \dot{m}_2 = 1$ kg/s.

Solving
$$\dot{m}_3 = \frac{x_2}{x_3} \tag{1}$$

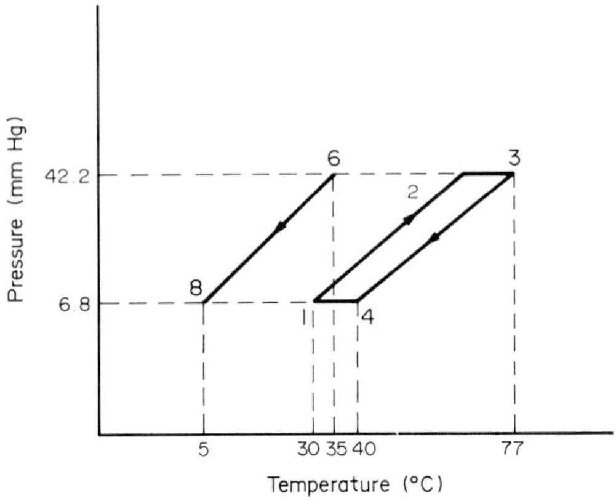

Fig. 7. Single stage water-lithium bromide cycle on p-t-x chart.

$$\dot{m}_5 = \frac{x_3 - x_2}{x_3} = \dot{m}_7 = \dot{m}_8 \tag{2}$$

$$\text{Relative ciculation} = \frac{\dot{m}_2}{\dot{m}_5}$$

$$= \frac{x_5}{x_3 - x_2} \tag{3}$$

Energy balances:

For the heat exchanger, $\dot{m}_3(h_3 - h_4) = \dot{m}_2(h_2 - h_1)$

$$h_2 = \frac{\dot{m}_3}{\dot{m}_2}(h_3 - h_4) + h_1 \tag{4}$$

For the generator, $$Q_G = \dot{m}_3 h_3 + \dot{m}_5 h_5 - \dot{m}_2 h_2 \tag{5}$$

For the evaporator, $$Q_E = \dot{m}_7(h_8 - h_7) = \dot{m}_5(h_8 - h_6) \tag{6}$$

Therefore $$CoP = \frac{Q_E}{Q_G} \tag{7}$$

The heat rejection and maximum supply temperatures are chosen so that crystallisation does not occur when the strong absorbent is cooled in the heat exchanger. For this basic cycle the following assumptions may be made:

temp. of heat abstraction in evap. (t_8) = 5 °C
temp. of weak solution leaving abs. (t_1) = 30 °C
temp. of strong sol. leaving gen. (t_3) = 77 °C
temp. of condensation (t_6) = 35 °C
temp. of strong sol. leaving heat exch. (t_4) = 37 °C

When the above information is plotted on the p-t-x chart, the weak and strong concentrations are respectively found to be 0.54 and 0.60 kg LiBr / kg solution.

Table 1. Calculation of CoP of H_2O-LiBr Cycle

State point	State	Pressure mmHg	Temp. °C	Liquid Concentration $\frac{\text{kg LiBr}}{\text{kg solution}}$	Enthalpy $\frac{\text{kJ}}{\text{kg}}$
1	solution	6.8	30	0.54	- 177
2	solution			0.54	- 102.3
3	solution	42.0	77	0.60	- 82.0
4	solution		37	0.60	- 165.0
5	H_2O sup. vap.				2633.6
6	H_2O sat. liq.	42.0	35		146.5
7					
8	H_2O sat. vap.	6.8	5		2510.8

It is convenient to commence the calculation of the CoP by completing Table 1. For H_2O-LiBr solution, pressure, temperature and concentration values may be read from Fig. 5, and enthalpy values from Fig. 6.

Enthalpy values at state points 5, 6 and 8 are obtained directly from, or calculated using steam tables. The values for h_6 and h_8 may be read from saturated steam tables. If enthalpy values for superheated steam at very low pressures are not available then h_5 may be calculated. If the temperature at which boiling commences in the generator is t_2' ($t_2' \neq t_2$) and the maximum solution temperature is t_3, the enthalpy of water and saturated steam at the temperature of Condensation (t_6) are h_f and h_g then

$$h_5 = h_g + C_{pv}\ 0.5(t_2' + t_3) - t_6 \quad (8)$$

where C_{pv} = 1.894 kJ/kg.°C.

From 1. $\dot{m}_3 = \frac{0.54}{0.60} = 0.9$ kg/s

2. $\dot{m}_5 = \frac{0.6 - 0.54}{0.6} = 0.1$ kg/s

4. $h_2 = 0.9(-82 + 165) - 177 = -102.3$ kJ/kg

8. $h_5 = 2565.4 + 1.894(0.5(65+77)-35)$
$= 2633.6$ kJ/kg

5. $Q_G = 0.9(-82) + 0.1(2633.6) - (-102.3)$
$= 291.9$ kW

6. $Q_E = 0.1(2510.8 - 146.5)$
$= 236.4$ kW

7 $CoP = \frac{236.4}{291.9} = 0.81$

The CoP may be improved by including a refrigerant pre-cooler so that the refrigerant from the expansion valve is cooled by vapour leaving the evaporator.

The variation in CoP with generator temperature is shown in Fig. 8. Further information may be obtained from refs. 5 and 6.

The work energy consumed by the pump can be estimated (see equation 8), but in the H_2O-LiBr cycle is negligibly small. Indeed in smaller units circulation can be maintained by bubble lift pumps. One important precaution is the operation of chiller with bubble lift pumps is to ensure that the chiller is not operated unless the minimum heat supply temperature is high enough to boil weak solution in the generator (65 °C in the above example of the basic cycle).

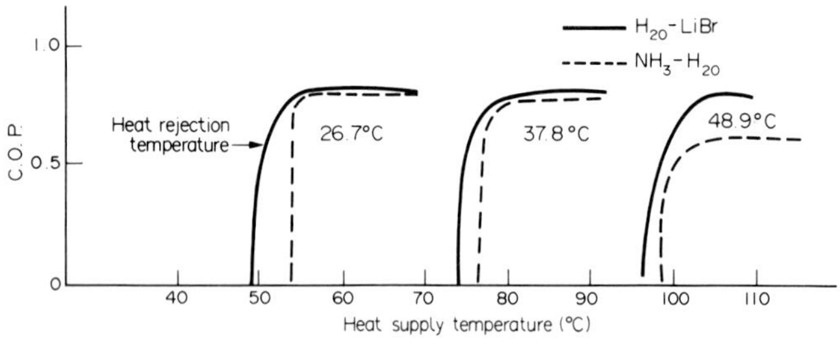

Fig. 8. Variation of CoP with heat supply and heat rejection temperatures for water-lithium bromide and ammonia-water single stage cycles with ideal solution heat exchangers. Evaporator temperature 4.4 °C (ref. 6).

In larger systems mechanical pumps maintain solution movement between generator and absorber, and recirculation in individual components. Precautions are taken to avoid crystallisation by appropriate design of the solution heat exchanger, and in some plant, mixing of solution streams.

4.2. Water-lithium bromide plant

Air conditioning plant operating on the H_2O-LiBr cycle are commercially available in many forms and capacities. Chillers are built as single vessels or double vessels.

Carrier Corporation, U.S.A. build a range of chillers varying in capacity from a few kW to several hundred kW operated by steam.

Arkla, U.S.A. and Yazaki, Japan build chillers operated by gas burners, and have modified these units to be operated by hot water from solar collectors. Yazaki provide a complete solar air-conditioning package including system design, solar collectors and hot loop, lithium bromide chiller, cooling water tower etc.

York Division, Borg Warner have also specified their ES absorption chillers for solar operation.

4.3. The ammonia-water cycle

In this system water is the absorbent and has a vapour pressure at operating generator temperatures. Hence the vapour that is boiled off in the generator is not pure NH_3 but a mixture of NH_3 and water vapour, the percentage of water vapour depending on the temperature of generation.

Therefore in plant incorporating this system, a rectifier and reflux condenser are attached to the generator to reduce the water vapour entering the condenser to an acceptable percentage. The design of the rectifier to achieve this aim is treated in standard text books (e.g. ref. 7).

While a rectifier is necessary in aqua-ammonia systems operated by gas burners or steam where generator temperatures will exceed 120 °C, in solar operated systems a rectifier may not be required because it is most unlikely that that temperature will be exceeded. A reflux condenser cooled by weak absorbent (rich in ammonia), or ambient air or cooling water, will suffice.

The flow circuit for an NH_3-H_2O cycle is shown in Fig. 9.

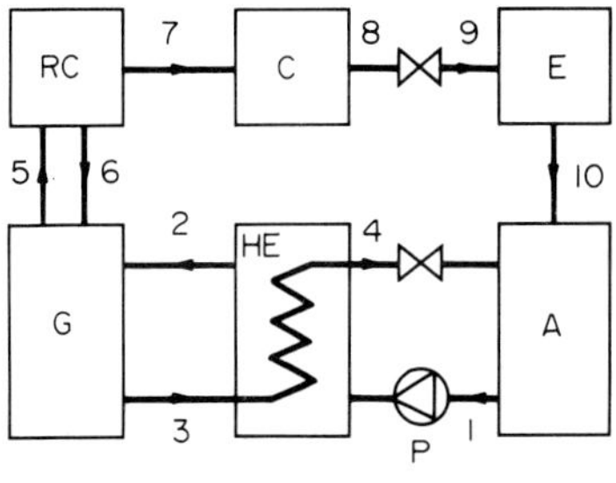

Fig. 9. Flow circuit for single-stage ammonia-water cycle. G, generator; RC, reflux condenser; C, condenser; E, evaporator; A, absorber; P, pump; HE, solution heat exchanger.

Heat is rejected to cooling water or ambient air in the absorber and condenser. Heat is supplied to the generator. The cooling effect occurs in the evaporator.

Assuming equilibrium conditions for the streams leaving the generator, absorber and the reflux condenser, the following mass and energy equations may be written.

Mass balances on generator-reflux condenser:

Total flow, $\dot{m}_2 = \dot{m}_7 + \dot{m}_3$

ammonia, $x_2\dot{m}_2 = x_7\dot{m}_7 + x_3\dot{m}_3$

Assume $\dot{m}_2 = 1$ kg/s; $x_7 = 1.0$

$$\dot{m}_3 = \frac{(1-x_2)}{(1-x_3)} \tag{4.9}$$

$$\dot{m}_7 = \frac{x_2-x_3}{1-x_3} \tag{4.10}$$

Mass balances on reflux condenser:

Total, $\dot{m}_5 = \dot{m}_6 + \dot{m}_7$

ammonia, $x_5\dot{m}_5 = x_6\dot{m}_6 + x_7\dot{m}_7$

Solving and substituting from previous eight lines,

$$\dot{m}_5 = \frac{(x_2-x_3)(1-x_6)}{(x_5-x_6)(1-x_3)} \tag{4.11}$$

$$\dot{m}_6 = \frac{(x_2-x_3)(1-x_5)}{(x_5-x_6)(1-x_3)} \tag{4.12}$$

Energy balances:

Solution heat exchanger, $h_2 = \frac{\dot{m}_3}{\dot{m}_2}(h_3-h_4) + h_1$ (4.13)

generator, $Q_G = \dot{m}_3h_3 + \dot{m}_5h_5 - \dot{m}_6h_6 - \dot{m}_2h_2$ (4.14)

evaporator, $Q_E = \dot{m}_7\ (h_{10}-h_9) = \dot{m}_7\ (h_{10}-h_8)$ (4.15)

Assuming pump operation to be isothermal, and flow incompressible then pump work

$$W_P = \dot{m}_2\ v_1\ (p_2-p_1) \tag{4.16}$$

where v = specific volume of solution (see Table 2)

Hence $CoP = Q_E/(Q_G+W_P)$ (4.17)

Table 2. Specific volume of aqua-ammonia. Multiply by 10^{-6} to obtain m^3/kg.

Concentration / Temp. °C	0.3	0.4	0.5	0.6	0.7
15	1116	1159	1207	1261	1325
30	1127	1174	1225	1285	1355
45	1141	1191	1248	1311	1390
60	1157	1209	1269	1340	1431
75	1174	1230	1295	1373	1479
90	1193	1253	1325	1415	1530
105	1213	1279	1358	1463	1591
120	1236	1308	1397	1516	1662
135	1262	1342	1444	-	-

As in the case of the H_2O-LiBr cycle, this cycle is determined by three (or four) temperatures: the heat abstraction temperature (t_{10}), the heat rejection temperatures in the absorber and condenser (which could be the same, or slightly different, as in this example, t_1 and t_8), the maximum heat supply temperature (t_3=t_5).

The following assumptions are made in this example: t_1 = 30 °C, t_3 = t_5 = 94 °C, t_6 = t_7 = 50°C, t_8 = 35 °C, t_{10} = -10 °C.

Depending on the design of the reflux condenser an alternative assumption is to assume $t_6 = t_3$. The net result of this assumption is a slightly smaller $\dot{m}_7$.

When these temperatures are plotted on the ln p vs $\frac{1}{T}$ (where T is the absolute temperature) for aqua ammonia (Fig. 10) the concentrations are obtained: 0.45 for the strong solution (weak absorbent) and 0.39 kg NH_3/(kg solution) for the weak solution (or strong absorbent). The operating pressures are also obtained from this chart: 2.92 kPa in the absorber and evaporator, 1350 kPa in the generator and condenser (Fig. 11).

The above equations and assumptions may be used to complete Table 3, using Figs. 12 and 13. All state points 5 and 7 enthalpy and concentration values are obtained from Fig. 13 using liquid

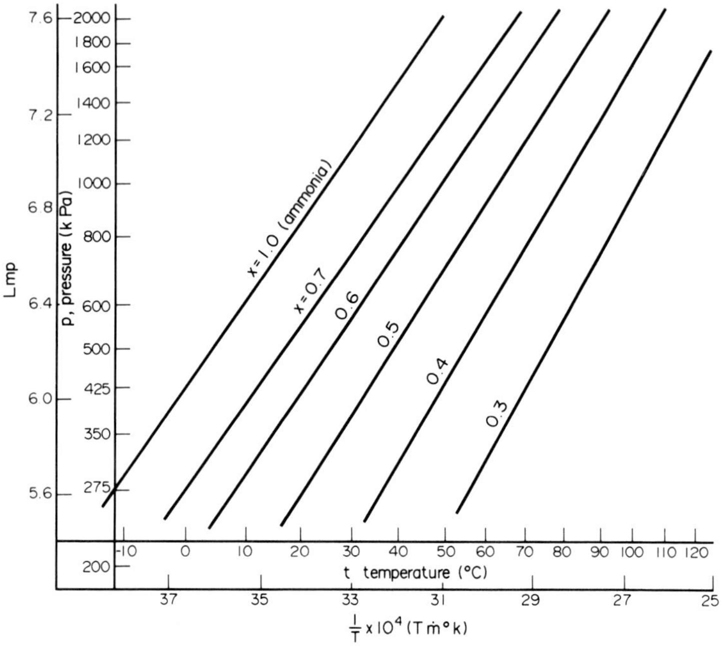

Fig. 10. Ln p vs $\frac{1}{T}$ plot for aqua-ammonia.

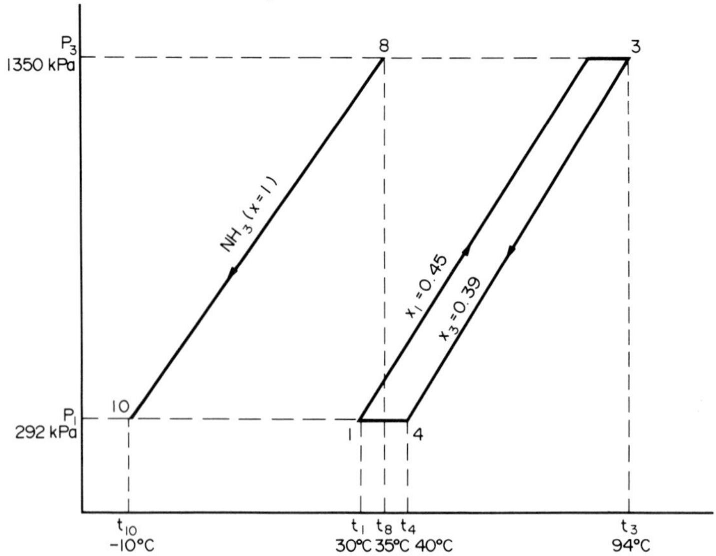

Fig. 11. Single-stage ammonia-water cycle plotted on ln p vs $\frac{1}{T}$ chart.

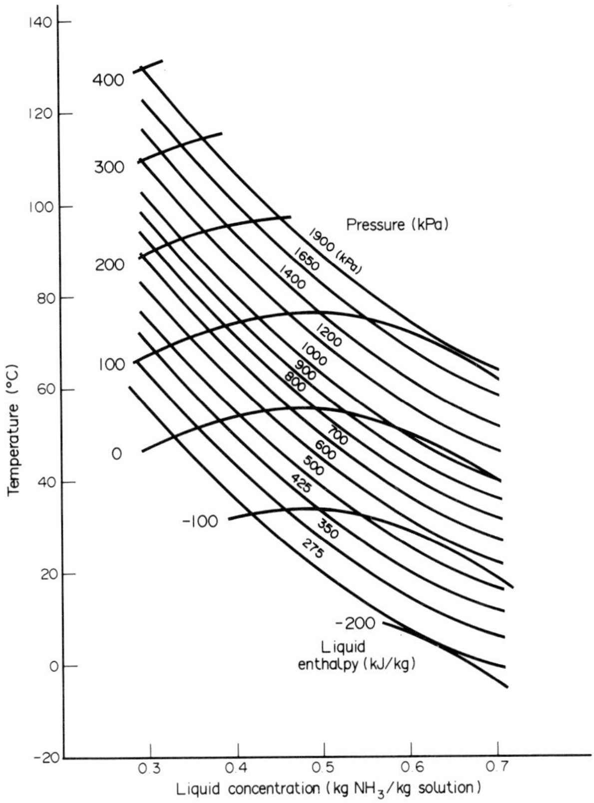

Fig. 12. Pressure-temperature-concentration-enthalpy properties for liquid ammonia-water solutions.

concentration values with which the vapour is in equilibrium. Tables for pure ammonia are used to obtain properties at state points 8 and 10.

From 9 $\dot{m}_3 = \frac{1-0.45}{1-0.39} = 0.9$

10 $\dot{m}_7 = \frac{0.45-0.39}{1-0.39} = 0.1$

11 $\dot{m}_5 = 0.1\left(\frac{1-0.71}{0.954-0.71}\right) = 0.12$

12 $\dot{m}_6 = \dot{m}_5 - \dot{m}_7 = 0.02$

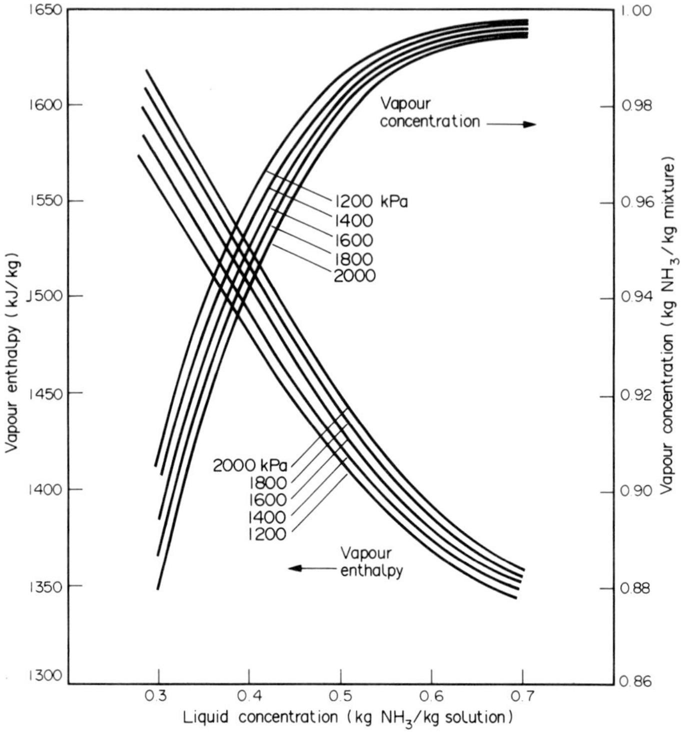

Fig. 13. Concentration and enthalpy of ammonia-water vapour mixtures.

13 $h_2 = 0.9(192 + 64) + (-117)$

$= 113.4$ kJ/kg

14 $Q_G = 0.9(192) + 0.12(1496) - 1(113.4) - 0.02(45)$

$= 238$ kW

15 $Q_E = 0.1(1433-347.5)$

$= 108.6$ kW

16 $W_P = 0.001197(1350-292)$

$= 1.27$ kW

17 $\text{CoP} = \dfrac{108.6}{238+1.27} = 0.45$

Table 3. Calculation of CoP for NH_3-H_2O cycle

State point	State	Pressure p kPa	temperature t °C	liq/vap Concent. x/xv (kgNH3)/(kgsol)	enthalpy h kJ/kg
1	liq	292	30	0.45	-117
2	liq	1350		0.45	
3	liq	1350	94	0.39	192
4	liq	1350	40	0.39	- 64
5	vap	1350	94	0.954	1496
6	liq	1350	50	0.71	45
7	NH_3 vap	1350	50	1.0	1346
8	NH_3 liq	1350	35		347.5
9					
10	NH_3 vap	292	-10		1433

There are two reasons why this CoP is less than the CoP of the water-lithium bromide cycle (section 1). Firstly the evaporator temperature is lower in the NH_3-H_2O cycle (-10 °C against 5 ° in the H_2O-LiBr cycle); and secondly, the significant vapour pressure of the absorbent (H_2O) in the NH_3-H_2O cycle at operating generator temperatures, and the consequent need for a reflux condenser, reduces the quantity of refrigerant NH_3 entering the condenser proper.

The H_2O-LiBr cycle cannot achieve sub-zero temperatures, and the NH_3-H_2O cycle is able to do this with generator temperatures in the region of 100 °C - which is within the range of high performance flat plate solar collectors.

LiBr units are usually water-cooled to avoid the possibility of crystallisation, thereby requiring a cooling water circuit with wet cooling tower, pump etc. Aqua-ammonia units for air conditioning can be air-cooled.

4.4 Aqua-ammonia refrigeration plant

In view of its toxicity NH_3 is not generally used in domestic applications - a notable exception being the triple fluid NH_3-H_2-H_2O refrigerator which is a sealed unit containing a relatively small mass of ammonia.

Arkla-Servel and Bryant in the U.S.A. have marketed aqua-ammonia air-cooled air-conditioning units which could be located outside the building.

Because of its cheapness and high latent heat, NH_3 is widely used in industry in vapour compression systems - but only occasionally in absorption systems, which are custom built for the applications in question.

Dornier in West Germany have announced the development of a solar operated NH_3-H_2O absorption cycle cold store. Apart from this unit it would seem that commercial development of solar operated aqua-ammonia systems has not occurred.

Solar operated experimental systems have been built and tested (refs. 8, 9, 10). The unit tested at Lawrence Berkeley Laboratories was an Arkla unit, the usual gas-fired generator replaced by a generator designed at LBL for hot water operation.

4.5. Actual and transient performance of lithium bromide and ammonia absorption systems

The analyses presented in Sections 4.1 and 4.3 are based primaril on the thermodynamics of steady state operation and equilibrium conditions. In actual plant heat transfer in the various components, as well as non-equilibrium conditions (e.g. when cycling will affect the coefficient of performance. These factors have been studied for the LiBr cycle is some detail (refs. 11, 12).

For the NH_3-H_2O system a method by which some of these factors may be taken into account is indicated in Section 2.7.5.

5. Multistage and complex vapour absorption cycles

In the single stage absorption cycle there is a single solution circuit including essentially a generator where heat is supplied, and an absorber where heat is rejected.

Absorption cycles which incorporate two or more solution circuits may be termed multistage cycles. These solution circuits can be arranged relative to each other in different ways, thereby altering the overall coefficient of performance.

The solution circuit (or circuits) can also be radically reorganised - in which case the cycle may be more appropriately termed complex.

The cycles may be classified as follows:

1. Double effect generation
2. Double effect cooling
3. Cascaded generation
4. Regenerative cycles
5. Mixed cycles.

5.1. The double-effect generation cycle

The flow circuits for the double-effect generation cycle are shown in Fig. 14. The solution is heated in the high temperature generator G_1 by an external source, and this heat represents the heat supplied to the cycle. The vapour from G_1 is condensed in the low temperature generator G_2, and boils off more refrigerant.

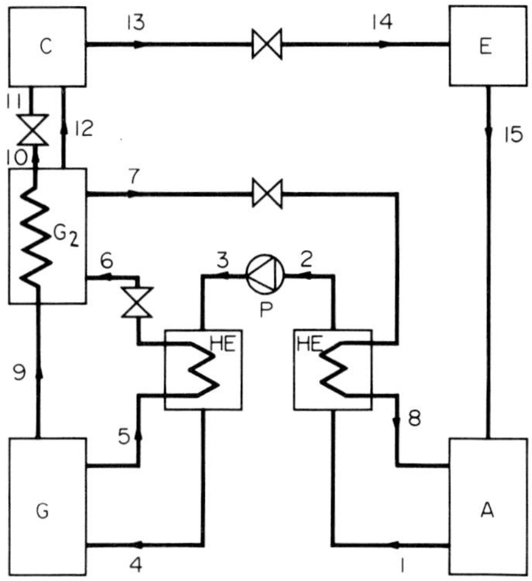

Fig. 14. Flow circuit for double-effect generation water-lithium bromide cycle. G_1, high temperature generator; G_2, low temperature generator; C, condenser; E, evaporator; A, absorber; HE, solution heat exchangers; P, pump. (ref. 13).

The conditions for an H_2O-LiBr cycle are shown plotted on the ln p vs 1/T chart in Fig. 15 (not to scale). The theoretical CoP may be calculated in the usual way assuming that the streams leaving a heat-mass exchanger (e.g. generator) are in equilibrium. For these conditions Whitlow (ref. 13) has calculated the net CoP to be 1.43 compared to 0.79 for the single-stage cycle.

A very much higher heat supply temperature is required for the operation of this cycle, upto 163 °C as compared to 80 - 85 °C. These higher temperatures, above 150 °C may be achieved by concentrating and evacuated tube collectors in solar operation.

Yazaki Corporation, Japan, are marketing an H_2O-LiBr chiller operating on this cycle.

Problems arise when aqua-ammonia is employed in a double-effect generation cycle. The pressure in the high temperature generator approaches 7600 kPa, and temperatures approach the critical temperature for ammonia. Phillips (ref. 14) concludes that the NH_3-H_2O pair is not well suited to air-cooled double-effect operation.

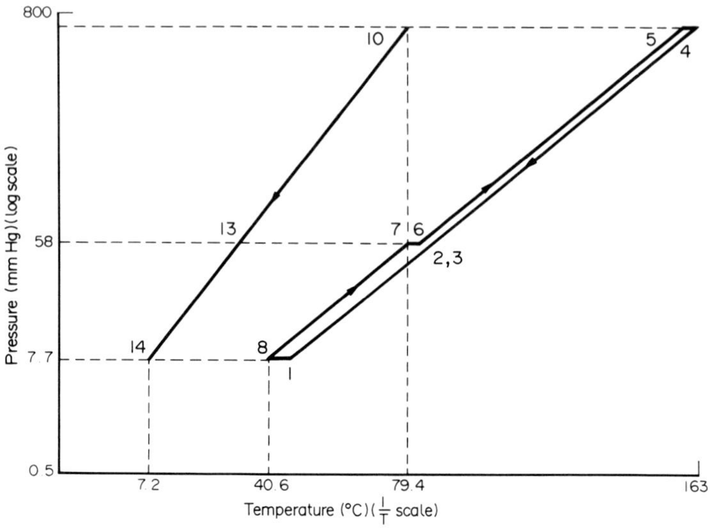

Fig. 15. Double-effect generation water-lithium bromide cycle (ref. 13).

Other pairs have been studied: R21 - dimethyl formamide, R22 - dimethyl formamide, R22 - dimethyl ether of tetra-ethylene glycol (ref. 15), and theoretical values of CoP upto 1.82 have been obtained.

5.2. The double-effect cooling cycle

The operation of this cycle is described below with NH_3-H_2O as the refrigerant-absorbent pair. Figure 16 shows the flow circuits. In the primary solution circuit, 1-2-3-4, the solution concentrations are less than in the secondary circuit, 11-12-13-14.

The refrigerant, NH_3, is generated in the primary circuit and enters the evaporator by way of the reflux condenser, condenser and expansion valve. The first cooling effect occurs in the evaporator.

Leaving the evaporator the NH_3 enters the secondary circuit to be absorbed in solution in the resorber. This solution boils in the desorber producing a second cooling effect. The NH_3 vapour from the desorber now dissolves in solution in the absorber - and so is returned to the primary circuit.

The following assumptions may be made regarding the operating temperature limits:

evaporator, $t_9 = 11$ °C
temperature range in resorber, t_{11}-t_{14} = 30 - 36 °C

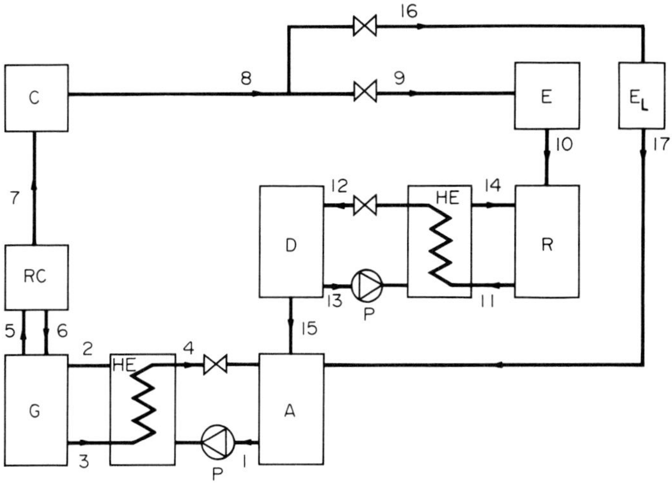

Fig. 16. Flow circuit for double-effect cooling ammonia-water cycle. G, generator; RC, reflux condenser; C, condenser; E_L, low temperature evaporator; E, evaporator; R, resorber; D, desorber; A, absorber; HE, solution heat exchanger; P, pump.

temperature range in desorber, t_{12}-t_{13}	= 7 - 13 °C
temperature range in absorber, t_4 - t_1	= 40 - 30 °C
maximum generator temperature, t_3	= 94 °C
streams leaving reflux condenser, t_6, t_7	= 50 °C
condenser, t_8	= 35 °C

The cycles can now be plotted on the log p vs $\frac{1}{T}$ chart (Fig. 17) and liquid concentrations in the generator, absorber, desorber and resorber obtained.

Assuming equilibrium conditions for streams leaving the generator, reflux condenser, evaporator, desorber: and also assuming that $\dot{m}_1$ = 1 kg/s, x_7 = 1.0, mass balances may be written for these components. From these relations expressions for the mass flow rates at various points in the flow circuits may be obtained (see section 4.4.3 for procedure).

Continuity considerations, i.e. a constant mass transfer rate for the refrigerant through the primary and secondary circuits results in a smaller change in concentration in the resorber-desorber circuit than in the generator-absorber circuit. The heat rejection temperature range in the resorber is also less than the range in the absorber.

Table 4 may now be completed using Figs. 4.12 and 4.13, and where those ranges are exceeded, the aqua-ammonia data charts in the ASHRAE Handbook of Fundamentals (ref. 16).

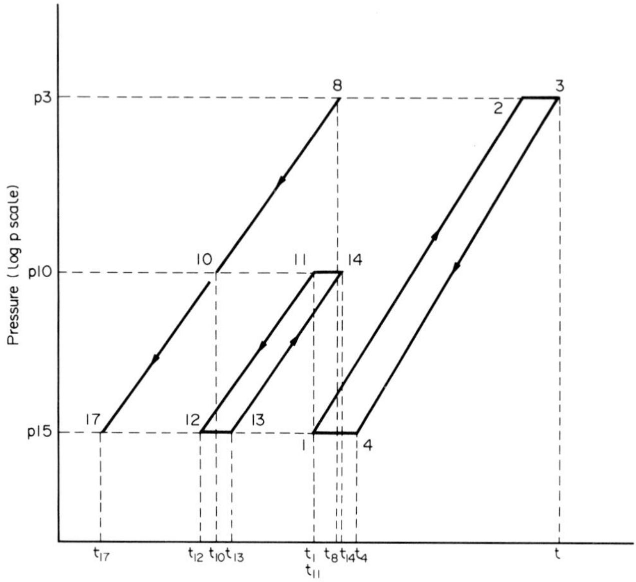

Fig. 17. Double-effect cooling ammonia-water cycle.

The following energy balances can now be evaluated:

For the heat exchangers,

$$h_2 = \frac{\dot{m}_3}{\dot{m}_2} (h_3 - h_4) + h_1$$

$$= 113.4 \text{ kJ/kg}$$

$$h_{12} = \frac{\dot{m}_{14}}{\dot{m}_{11}} (h_{13} - h_{14}) + h_{11}$$

$$= -117 \text{ kJ/kg}$$

For the generator, $Q_g = \dot{m}_3 h_3 + \dot{m}_5 h_5 - m_2 h_2 - m_6 h_6$

$$= 238 \text{ kW}$$

For the evaporator, $Q_e = \dot{m}_8 (h_{10} - h_8)$

$$= 110.75 \text{ kW}$$

For the desorber, $Q_d = m_{15} h_{15} + m_{13} h_{13} - m_{12} h_{12}$

$$= 141.2 \text{ kW}$$

Table 4. Property values for NH_3-H_2O double cooling effect cycle

State point	State	Pressure p kPa	temp. t °C	conc. x/xv $\frac{kg\ NH_3}{kg\ sol}$ or $\frac{kg\ NH_3}{kg\ vap}$	enthalpy h $\frac{kJ}{kg}$	mass flow rate $\dot{m}$ kg/s
1	liq	292	30	0.45	-117	1
2	liq	1350				1
3	liq	1350	94	0.39	192	0.9
4	liq		40	0.39	- 64	
5	vap	1350		0.954	1496	0.12
6	liq	1350	50	0.71	45	0.02
7	vap	1350	50	0.998=1.0	1346	0.10
8	liq	1350	35	1.0	347.5	0.10
9						
10	vap	627	11	1.0	1455	0.10
11	liq	627	30	0.625	- 89	
12	liq	292	7	0.625		0.78
13	liq	292	13	0.57	-188	0.68
14	liq	627	36	0.57	- 81	
15	vap	292	13	1.00	1310	0.1
17	vap	292	-10	1.00	1433	

Hence $$\text{CoP} = \frac{Q_e + Q_d}{Q_g}$$

$$= \frac{110.75 + 141.2}{236}$$

$$= 1.06$$

Pump work has been ignored in these calculations.

This value of the CoP is a substantial improvement on the value of the CoP for the NH_3-H_2O single stage cycle operating between the same limits.

Other refrigerant-absorbent pairs R21-DMF, R22-DMF, R22-DMETEG have been assessed while operating on this cycle with a similar improvement in the theoretical CoP (ref. 17).

The cycle as described above is suitable for air-conditioning as the cooling temperature range is 7 ° to 12 °C. However the circuit as shown in Fig. 16 also includes a second low temperature evaporator E_L. Refrigerant to this evaporator bypasses the resorption circuit. The temperature in this evaporator is -10 °C, and is suitable for food preservation.

This flow circuit (1-2-3-4-5-6-7-8-16-17) is the circuit for a single stage NH_3-H_2O cycle with the same operating temperature limits as the cycle described in section 4.3. The theoretical CoP of this cycle is therefore 0.45.

This second evaporator draws attention to the many possibilities inherent in vapour absorption refrigeration cycles when the number of solution circuits is increased. Another mode in which these circuits can be operated is discussed in ref. 18.

The realisation of these cycles is dependent on the refrigerant-absorbent pair. Aqua-ammonia lends itself more easily to these adaptations than most other refrigerant-absorbent combinations.

5.3. The cascaded generation cycle

In this cycle (Fig. 2.18) there are two solution circuits or stages. In the low pressure stage (G_1-A_1) the solution concentrations are less than in the high pressure stage (G_2-A_2).

The refrigerant produced in the low pressure stage (G_1) is absorbed in the absorber of the high pressure stage (A_2).

Heat is supplied in both generators (G_1 and G_2) and there is a single cooling effect in the evaporator.

The CoP of this cycle has been calculated for the NH_3-H_2O pair, for two evaporator temperatures: - 10 °C (a freezing cycle) and 5 °C (an air-conditioning cycle). The operating temperature limits are given in Table 5.

The CoP is calculated using the same procedure as described in preceding sections for aqua-ammonia cycles; and tables 6 and 7 completed for the two values of the evaporator temperature, -10 °C and 5 °C. Figure 19 is a sketch of the cycle on the ln p vs $\frac{1}{T}$ plot.

Finally energy balances on the heat exchangers, generators and evaporators permit the evaluation of the CoP.

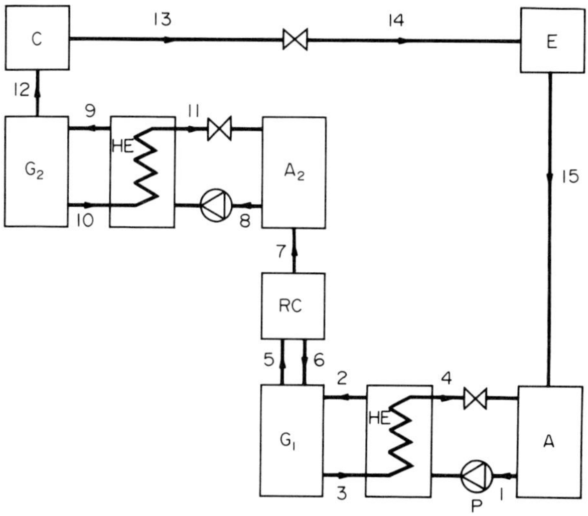

Fig. 18. Flow circuit for cascaded generation ammonia-water cycle. G_1,G_2, low and high pressure generators respectively; A_1,A_2, low and high pressure absorbers respectively; HE, solution heat exchanger; RC, reflux condenser; C, condenser; E, evaporator; P, pump.

Table 5. Cascaded Generation cycle operating temperature limits and Coefficient of Performance

			Freezing cycle	Air-conditioning cycle
Evaporator temp.,	t_{15}	°C	-10	5
Absorber temp. ranges,				
	(t_1-t_4)	°C	30-40	30-40
	(t_8-t_{11})	°C	30-37	30-37
Condenser temp.,	t_{13}	°C	35	35
Max. generator temp.,	t_3,t_9	°C	65	57
Q_{G1}		kW	212.6	245.9
Q_{G2}		kW	157	219.2
Q_e		kW	108.6	164.3
CoP			0.29	0.35

Table 6. Cascaded generation freezing cycle (evaporated temperature -10 °C)

State point	State	pressure p kPa	temp. t °C	liq/vap conc. x/xv $\frac{kg\ NH_3}{kg\ sol}$	enthalpy h $\frac{kJ}{kg}$	mass flow rate $\dot{m}$ $\frac{kg}{s}$
1	liq	292	30	0.45	-117	1.0
2				0.45	- 10	1.0
3		627	65	0.39	55	0.9
4	liq		40	0.39	- 64	
5	vap	627	65	0.979	1438	0.106
6	liq	627	30	0.625	- 89	0.006
7	vap	627	30	1.0		0.10
8	liq	627	30	0.625	- 89	0.725
9			54	0.625	22	0.725
10		1350	65	0.565	56	0.625
11	liq		37	0.565	- 74	0.625
12	vap	1350	65	1.0	1387	0.1
13	liq	1350	35		347.5	0.1
14		292			347.5	0.1
15	vap	292	-10		1433	0.1

Heat supplied to G_1, $Q_{G1} = m_3h_3 + m_5h_5 - m_2h_2 - m_6h_6$

Heat supplied to G_2, $Q_{G2} = m_{10}h_{10} + m_{12}h_{12} - m_9h_9$

Heat abstracted in evaporator, $Q_e = m_{13}(h_{15} - h_{13})$

$$CoP = \frac{Q_e}{(Q_{G1} + Q_{G2})}$$

Values are given in Table 5.

5.4 Comparison of CoP of absorption cycles

The operating temperature limits and the values of the theoretic CoP of the cycles discussed above are presented in Table 8.

Table 7. Cascaded generation air-conditioning cycle (evaporated temperature 5 °C)

State point	State	pressure p kPa	temp. t °C	liq/vap conc. x/xv $\frac{kg\ NH_3}{kg\ sol}$	enthalpy h $\frac{kJ}{kg}$	mass flow rate $\dot{m}$ kg/s
1	liq	516	30	0.57	-103	1
2	liq			0.57	- 25	1
3	liq	835	57	0.495	9	0.85
4	liq		37	0.495	- 83	
5	vap	835	57	0.992	1393	0.153
6	liq	835	36	0.665	- 40	0.004
7	vap	835	36	0.9985		0.149
8	liq	835	30	0.74	- 70	0.486
9	liq					0.486
10	liq	1350	57	0.625	45	0.387
11	liq		37	0.625	- 54	0.387
12	vap	1350	57	0.996	1365	0.149
13	liq	1350	35		347.5	0.149
14		516			347.5	0.149
15	vap	516	5		1450	0.149

Table 8. Comparison of theoretical coefficients of performance

Cycle	Evap. temp. °C	Min. heat rejection temp. °C	Max. heat supply temp. °C	Theor. CoP
Single stage H_2O-LiBr	5	30	77	0.81
Single stage NH_3-H_2O	-10	30	94	0.45
Double effect gener. H_2O-LiBr	7.2	40.6	163	1.43
Double effect cooling NH_3-H_2O	7.13	30	94	1.06
Cascaded gener. NH_3-H_2O:				
(i) freezing	-10	30	65	0.29
(ii) air-conditioning	5	30	57	0.35

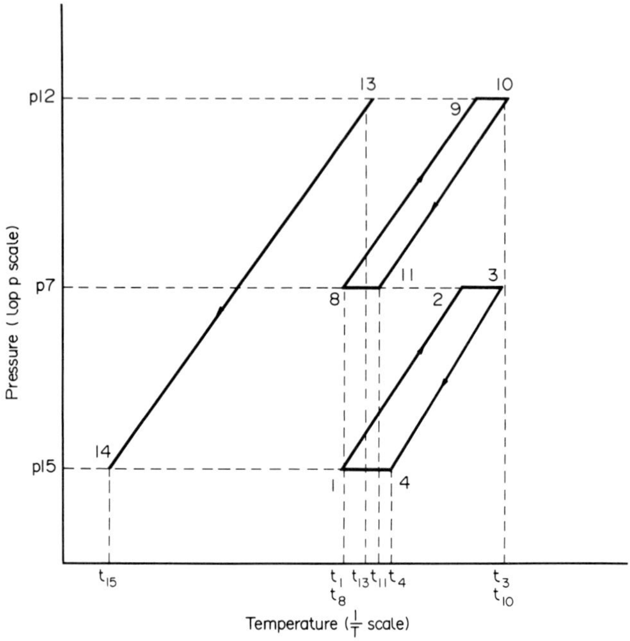

Fig. 19. Cascaded generation ammonia-water cycle.

When considering solar operation the maximum heat supply temperature is significant, as the efficiency of solar heat collection decreases with increasing collector temperatures. This effect is discussed in relation to the theoretical CoP in section 7.1.

5.5. Regenerative cycles

If the concentration change in the solution in the cycle is large enough (say from 0.15 to 0.70 kg NH_3/kg sol), then the temperatures at which absorption occurs at lower NH_3 concentrations will be higher than the generation temperatures at the higher concentrations.

It is then possible to use the heat rejected in the absorber at lower NH_3 concentrations to generate NH_3 from the solution at higher concentrations.

Two advanced cycles based on this regenerative principle have been devised - the single-effect and double-effect regenerative cycles by Dao et al (refs. 19 and 20). These are complex cycles in which there are a number of heat exchangers and circuits requiring multistage pumps. High generating temperatures are also required. At maximum generating temperatures over 140 °C, the theoretical CoP is twice that for the single stage NH_3-H_2O cycle.

5.6. Mixed cycles

The flexibility inherent in absorption refrigeration cycles enables cycles operating on two different refrigerant-absorbent pairs to be coupled together - thereby taking advantage of the properties of both refrigerant-absorbent systems. Kaushik and Kumar (ref. 21) have assessed the capability of a two-stage cycle in which the absorber of a NH_3-H_2O system is cooled by the evaporator of an H_2O-LiBr system - an arrangement which may be termed 'cascaded cooling'.

Both generators are separately supplied with solar heat, and an overall CoP of 0.3 is realised with maximum generator temperatures of 70 °C and 110 °C in the two generators. The evaporator temperature is -10 °C.

This cycle can be compared to the cascaded generation freezing cycle (section 5.3).

6. Hybrid systems

When a vapour absorption system (VAS) is coupled to another type of refrigeration cycle, for example, a vapour compression system (VCS) then the combination may be termed a hybrid system.

A number of combination are possible. Some examples are:

(i) a VAS - VCS cascaded hybrid system (more fully described below in section 6.1).

(ii) a VAS - VCS combination where the VAS evaporator and the VCS evaporator are in series, with the chilled water passing through the VAS evaporator first.

(iii) The VAS and a vapour jet refrigeration cycle in cascaded hybrid mode (ref. 22).

(iv) an integrated VAS - VCS combination where the compressor is included within the VAS refrigerant flow circuit between the evaporator and the absorber (see section 7.5).

It will be demonstrated later (see section 7.3) that solar operated hybrid systems are more likely to be economically viable than the VAS systems with auxiliary heating.

6.1. The vapour absorption - vapour compression hybrid system

The flow circuit for the VAS - VCS cascaded hybrid combination is shown in Fig. 20.

In this arrangement the evaporator of the VAS serves as a second (or bypass) condenser for the VCS. The objective of the arrangement is to reduce the power consumption of the compressor by lowering the compressor discharge pressure.

The VCS can therefore work in two modes: in the conventional

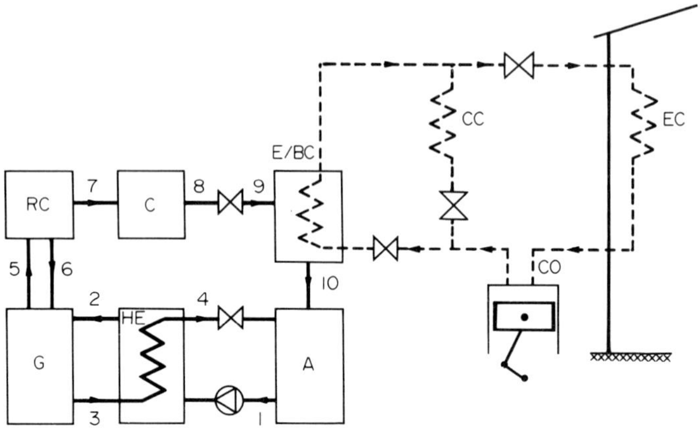

Fig. 20. Basic flow circuit for cascaded hybrid ammonia-water vapour absorption (VAS)/ vapour compression (VCS) air-conditioning system. G, generator; HE, solution heat exchanger; A, absorber; RC, reflux condenser; C, condenser; E/BC, ammonia evaporator and bypass condenser for VCS; CC, air-cooled condenser for VCS; EC, evaporator for VCS; CO, compressor.

mode with an air-cooled (or water-cooled) condenser, or in the hybrid mode with an ammonia-cooled condenser.

To calculate the compressor work per unit of cooling (w_c) the following expression may be used (ref. 7):

$$w_c = \frac{0.00606(n)(p_b v_b)}{\zeta_0(n-1)(h_3-h_2)} \left[\left(\frac{p_c}{p_b}\right)^{\frac{n-1}{n}} - 1\right] \frac{\text{hp}}{\text{ref.ton}} \tag{4.18}$$

$$\text{or in SI units} = \frac{n.p_b.v_b}{\zeta_0(n-1)(h_3-h_2)} \left[\left(\frac{p_c}{p_b}\right)^{\frac{n-1}{n}} - 1\right] \frac{\text{kW}}{\text{kW cooling}} \tag{4.19}$$

where n = index of compression

p_b, v_b = pressure and volume at the commencement of compression

ζ_0 = overall mechanical-electrical efficiency of compressor and electric motor

h_2, h_3 = enthalpy of refrigerant at entrance and exit of the evaporator

p_c = pressure at the end of compression.

This expression allows for heat gain by refrigerant from evaporator exit to compressor cylinder (increasing v_b), and pressure drop at suction and delivery valves (affecting p_b and p_c).

The absorption system may be analysed as in previous sections.

The operation of an NH_3-H_2O VAS and an R22 VCS cascaded hybrid system is illustrated by the following example.

Example:

The following particulars refer to a VAS - VCS cascaded hybrid air conditioning system.

R22 is the refrigerant in the VCS. When the VCS is operating conventionally with an air cooled condenser, the pressure in the condenser is 1533 kPa, and there is no subcooling of refrigerant. The temperature and pressure in the evaporator are 7 °C and 625 kPa; and the enthalpy of the superheated vapour at the evaporator outlet is 257 kJ/kg. At the end of the compressor suction stroke the pressure of the vapour is 598 kPa (i.e. pressure drop across suction valve is 27 kPa); and its specific volume 0.0462 m³/kg. At the end of compression the pressure is 1575 kPa; and there is a drop of 42 kPa across the delivery valve. Assume the index compression = 1.16.

When the VCS is operating in hybrid mode with its bypass condenser cooled by the evaporator of the VAS, the condenser pressure is 801 kPa with no subcooling of the refrigerant. There are no other changes in the VCS operating cycle.

In the NH_3-H_2O VAS operating in the hybrid mode the evaporator temperature is 15.5 °C, and the heat rejection temperature 40 °C. The change in solution concentration in the generator is 0.04.

Calculate (i) the saving in compressor indicated work per kW of cooling, w_{ci}, when operating in the hybrid mode;
(ii) the CoP of the VAS in the hybrid mode;
(iii) the heat supplied to the VAS generator per kJ of compressor indicated work saving.

Solution:

With the VCS operating in conventional mode, $n = 1.16$, $p_b = 598$ kPa, $v_b = 0.0462$ m³/kg, $\zeta_0 = 1$ (for indicated work), $h_3 = 257$ kJ/kg, $p_c = 1575$ kPa, and from R22 tables, $h_2 = 94.3$ kJ/kg.

From equation 19

$$\text{conventional } w_{ci} = \frac{1.16(598)(0.0462)}{0.16(257-94.3)}\left[\left(\frac{1575}{598}\right)^{\frac{0.16}{1.16}} - 1\right]$$

$$= 0.176 \text{ kJ/kJ cooling}$$

With the VCS operating in hybrid mode, $p_c = 801 + 42 = 843$ kPa, $h_2 = 63.2$ kJ/kg.

$$\text{hybrid } w_{ci} = \frac{1.16(598)(0.0462)}{0.16(257-63.2)}\left[\left(\frac{843}{598}\right)^{\frac{0.16}{1.16}} - 1\right]$$

$$= 0.050 \text{ kJ/kJ cooling}$$

(i) Saving in compressor indicated work = 0.176 - 0.050

= 0.126 kJ/kJ cooling

or 71.5 per cent

For the VAS it is possible to obtain solution concentrations and to complete table 2.9 employing the procedure described in Section 2.4.3 for the single stage NH_3-H_2O cycle.

Table 9. Property values for NH_3-H_2O absorption cycle in cascaded hybrid system

State point	State	pressure kPa	temp. °C	liq/vap conc. x/xv $\frac{kg\ NH_3}{kg\ sol}$	enthalpy kJ/kg	mass flow rate kg/s
1	liq	742	40	0.59	-52.5	1.0
2	liq	1554		0.59	62.6	1.0
3	liq	1554	73	0.55	90.5	0.91
4	liq	742	45	0.55	-36	
5	vap	1554	73	0.99	1402	0.093
6	liq	1554	55	0.71	73	0.003
7	vap	1554	55	1.0		0.09
8	liq	1554	40		371.5	0.09
9	-		15.5		371.5	0.09
10	vap	742	15.5		1459.5	0.09

From energy balances on the generator and evaporator,

$$Q_G = 0.91(90.5) + 0.093(1402) - (62.6) - 0.003(73)$$

$$= 149.9 \text{ kJ/s}$$

$$Q_E = 0.09(1459.5 - 371.5)$$

$$= 97.9 \text{ kJ/s}$$

(ii) CoP of VAS $= \frac{97.9}{149.9} = \underline{0.65}$

During hybrid operation the heat transferred in the NH3 evaporator R-22 condenser is the sum of the compressor work (0.05 kW) and the cooling effect in the R-22 evaporator (1.00 kW), namely 1.05 kW/kW cooling.

The heat supplied in the generator $= \frac{1.05}{0.65} = 1.615 \frac{\text{kW heat}}{\text{kW cooling}}$

This heat supply has effected a saving in compressor indicated work of 0.176 - 0.05

$$= 0.126 \frac{\text{kW work}}{\text{kW cooling}}$$

Therefore heat supplied to the generator per kW compressor indicated work saving $= \frac{1.615}{0.126}$

$$= 12.8 \frac{\text{kW heat}}{\text{kW work}}$$

While these are theoretical values it is clear that there are possibilities in this arrangement for solar operation. The maximum heat supply temperature of around 73 °C is possible with flat-plate collectors. The CoP of this single stage NH_3-H_2O cycle is acceptable when compared to other cycles (Table 8).

The cascaded hybrid system is currently under active investigation (refs. 23 and 24).

7. Solar operation of continuous absorption cycles

Solar energy is variable and intermittent. These characteristics, together with the means adopted to collect solar radiation, all impact on procedures which use vapour absorption cycles to supply a cooling load.

This section will discuss these matters, and also suggest methods of relating solar energy inputs to satisfy specified cooling requirements.

7.1. Temperature of collection

The collectors best suited for use in solar cooling are the high performance flat plate collector and the evacuated tube collector. However it would appear that the flat-plate collector is currently the more likely choice on economic grounds, and so its performance characteristics have been used in the discussion which follows.

The performance of the flat-plate collector with a liquid heating medium is represented by

$$Q_U = F'(\tau\alpha)I - F'U_C[\frac{T_{fo}-T_{fi}}{2} - T_a] \qquad (20)$$

where Q_U = useful heat collection

F' = collector efficiency factor

τ = transmittance of glass cover system for incident solar radiation I

α = absorbtance of absorber plate

I = incident solar radiation

U_C = overall heat loss coefficient for collector

T_{fi}, T_{fo} = inlet and outlet temperatures of liquid heating medium

T_a = ambient temperature.

This expression assumes U_C is constant over the operating temperature range. In reality U_C will vary slightly. The derivation of this equation may be found in ref. 25 and 26.

Standard tests of the performance of flat-plate collectors enable a plot to be obtained as in Fig. 21, which is the curve for a collector with a black chrome selective surface on the absorber plate, and a single low-iron glass cover.

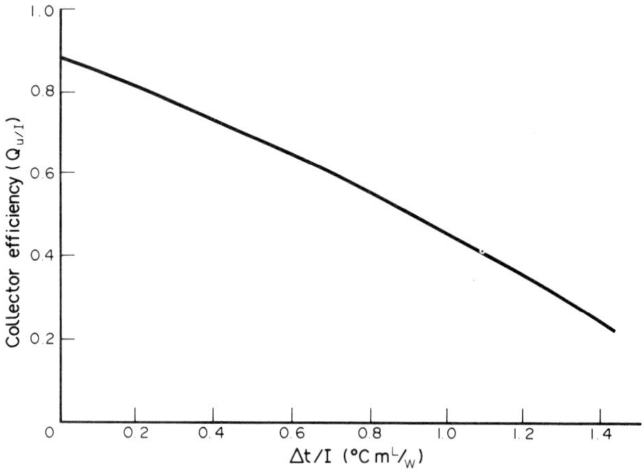

Fig. 21. Performance curve for a flat plate collector under standard conditions. Collector specifications: one low iron glass cover, black chrome selective surface, water heating (equation 7.1.2).

The equation to the curve is

$$\frac{Q_U}{I} = 0.882 - 3.419\left(\frac{\Delta T}{I}\right) - 0.00817\left(\frac{\Delta T}{I}\right)^2 .I \qquad (4.21)$$

where $\Delta T = \left(\frac{T_{fo}-T_{fi}}{2} - T_a\right)$

This curve indicates the slight variation in U_C as the temperature increases.

When $I = 1000$ W/m² and $\Delta T = 0$, then $Q_U = 882$ W/m². Also when $I = 1000$ W/m² and $\Delta T = 70$ °C, then $Q_U = 603$ W/m².

Introducing these sets of values in equation 4.20 we have

$$882 = F'\tau\alpha(1000), \text{ or } F'\tau\alpha = 0.882$$

and $$603 = 0.882(1000) - F'U_C(70),$$

i.e. $F'U_C = 4.0$

So equation 4.20 becomes

$$Q_U = 0.882\ I - 4.0\ (\Delta T) \qquad (22)$$

which is the straight line approximation to the collector performance curve, equation 21.

When $Q_U = 0$, the critical value of incident radiation

$$I_{min} = \frac{F'U_C}{F'\tau\alpha}\ (\Delta T).$$

For the collector under examination, $I_{min} = \frac{4.0}{0.882}(\Delta T)$.

The implications of this critical value, I_{min}, may be explained by a simplified example.

Consider a hypothetical day where I varies as shown in Fig. 22.

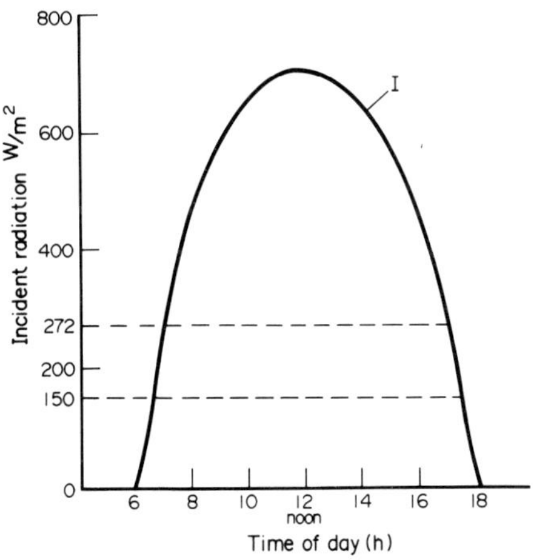

Fig. 22. Radiation incident on collector on hypothetical day.

If the mean heat supply temperature $(\frac{T_{fo}+T_{fi}}{2})$ is 90 °C, and the ambient temperature, T_a = 30 °C, then

$$I_{min} = \frac{4.0}{0.882}\ (90-30) = 272\ W/m^2.$$

The useful heat collection during the day is the area enclosed above this value of I_{min} and the I-curve in Fig. 22. The efficiency of heat collection

$$\zeta_c = \frac{\sum^{day} Qu}{\sum^{day} I} = 0.50 \text{ approx.}$$

If the mean heat supply temperature is reduced to 63 °C, then I_{min} = 150 W/m²; and the efficiency of heat collection is 0.71 approximately.

In other words when the temperature of heat supply is less, a smaller area of solar collector can provide the same heat supply. Clearly this has consequences for vapour absorption cooling systems.

From Table 8 compare the single stage NH_3-H_2O cycle and the cascaded generation NH_3-H_2O cycle with the same evaporator (-10 °C) and heat rejection temperatures. This comparison is presented in table 4.10.

Table 10. Comparison of CoP and solar collection efficiency for single stage and cascaded generation NH_3-H_2O cycles.

Cycle	Max. heat supply temp. °C	CoP (theor.)	Estimated solar collection eff. for hypothetical day per cent
single stage NH_3H_2O	94	0.45	50
cascaded generation NH_3-H_2O	65	0.29	71

The lower CoP of the cascaded generation cycle is offset by a higher solar collection efficiency, and hence a smaller and cheaper solar collector array.

Actual insolation on the collector can vary every few minutes, and from hour to hour depending upon cloud cover and time of day. It also varies from day to day, and with the season of the year. The mean temperature of heat collection varies with the load and with the heat capacity of the hot water circuit. Nevertheless the procedure described in the simplified example is essentially valid; and this forms the basis for more exact procedures described in later sections.

7.2. Thermal storage

Insolation above the critical value would be available at best for 7 or 8 hours a day. Where cooling is required outside these hours, thermal storage is necessary either as hot water or as refrigerant, the collector area being increased sufficiently to collect this heat.

Collector area and thermal storage capacity are therefore important parameters in the design of solar cooling plant.

7.3. Energy usage in solar cooling systems and the critical solar fraction

In order to drive a complete solar operated vapour absorption cooling plant two sources of energy are required: a supply of heat to the generator and a supply of work energy to drive fans, circulating pumps etc. (termed 'parasitic power').

It is quite uneconomical to supply all the heat required from solar collectors. Therefore auxiliary energy from a source (or sources) other than solar is necessary. Auxiliary energy can be in the form of heat, or in the form of work (the work being in magnitude equal to the work output from an engine cycle to which the auxiliary heat is supplied).

> *The solar fraction, f, is defined as that fraction of the useful energy input (excluding parasitic power) which derives from solar sources.*

The basis of economic evaluation of solar systems is the cost-benefit ratio. The benefit is obtained by a comparison of the capital and operational costs of a conventional system and the solar system which replaces it.

Consider first the vapour compression system. The conventional VCS consists essentially of a compressor, condenser and evaporator. The compressor is usually driven by an electric motor. There would also be one or more pumps/fans for cooling and circulation.

The electrical energy consumption of the motor driving the compressor

$$W_C = \frac{C_L}{(CoP)_C \cdot E_m}$$

where C_L = cooling load

$(CoP)_C$ = coefficient of performance of VCS

E_m = efficiency of electric motor.

Assuming parasitic power consumption = $p_C \cdot C_L$

the total power consumption of the VCS

$$W_{ct} = \frac{C_L}{(CoP)_c . E_m} + p_c . C_L$$

Consider next a solar vapour absorption system, with an auxiliary heat source (e.g. a fuel-fired boiler).

$$\text{Heat energy required to operate VAS} = \frac{C_L}{(CoP)_a}$$

If the solar fraction is f, then heat energy supplied from auxiliary boiler

$$Q_x = \frac{(1-f)C_L}{(CoP)_a . E_b}$$

where

$(CoP)_a$ = CoP of VAS

E_b = efficiency of the boiler

If the parasitic power consumption of the VAS = $p_a . C_L$ then total auxiliary energy consumption of the VAS (i.e. from non-solar sources)

$$Q_{xt} = \frac{(1-f)C_L}{(CoP)_a . E_b} + p_a . C_L$$

Therefore energy savings = $W_{ct} - Q_{xt}$

$$= \frac{C_L}{(CoP)_c . E_m} + p_c . C_L - \frac{(1-f)C_L}{(CoP)_a . E_b} + p_a . C_L \quad (23)$$

The critical value of the solar fraction, f_c, is obtained when the energy savings are zero.

$$0 = \frac{C_L}{(CoP)_c . E_m} + p_c . C_L - \frac{(1-f_c)C_L}{(CoP)_a . E_b} + p_c . C_L$$

$$f_c = 1 - (CoP)_a . E_b \quad \frac{1}{(CoP)_c . E_m} \quad - (p_a - p_c)$$

For representative values of $(CoP)_a = 0.6$, $E_b = 0.8$, $(CoP)_c = 2.5$, $E_m = 0.9$, $p_a = 0.2$, $p_c = 0.1$

$$f_c = 0.835.$$

This means that when the operating solar fraction is less than 0.835 (approximately) the solar operated VAS will be consuming more auxiliary energy than the electrical energy required by the conventional VCS.

Equation 23 can be modified to indicate cost savings:

$$\text{Cost savings in VAS} = \frac{u_e \cdot C_L}{(CoP)_c \cdot E_m} + u_e \cdot p_c \cdot C_L - \frac{u_h(1-f')C_L}{(CoP)_a \cdot E_b} + u_e \cdot p_a \cdot C_L \qquad (24)$$

where u_e, u_h are unit costs of electricity and auxiliary heat respectively.

The critical cost solar fraction f'_c is obtained when the cost savings are zero.

$$0 = \frac{u_e \cdot C_L}{(CoP)_c \cdot E_m} + u_e \cdot p_c \cdot C_L - \frac{u_h(1-f'_c) \cdot C_L}{(CoP)_a \cdot E_b} + u_e \cdot p_a \cdot C_L$$

$$f'_c = 1 - \frac{u_e}{u_h} \cdot (CoP)_a \cdot E_b \quad \frac{1}{(CoP)_c \cdot E_m} - (p_a - p_c)$$

For representative values of $(CoP)_a = 0.6$, $E_b = 0.8$, $(CoP)_c = 2.5$, $E_m = 0.9$, $p_a = 0.2$, $p_c = 0.1$, f'_c has been calculated for a range of values of u_e/u_h:

$\frac{u_e}{u_h}$	f'_c
1.0	0.835
1.5	0.752
2.0	0.670
2.5	0.588

Cost savings accrue when the operating solar fraction is greater than the critical value f'_c. Even when auxiliary heat is half the cost of electricity, cost savings only begin to accrue when the operating f exceeds 0.67.

In practice the values of f_c and f'_c stated above may be exceeded for short periods during clear days, but are rarely exceeded on a seasonal or annual basis. Therefore for VAS with auxiliary boilers cost-effectiveness is virtually impossible to achieve. This conclusion has been borne out by the actual performance of such systems - except where the solar system also supplies a substantial heating load.

The reduction in f_c and f'_c is crucial to the economic viability of solar cooling systems.

This reduction in f_c and f'_c may be achieved by incorporating electrically driven vapour compression plant with the solar driven VAS (hybridisation) so that auxiliary energy is supplied to drive the VCS. By these arrangements f_c can be reduced to 0.12 or less (ref. 27). Economic viability is enhanced, even though initial capital costs of hybrid plant are somewhat greater than a VAS with an auxiliary boiler.

Hybrid systems are described in sections 2.6 and 2.7.5.

7.4. The $\bar{\phi}$, f relation

For the system shown in Fig. 23 the $\bar{\phi}$, f method is a procedure which estimates the solar fraction, f over a year of operation at a specified location given the load, the minimum temperature of operation, and the climatic data at site.

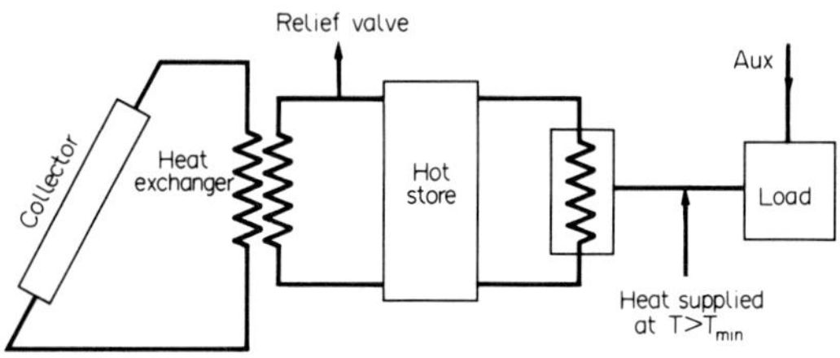

Fig. 23. Solar system for derivation of ϕ, f relation.

The importance of this design method for solar operated vapour absorption cooling systems is that these systems have a critical minimum heat supply temperature, and this method takes this into account.

The derivation of this method is outside the scope of this book, and may be found in ref. 28.

The fundamental principle basic to the method is that explained in the simplified hypothetical example in section 7.1, but the method itself takes into account all the variations in climatic data, flat-plate collector characteristics, hot store capacity and losses, and minimum operating temperature over a whole year by computing the solar fraction month by month.

The solar fraction for any month is given by

$$f = \bar{\phi}_{max}.Y - a \quad \exp(bf)-1 \quad 1-\exp(cX) \; .$$

In this expression $\bar{\ }_{max}$ is the quantity of solar radiation above the critical minimum received during the month. Y is a non-dimensional number relating the energy absorbed by the solar collector to the load; while X is a non-dimensional number including heat lost from the collector. The value of a depends on the capacity of the hot water (thermal) store. b and c are constants.

A full explanation of this procedure is given in ref. 25; and software for computer operation are commercially available.

7.5. Simulation of the integrated VAS-VCS hybrid system

The flow circuit for an NH_3-H_2O integrated hybrid system is shown in Fig. 24.

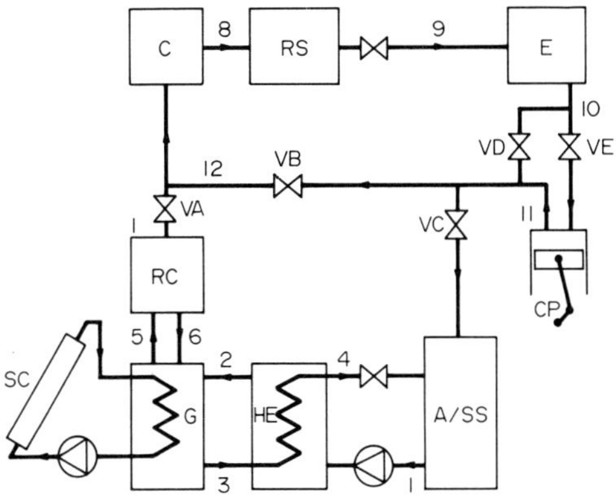

Fig. 24. Flow circuit for ammonia-water integrated hybrid system with refrigerant storage. SC, solar collector; G, generator; RC, reflux condenser; C, condenser; RS, refrigerant storage; E, evaporator; CP, compressor; A/SS, absorber and solution store; R, recuperator or solution heat exchanger.

This system can be operated in 3 modes:

(i) as a VAS with valves VA, VC, VD open and VB, VE closed;
(ii) as a VCS with VB, VE open and VA, VC, VD closed;
(iii) as a hybrid system with VA, VC, VE open and VB, VD closed.

This system includes a refrigerant store, which stores refrigerant during the solar generation phase for use as coolant after generation ceases (during late evening and at night). When generation does not occur due to insufficient solar radiation and when there is no refrigerant in the store, the system operates conventionally in mode (ii).

When the VAS absorber is functioning, the pressure difference across the compressor is reduced; and hence the power consumption of the compressor (W_{input}) is less than the power consumption in conventional operation in mode (ii) ($W_{conv.}$).

If $$W_{saving} = W_{conv.} - W_{input}$$

then the solar fraction, $f = W_{saving}/W_{conv.}$.

The operation of this system in mode (i) has not been included in this simulation.

The different components of this system have been modelled as follows to study the transient performance over a long period (refs. 29, 30) based on previous work on the transient model of the cascaded hybrid system (ref. 23).

For the generator-reflux condenser,

$$\dot{m}_2 = 1, \; x_7 = 1.0$$

$$\dot{m}_3 = \frac{1-x_2}{1-x_3}$$

$$\dot{m}_5 = \frac{(x_2-x_3)(1-x_6)}{(1-x_3)(x_5-x_6)}$$

$$\dot{m}_7 = \frac{(x_2-x_3)}{(1-x_3)}$$

For the recuperator,

$\dot{m}_{SS}$ = strong solution flow rate = $\dot{m}_2$

$\dot{m}_{WS}$ = weak solution flow rate = $\dot{m}_3$

C_{psol} = specific heat capacity of solution = 4.78 kJ/kg.°C

$$N_R = NTU = \frac{(UA)_R}{\dot{m}_{WS} \cdot C_{psol}}$$

$(UA)_R$ = product of heat transfer coefficient and heat transfer surface

$$C_R = \frac{C_{min}}{C_{max}} = \frac{1-x_2}{1-x_3}$$

Effectiveness $$E_R = \frac{1-\exp[-N_R(1-C_R)]}{1-C_R.\exp[-N_R(1-C_R)]}$$

Also $$E_R = \frac{\dot{m}_{SS} \cdot C_{psol}(t_2-t_1)}{\dot{m}_{WS} \cdot C_{psol}(t_3-t_1)}$$

For the generator,

Heat from hot water = Energy stored in generator + Heat loss to ambient + Heat to solution

$$\dot{m}_d \, C_{pw}(t_{ho}-t_{hi}) = C_{gen}\left(\frac{\Delta t_3}{\Delta \tau}\right) + (UA)_{go}\left(\frac{t_3+t_2}{2} - t_a\right) + Q_S$$

Where $\dot{m}_d$ = mass flow rate of hot water

t_{ho}, t_{hi} = outlet and inlet temperatures at collector,

C_{pw} = specific heat capacity of water

C_{gen} = heat capacity of generator

Δt_3 = change in t_3 in time interval $\Delta\tau$

$(UA)_{go}$ = product of heat transfer coefficient and external surface area

t_a = ambient temperature,

Q_S = $\dot{m}_3 h_3 + \dot{m}_5 h_5 - \dot{m}_2 h_2$

Also for generator as heat exchanger,

$$E_g = \frac{1-\exp[-Ng(1-C_g)]}{1-C_g.\exp[-N_g(1-C_g)]}$$

where $N_g = (UA)_g / C_{gmin}$

$(UA)_g$ = product of heat transfer coefficient and surface area for heat transfer from hot water to solution

$C_{gmin} = \text{Min } [C_{gsol} \; ; \; C_{gw}]$

$C_{gsol} = \dot{m}_{SS}(-172.63 + 1168.64x_1 - 2407.27x_1^2 + 1703.87\; x_1^3)$

$C_{gw} = 4.187\; \dot{m}_d$

$C_g = C_{gmax}/C_{gmin}$

also $$E_g = \frac{C_{gsol}(t_3 - t_2)}{C_{gmin}(t_{ho} - t_2)}$$

For the collector, from equation 4.20

$$\dot{m}_d(4.187)(t_{ho} - t_{hi}) = A_C F'[(\tau\alpha)I - U_C(\frac{t_{ho}+t_{hi}}{2} - t_a)]$$

where A_c = area of collector field.

For refrigerant store and evaporator,

In the time interval $\Delta\tau$,

$$\text{Mass stored in refrigerant store, } \delta M_s = \text{Mass condensed } \delta M_g + \text{Mass evaporated } \delta M_e$$

where $\delta M_g = \dot{m}_{SS} \dfrac{(x_1 - x_3)}{(1-x_3)} \cdot \Delta\tau$

$$\delta M_e = \frac{Q_{ev}}{(h_{10} - h_8)} \cdot \Delta\tau$$

Q_{ev} = refrigeration load (in kW)

For the solution store,

Solution concentration at the commencement of time interval, $\Delta\tau$

$$x_1 = M_a/(M_a + M_w)$$

where M_a = mass of ammonia

M_w = mass of water in solution.

Solution concentration at the end of time interval, $\Delta\tau$

$$x_1' = \frac{x_1(M_a + M_w) + (\delta M_e - \delta M_g)}{(M_a + M_w) + (\delta M_e - \delta M_g)}$$

For the ammonia compressor, based on an equation in Threlkeld (ref. 7), a performance relation is obtained as follows:

$$\frac{W_c}{Q_{evap}} = 0.18441(p_c)^{0.38419} - 0.4902(p_c)^{0.14755}$$

where p_c is in psia.

For transient operation in hybrid mode

$$p_c = \frac{p_1 + p_1'}{2} + \Delta p_L$$

where p_1 and p_1' are the pressures in the absorber at the commencement and end of the time interval. $\Delta\tau$, and Δp_L is the pressure drop between the compressor and the absorber.

In conventional operation, the condenser pressure p_c is equal to p_3.

A computer simulation incorporating the above models for the system sub-components uses isolation on the collector (I), ambient temperature (t_a), and cooling load (Q_{ev}) as the driving inputs. System specifications include collector area (A_C) and collector specifications, heat transfer coefficients and areas, etc all as required in the above relations.

Preliminary results using this simulation procedure have been obtained for one year using meteorological data for Singapore (Fig. 25).

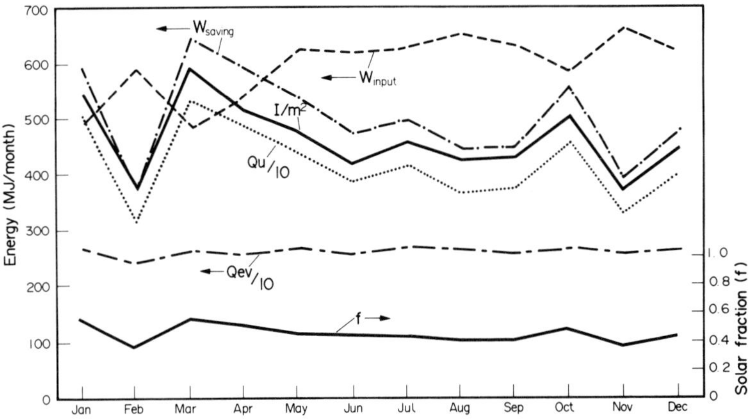

Fig. 25. Preliminary results of simulated integrated hybrid system performance at Singapore. A_C = 36 m², $F_R(\tau\alpha)$ = 0.75, F_RU_C = 5.0 W/m².°C (ref. 31).

8. The intermittent vapour absorption cycle and solar operation

The principle of the intermittent VAS has been described in section 1, and possible refrigerant-absorbent pairs in section 4.3. In this section two solar applications of the intermittent cycle are described.

8.1. The intermittent NH_3-H_2O vapour absorption cycle

8.1.1. The theoretical cycle

The flow circuit is shown in Fig. 26. The arrows designated G show the direction of flow of solution and refrigerant during the generation phase.

The arrows designated A show the solution and refrigerant flow directions during the absorption phase.

The cycle is shows sketched on the ln p vs 1/T chart in Fig. 27.

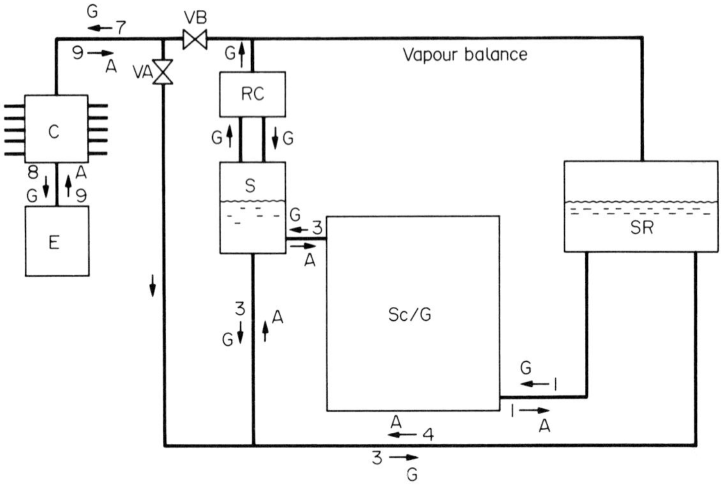

Fig. 26. Flow circuit for solar intermittent ammonia-water refrigerator. SC/G, solar collector/generator; S, separator; SR, solution reservoir; RC, reflux condenser; C, condenser; E, evaporator; arrows G and A indicate flow during generation and absorption respectively.

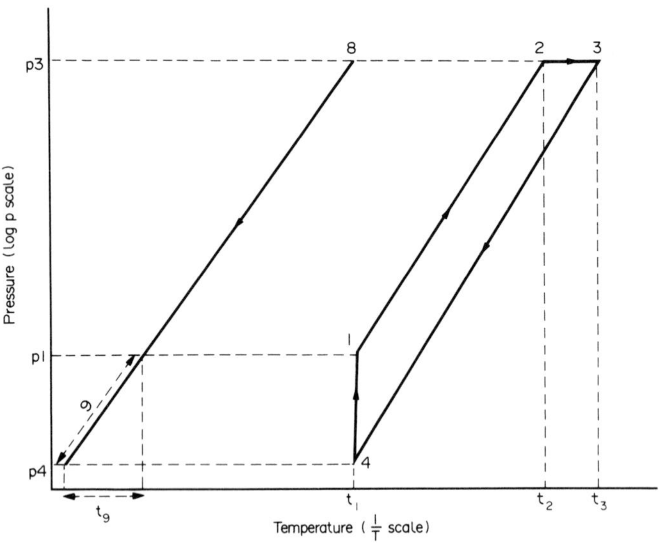

Fig. 27. Intermittent ammonia-water cycle with constant temperature absorption.

8.1.2. Analysis of theoretical cycle

The generation phase takes place over a period of time, depending upon the mass of solution in the store, the collector area, and the insolation.

The heating consists of two processes. In the first, process 1 - 2, the temperature of the bulk of the solution is raised from t_1 to t_2 while the mass remains unchanged ($m_1 = m_2$).

Therefore

$$Q_{1-2} = m_1 \ (h_2-h_1)$$

In the second process, process 2 - 3, boiling begins and continues till the temperature reaches the maximum temperature t_3, with a decrease in concentration ($x_3<x_2$), and a decrease in mass ($m_3<m_2$).

During process 2 - 3, assume a small decrease in mass, dm, due to boiling, while there is a small increase in h,

i.e. $$dQ = (m-dm)(h+dh-h) - dm(h_{fg}+h+\frac{dh}{2}-h)$$

where h_{fg} is the heat of evaporation.

$$dQ = mdh - h_{fg} \ dm$$

$$= mdh + hdm - (h_{fg}+h)dm$$

$$= d(mh) - h_v \ dm$$

So for process 2 - 3, $$Q_{2-3} = \int_{m_2h_2}^{m_3h_3} d(mh) - \int_{m_2}^{m_3} h_v \ dm$$

$$= m_3h_3 - m_2h_2 + \int_{m_3}^{m_2} h_v \ dm$$

Therefore $$Q_g = Q_{1-2} + Q_{2-3}$$

$$= m_3h_3 - m_1h_1 + \int_{m_3}^{m_2} h_v \ dm \qquad (25)$$

This expression assumes heat returned to the generator by the reflux liquid is negligible.

The mass of vapour dm leaving the generator as S is related to the mass of vapour dm_r leaving the reflux condenser at 7 by the following expression:

$$dm = \frac{(1-x)}{(x_v-x)} \ dm_r \qquad (26)$$

where x_v = vapour concentration leaving generator at the solution temperature at which dm was evaporated,

x = solution concentration of reflux liquid assumed to be at t_3.

Equations 25 and 26 are evaluated numerically introducing values from aqua-ammonia data charts and tables.

Refrigerant vapour condensing in the condenser at t_8 collects in the evaporator at the same temperature. The final mass $m_{r8} = m_1 - m_3$.

The solution is now allowed to cool for some time while the pressure in the system falls. The refrigeration phase is commenced by closing valve VA, and gradually opening VB.

As VB is opened the mass of refrigerant in the evaporator, m_{r8}, initially at the temperature of condensation (t_8) cools adiabatically till its vapour pressure is the same as the system pressure. Some of the refrigerant is evaporated to achieve this, the final mass being m_{r9}.

During the adiabatic cooling of M_r, let a small mass of refrigerant dm_r be evaporated. So

$$-0.5\ \{m_r + (m_r - dm_r)\}\ c_{pv}\ dt = -\ h_{fg}.dm_r$$

where dt = drop in temperature of refrigerant

C_{pr}, h_{fg} = specific heat and latent enthalpy of refrigerant

So
$$\int_{m_{r9}}^{m_{r8}} \frac{dm_r}{m_r} = \int_{t_9}^{t_8} \frac{C_{pv}.dt}{h_{fg}}$$

$$\frac{m_{r9}}{m_{r8}} = \exp\ \{(h_{f8} - h_{f9})/h_{fgm}\} \qquad (27)$$

where h_{f8}, h_{f9} = liquid enthalpy of refrigerant at 8 and 9 respectively,

h_{fgm} = mean latent enthalpy of refrigerant during process 8-9.

Thereafter further evaporation produces a cooling effect Q_e. During this process 4-1, the pressure in the system rises from p_4 to p_1 while the solution concentration increases from x_3 to x_1.

$$\text{The cooling effect, } Q_e = m_{r9}\ (h_{fgm}) \qquad (28)$$

Where h_{fgm} is the mean h_{fg} over the pressure range p_4 to p_1.

During this absorption phase the solution circulation is reversed (arrows with A). The backs of the solar collector cases are opened to the atmosphere to facilitate cooling of the solution.

Following the procedure described for the analysis of the single stage continuous NH_3-H_2O cycle, completing a table of properties, and applying mass and energy balances to components it is possible to evaluate equations 4.25 to 4.28 and obtain Q_g and Q_e.

So CoP = Q_g/Q_e.

The CoP varies with the operating temperature limits as shown in Fig. 28.

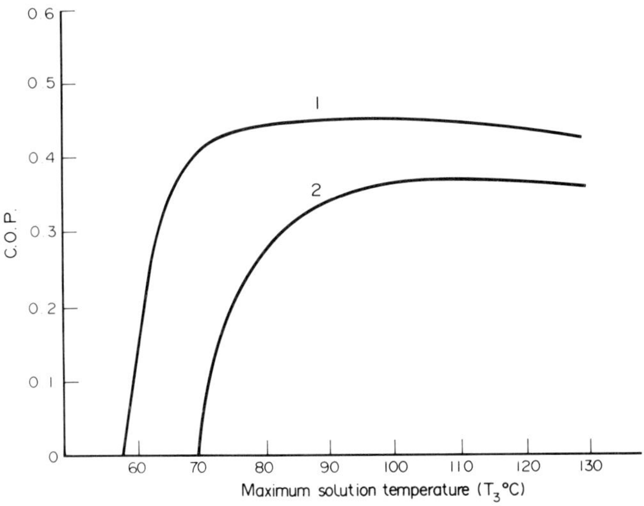

Fig. 28. Theoretical coefficient of performance of intermittent ammonia-water cycle with constant temperature absorption. Curve 1, x_1 = 0.55, t_1 = t_8 = 26.7 °C, max t_9 = 0 °C. Curve 2, x_1 = 0.48, t_1 = t_8 = 26.7 °C, max t_9 = - 8.3 °C.

The results of a test on an experimental solar-operated intermittent NH_3-H_2O refrigerator have been abstracted from ref. 31 and are presented in Table 4.11.

This cycle has received much attention, and larger NH_3-H_2O refrigerators have been built and tested, notably the 'village size' refrigerator at the Asian Institute of Technology (ref. 32) described later in this book.

The cooling capacity of an intermittent system is dependent on the initial mass of solution: the solution store is therefore a constraint on the size of the plant.

Table 11. Test results for NH_3-H_2O intermittent solar refrigerator. Collector area 1.63 m². (ref. 31)

Generation phase.	
Time of starting	9.41 am
Duration	6h 40min
Initial mass of solution (m_1)	8.88 kg
Initial solution temp. (t_1)	30 °C
Initial concentration (x_1)	0.464
Max. solution temp. (t_3)	100 °C
Max. solution pressure (p_3)	1413 kPa
Final solution conc. (x_3)	0.370
Mass of distillate, NH_3 (m_{r8})	1.32 kg
Temp. of condenser cooling water	30 - 35 °C
Heat received by solution and distillate	5137 kJ
Total incident solar radiation	24872 kJ
Absorption phase	
Duration (approximate)	6h
Temp. of absorbing solution	33 - 35 °C
Minimum evaporator temp.	- 12.5 °C
Total effective cooling (Q_c)	1435 kJ
Effective cooling per m² of collecting surface per day	882.5 kJ/m²
Performance ratios	
Actual COP	0.279
Theoretical COP	0.316
Ratio of total effective cooling cooling to total radiation	5.4 %

This cycle can also be multistaged with cascaded generation, thereby operating at lower maximum temperatures. Such a two-stage cycle has been investigated (ref. 33).

8.2. Intermittent systems using salts

Ammonia is absorbed by some salts, and the following have been identified as promising absorbents in solar operated cooling cycles: $CaCl_2$, $SrCl_2$, NaSCN, $LiNO_3$.

Of these ammonium thiocyanate (NaSCN) and lithium nitrate ($LiNO_3$) from liquids with NH_3: and data for these solutions are available (ref. 36).

Calcium chloride ($CaCl_2$) and strontium chloride ($SrCl_2$) are solid absorbents and are used in so called 'dry absorption machines'.

Eggers Lura et al (ref. 34) have theoretically assessed the performance of these combinations for solar operation for the following conditions:

Total solar radiation during generating period of 5 hours	1040 W/m^2
Efficiency of solar collector	0.55 to 0.45
Ambient temp.	31 °C
Condensing temp.	40 °C
Absorbing temp.	40 °C
Duration of absorption phase	12 hours
Evaporation temp.	- 10 °C
Estimated spec. mass of generator/-absorber vessel (kg/kg refrig. evaporated	10 kg

The results are shown in Fig. 29.

Using an experimental rig with a flat-plate collector (1.5 m^2), Swartman et al (ref. 35) have obtained higher values of solar COP for the NH_3-NaSCN combination.

With solid absorbents many problems arise in practice: there is poor heat transfer to and from the solid absorbent, and the solid absorbent swells very considerably to about ten times its original volume as ammonia is absorbed. During generation the absorbent tends to lose its crystalline structure, and break up into a powder which makes subsequent absorption more difficult. Absorbing capacity also decreases with cycling as the powdered absorbent is packed into regions less accessible to NH_3.

Progress has been made in overcoming these problems by Iloeje (ref. 3) who mixed Portland cement and $CaSO_4$ with the $CaCl_2$ absorbent to obtain hard porous granules.

The equations governing the operation of the NH_3-$CaCl_2$ refrigerator are as follows:

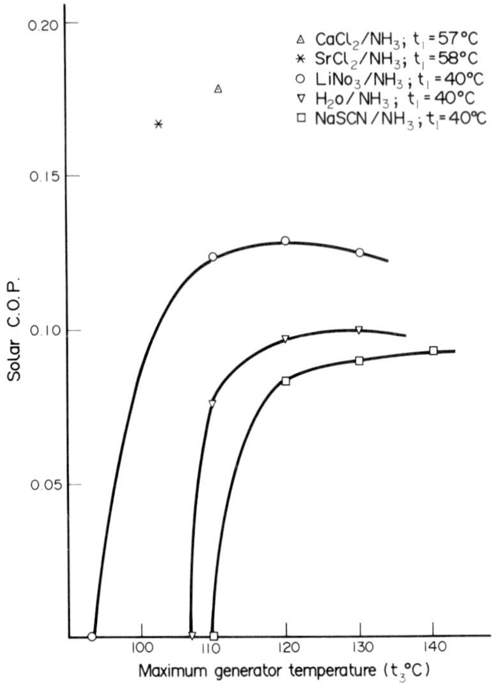

Fig. 29. Coefficient of performance of solar operated theoretical intermittent cycle with constant pressure absorption for various working pairs. t_8 = 40 °C, t_9 = -10 °C (ref. 35).

Generation at 13.5 bar:

$$Q_{G1} + CaCl_2.8NH_3 \quad CaCl_2.4NH_3 + 4NH_2 \text{ at } 90\ °C$$

$$Q_{G2} + CaCl_2.4NH_3 \quad CaCl_2.2NH_3 + 2NH_3 \text{ at } 101\ °C$$

Absorption at 3 bar, -10 °C in the evaporator, heat rejection temps. 53 - 63 °C:

$$CaCl_2.2NH_3 + 6NH_3 \quad CaCl_2.8MH_3 + Q_A$$

Iloeje has constructed a solar powered refrigerator using $CaCl_2$ mixed with 20 percent by mass $CaSO_4$. To quote from ref. 38:

"The double glazed collector/absorber/generator unit used clear PVC and plane glass sheets, with the former as the outer cover. Overall collector plate exposed area was 1.41 m². A stagnant water evaporative condenser was designed and constructed of re-inforced sandcrete, with steel condenser tubes, and was coupled to the system. The evaporator was a spirally coiled steel tube immersed in a pool of stagnant water. Absorbent cooling during absorption was by natural convection of air over the collector plate and tubes, coupled with sky radiation. All construction, except the collector covers and porous condenser walls, were of steel sheets and tubes.

Ambient temperatures during absorption and generation ranged over (25 ° - 35 °C). Tests indicated that cooling capacities of the NH_3 condensed were around 714 KJm^{-2}, while effective cooling obtained was equivalent to an ice production of 1 kg m^{-2}, per day".

9. Conclusions

The use of solar energy in the form of heat to produce cooling is subject to the second law of thermodynamics and so is inherently more complex than the use of solar energy directly for a heating application, e.g. house heating.

The problem is not made any easier by the manner in which an economic assessment of a solar thermal cooling system is made - namely a comparison of the capital and operational costs of a vapour compression refrigerator vis a vis those of a solar cooling plant. For the vapour compression plant handles only the second of the two processes required for the production of cooling from heat (section 2); the first process, the heat-to-work cycle, occurs at the thermal power station. The solar cooling plant, on the other hand, handles both, and so is necessarily more expensive in terms of capital outlay on plant.

Nevertheless, the economic evaluation is the ultimate and crucial determinant of success. Methods of economic assessment are described in refs. 25 and 26.

The foregoing chapter has attempted to address the primary factors affecting economic viability of vapour absorption systems - the coefficient of performance of these systems in relation to plant complexity, associated solar collector performance and auxiliary energy supply; and indicates directions which future investigations might follow.

10. References

1. R.T. Ellington, G. Kunst, R.E. Peck, J.F. Reed. The absorption cooling process. Inst. Gas. Technology, Res. Bull. No. 14, 1957.
2. J.E. Aker, R.G. Squires, L.F. Albright. An evaluation of alcohol-salt mixtures. Trans. ASHRAE 71, II, 14, 1965.
3. J.P. Robertson, L. Chien-Yung, R.G. Squires, L.F. Albright. Trans. ASHRAE 72, I, 198, 1966.
4. P.S. Asha Iyer, S. Srinivasa Murthy, M.V. Krishna Murthy. Thermodynamic properties of some VAS refrigerant-absorbent pairs Active Cooling Systems. Procs. Solar Cooling Workshop. ed. Chinnappa et al. James Cook Uni. Queensland 4811 1984.
5. R.W. Allen, F.H. Morse, S.L. Sargent. Optimisation studies of solar absorption air-conditioning systems. Report No. NSF/RANN/SE/GI - 39117/PR/74/1. Uni. of Maryland, 1974.

6. R.W. Allen, F.H. Morse, S.L. Sargent. Optimisation studies of solar absorption air-conditioning systems (Final progress report). No NSF/RANN/SE/GI - 39117/PR/76/2. Uni. of Maryland, 1976.
7. J.L. Threlkeld. Thermal environmental engineering. Prentice-Hall.
8. J.C.V. Chinnappa, P.L. Kok. A continuous cycle solar ice-plant. Procs. 8th Biennial Solar World Congress, Perth. Vol. 1, 640-644, 1983.
9. K. Dao, M. Simmons, R. Wolgast, M. Wahlig. Performance of an experimental solar driven absorption air conditioner. Report No: LBL-5911. L.B. Labs, U.C./Berkeley 1976.
10. E.J. Cairns, M.A. Wahlig. Applied Science Division Annual Report. Solar Energy Program, FY 1984. Report No. LBL-18752/U.C.-13, p. 2-23. L.B. Labs/Berkeley, 1985.
11. D. Miller. Performance of water-cooled lithium bromide absorption units for solar energy applications. Procs. 2nd ERDA solar cooling workshop. (Ed. de Winter and de Winter), N.T.I.S., VA 22161, p. 186, 1976.
12. D.K. Anand, R.W. Allen, B. Kumar. Transient Simulation of Absorption Machines. J. Solar Energy Eng., Vol. 104, p. 197, 1982.
13. E.P Whitlow. Trends of efficiencies in absorption refrigeration machines. ASHRAE Jour. Dec. 1966. p. 44.
14. B.A. Phillips. Absorption cycles for air cooled solar air-conditioning. ASHRAE symp. Dallas Vol 4/15. 966-974. 1976.
15. P.S. Asha Iyer, S. Srinivasa Murthy, M.V. Krishna Murthy. Comparative thermodynamic analysis of double-effect vapour absorption refrigeration systems with different working fluids. Active Cooling Systems. Procs. Solar Cooling Workshop. Op cit. p. 68. 1984
16. ASHRAE Handbook of Fundamentals, 1981.
17. P.S. Asha Iyer, S. Srinivasa Murthy, M.V. Krishna Murthy. Comparative thermodynamic analysis of dual-stage vapour absorption refrigeration systems with different working fluid pairs. Active Cooling Systems: Procs. Solar Cooling Workshop. Op. cit. p. 56, 1984.
18. J.C.V. Chinnappa, H.A. Martin. Dual-mode operation by solar energy of an ammonia-water two stage cycle Solar Energy 18, 4, 337-342, 1976.
19. K. Dao. A new absorption cycle. The single effect regenerative absorption refrigeration cycle. Report No. LBL-6879. L.B. Labs U.C./Berkeley, 1978.
20. K. Dao. Conceptual design of an advanced absorption cycle. The double-effect regenerative absorption refrigeration cycle. Report No: LBL-8405. Lawrence Berkeley Labs, 1978.
21. S.C. Kaushik, Rajesh Kumar. Computer aided conceptual thermodynamic design of a two-stage dual-fluid absorption cycle for solar refrigeration. Solar Energy, 35, 5, p. 401-407, 1985.
22. R. Balasubramaniam, S. Srinivasa Murthy, M.V. Krishna Murthy. Analytical studies of the solar operated vapour jet - vapour compression cascaded hybrid systems for cogeneration of cold and heat. Active cooling systems: Procs. Solar Cooling Workshop. Op cit. p. 107. 1984.
23. J.C.V. Chinnappa, M.R. Crees, K.P. Stark. A hybrid system for solar air-conditioning. Procs. ISES (ANZ Sect.) Conference, Vol. 2, Sydney 1981.

24. J.C.V. Chinnappa, M.R. Crees, K.P. Stark. Ammonia-water vapour absorption - vapour compression cascaded hybrid systems for heating and cooling. Active Cooling Systems: Procs. Solar Cooling Workshop. Op. cit. p. 96, 1984.
25. J.A. Duffie, W.A. Beckman. Solar Engineering of Thermal Processes. Wiley.
26. F. Kreith, J.F. Kreider. Principles of Solar Engineering. Hemisphere/McGraw-Hill.
27. J.C.V. Chinnappa. Energy usage by solar cooling systems Active Cooling Systems: Procs. of Solar Cooling Workshop. Op. cit. p. 40, 1984.
28. S.A. Klein, W.A. Beckman. A general design method for closed loop solar energy systems. Solar Energy, 22, 269-282, 1979.
29. J.C.V. Chinnappa. Preliminary assessment of an integrated hybrid ammonia-water solar air-conditioning system. Active Cooling Systems: Procs. Solar Cooling Workshop. Op. cit. p. 131, 1984.
30. N.E. Wijesundera (National Univ. of Singapore). Private communication, 1987.
31. J.C.V. Chinnappa. Performance of an intermittent refrigerator operated by a flat-plate collector. Solar Energy 6, 4, 143-150, 1962.
32. R.H.B. Exell, S. Kornsakoo, S. Oeapipatanakul, S. Chanchaona. A village-size solar refrigerator. Res. Report No. 172. Asian Inst. Technology, Bangkok 10501, 1984.
33. A. Mani, A. Venkatesh. A two-stage intermittent solar refrigeration system: evaluation of salient parameters. Energy conv. and management (GB), 26, 3-4, 271-276, 1986.
34. A. Eggers-Lura, P. Bechtoft Nielson, B. Stubkier, P. Worsøe-Schmidt. Potential use of solar powered refrigeration by an intermittent solid absorption system. COMPLES Inter. meeting, Univ. Petr. Minerals, Dhahran, Saudi Arabia, 1975.
35. R.K. Swartman, V.Ha, C. Swaminathan. Comparison of ammonia-water and ammonia-sodium thocyanate as the refrigerant-absorbent in a solar refrigeration system. Solar Energy, 17, p. 123-127, 1975.
36. C.A. Infante Ferreira. Thermodynamic and physical property data equations for NH_3-$LiNO_3$ and NH_3-NaSCN solutions. Solar Energy, 32, 2, 231-236, 1984.
37. O.C. Iloeje. Quantitative comparisons for treated $CaCl_2$ absorbent for solar refrigeration. Solar Energy, 37, 4, 253-260, 1986.
38. O.C. Iloeje. Design, construction and test run of a solar powered solid absorption refrigerator. Solar Energy, 35, 5, 447-455, 1985.

CHAPTER 3

SOLAR ABSORPTION COOLING

Professor Ken-ichi Kimura

Department of Architecture
Waseda University, Japan

Solar cooling is one of the wonders of solar energy utilization. It is particularly interesting to appreciate how solar heat can be converted into colder energy flows in the midst of the hot season. The principles, equipment, the system and the applications of solar cooling with absorption refrigeration machines are described below.

1. The Principles of the Absorption Refrigeration Machine

When salt is exposed to humid air, it naturally absorbs moisture and appears to form a molten solution. Then the salt may be called absorbent. There are other substances with stronger powers of absorption than salt, and both lithium bromide and lithium chloride are the most commonly used substances for absorption refrigeration machines. For example, a strong solution of water-LiBr can absorb moisture in the air and it becomes a weak solution.

The strong LiBr solution was put into the container on the left as shown in Fig. 1 and water into the right-hand container. The containers are connected by a tube. Water is evaporated if the containers and tube are evacuated, as water vapour becomes absorbed in the LiBr solution. The water is cooled by the heat lost be evaporation, and is reduced to "chilled water". In this case, the water is called the refrigerant and the container of the refrigerant is called the evaporator. Solar energy is used to boil the diluted solution of LiBr, which must be strengthened so that it can absorb moisture again. This process is called regeneration.

The basic cycle of the single effect absorption machine consists of the four processes of absorption, evaporation, regeneration and condensation. Figure 2 shows the principal features of the systems. [2]

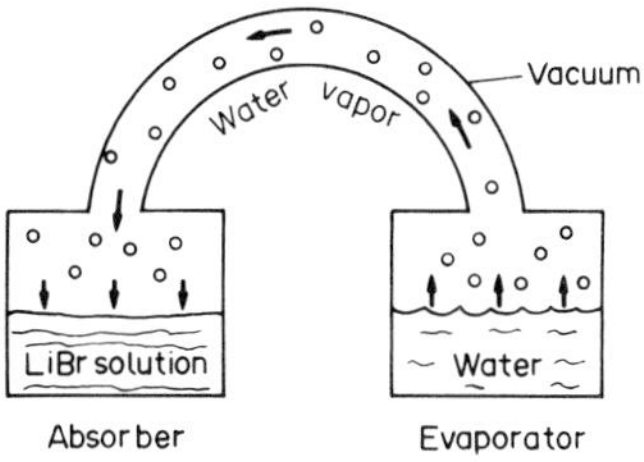

Fig. 1. Absorber and evaporator.

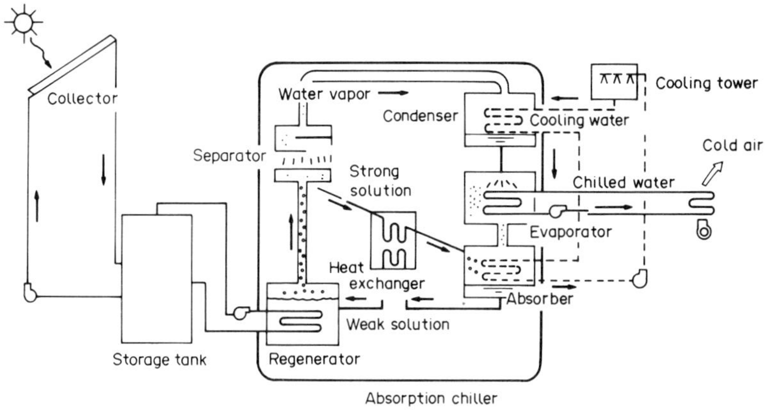

Fig. 2. Solar absorption air-conditioning system.

In the evaporator the water is evaporated under vacuum pressure as it is sprayed on the tubes through which chilled water flows. The latent heat of evaporation is transferred from the chilled water to the water on the outside surface of the tubes, and the chilled water temperature is lowered. This is the cooling effect. The chilled water is then delivered to the heat exchanger coil where the air is cooled to be discharged to the room space for air-conditioning as it is needed. The water to be evaporated and to cool the chilled water is called the refrigerant.

In the absorber the concentrated solution of the absorbent LiBr absorbs water vapour from the evaporator under vacuum pressure. As the solution is sprayed on the cooling tubes within the absorber container, it is cooled down and the vapour pressure is lowered to be capable of absorbing more water vapour. The absorption heat produced there is to be carried away by the cooling water, the temperature rise of the absorber solution being suppressed. The solution is diluted by absorbing water vapour and reduced to a weak solution, which has less capacity for absorbing water vapour.

In the regenerator, the diluted solution led from the absorber is heated and concentrated by dispelling some water vapour, so that the solution may again be capable of absorbing water vapour. Thus this process is called regeneration. The heat must be given from outside of the system and may come from the solar collector, if the temperature of the hot water from solar collector is high enough to let the weak solution boil and then be strengthened. This process produces high temperature water vapour and a strong LiBr solution. Before being brought to the absorber, the strong solution at high temperature is cooled through the heat exchanger with the weak solution of lower temperature from the absorber. This process of heat exchange is also beneficial to the weak solution, because it is afterwards to be heated in the regenerator.

In the condenser, the high temperature water vapour produced in the regenerator is condensed on the surface of the cooling tubes through which the cooling water flows from the cooling tower, thus being changed into water to work as a liquid refrigerant. The liquid refrigerant is to enter into the evaporator by pressure difference, being evaporated to take heat out of the chilled water. The cycle is thus completed.

The cooling water in the absorber and in the condenser enables the water to cool the absorbent in the absorber and the refrigerant in the condenser. While taking heat out of the condenser and absorber, the cooling water is heated and then must be cooled by the cooling tower so that the process can re-cycle. The role of the cooling tower is to cool the cooling water down to the vicinity of the wet bulb temperature of the outside air.

The machine in which the above basic cycle is completed is called a LiBr-water type of absorption refrigeration machine or absorption chiller with LiBr as absorbent and water as refrigerant.

2. Temperature vs. Vapour Pressure in the Cycle

The above processes may be shown on the temperature-pressure diagram with a parameter of concentration of the LiBr solution as shown in Fig. 3.

The lines from A to B and A to C are identified as heated and concentrated processes of the LiBr solution in the regenerator. The diluted solution of 58 % concentration with an entering temperature of 71 °C is heated up to 83 °C and boiled to be concentrated to a 62 % strong solution until it is heated to reach point C at 93 °C at the same vapour pressure of 65 mmHg. The line from C to D shows the process in the heat exchanger. The strong solution at 93 °C is sensibly cooled down to exchange heat with the weak solution of lower temperature.

The lines from D to E and E to F show the process in the absorber. The strong solution of 60 °C enters the absorber and is cooled down to 46 °C by the cooling water with a vapour pressure of 6.5 mmHg. The strong solution absorbs the refrigerant water vapour from the evaporator at a constant pressure of 6.5 mmHg until the solution reaches a concentration of 58 % and a temperature of 42 °C. The line from F to A corresponds to the heating process in which the weak solution rises in temperature from 42 °C to 71 °C by a heat exchange with the strong solution.

The cycle must be completed so that the LiBr solution never crystallizes and the closed loop of A-B-C-D-E-F-A must be drawn above the line of crystallization in Fig. 3.

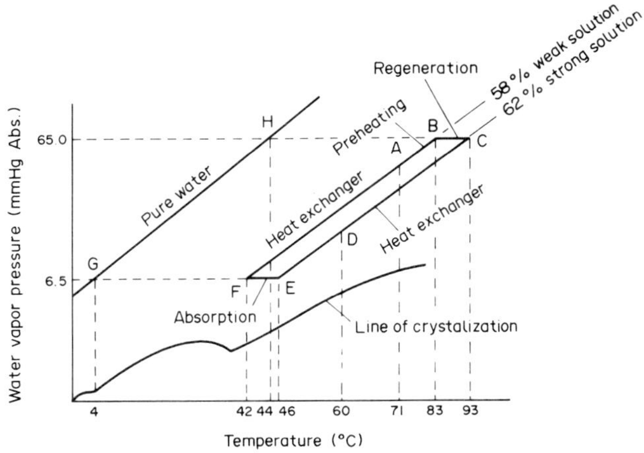

Fig. 3. Pressure-concentration-vapour pressure diagram of single effect absorption cycle.

The line from H to G represents the self cooling process of the refrigerant by evaporation through orifices; the point G is the state of the liquid refrigerant being evaporated at 6.5 mmHg and 4 °C, by which the chilled water is produced.

3. Coefficient of Performance

From the viewpoint of an energy balance, a solar absorption refrigeration machine is one which takes heat out of the room spaces and takes solar energy into the system to reject heat off to the surroundings through the cooling tower. The performance of any refrigeration machine is usually expressed by the Coefficient of Performance, COP. The COP of an absorption chiller is defined as the ratio of the cooling effect to the input energy to the regenerator. The value of the COP would increase with a higher temperature of the input energy to the regenerator, and/or a lower temperature of the cooling water and/or a higher temperature of the chilled water.

Theoretically, the COP may be divided into two components and expressed by the product of Carnot efficiency and refrigeration efficiency with reference to the temperature-entropy chart of Fig. 4 [1]. The upper rectangle of Fig. 4 represents the absorber cycle where the amount of solar energy input to regenerator Q_G[J] at a temperature of T_G[K] minus the heat discharged to environment through the cooling tower Q_C'[J] at a temperature of T'_C[K] is equal to the work L[J] to drive the refrigeration cycle. Thus the Carnot efficiency η_C can be expressed as

$$\eta_C = \frac{L}{Q_G} = \frac{Q_G - Q'_C}{Q_G} = \frac{T_G - T'_C}{T_G} \tag{1}$$

The lower rectangle of Fig. 4 represents the refrigeration cycle where the amount of cooling effect Q_E[J] at the evaporation temperature T_E[K] plus the work L[J] from the absorber cycle is discharged at the condensing temperature T_C[K]. Thus the refrigeration efficiency η_R can be expressed as

$$\eta_R = \frac{Q_E}{L} = \frac{Q_E}{Q_C - Q_E} = \frac{T_E}{T_C - T_E} \tag{2}$$

Then the overall COP of the absorption refrigeration machine may be expressed by the following:

$$COP = \frac{L}{Q_G} \cdot \frac{Q_E}{L} = \eta_C \cdot \eta_R = \frac{T_G - T'_C}{T_G} \cdot \frac{T_E}{T_C - T_E} \tag{3}$$

Substituting the values in Fig. 2 into equation (3), it gives

$$COP = \frac{(273+93) - (273+32)}{273+93} \times \frac{273+11}{(273+32) - (278+11)}$$

$$= \frac{61}{366} \times \frac{284}{21} = 0.166 \times 13.52 = 2.25.$$

This represents an ideal value of COP. In practice, however, the heat exchange losses at every component and other technical difficulties would reduce the COP value of a currently commercialized single effect solar refrigeration machine to the range of 0.65 to 0.5.

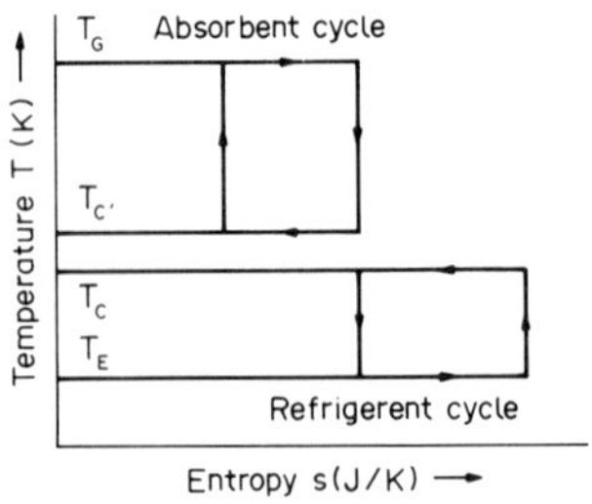

Fig. 4. T-s chart of absorption chiller.

4. Performance Characteristics

In order to raise the COP value of an absorption chiller it is considered appropriate to take the following measures especially for solar air-conditioning. As shown in Fig. 5 [5] it is generally effective to lower the condensing temperature both for absorber and condenser by letting cooling water flow through both components in parallel. It is also important to raise the evaporating temperature by trying to make use of the higher temperature of the chilled water as shown in Fig. 6 [5].

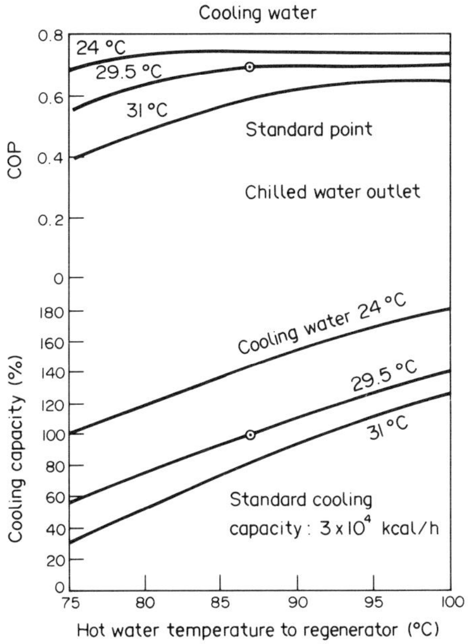

Fig. 5. Performance of small size absorption chiller of water-LiBr type.

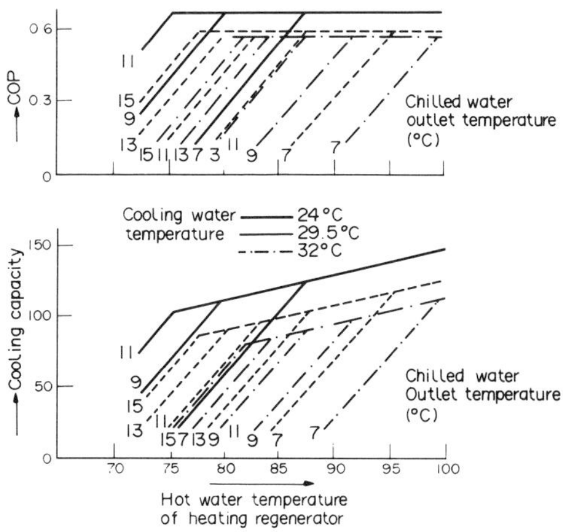

Fig. 6. Performance of absorption chiller with variable concentration of LiBr solution.

Although the COP value increases with higher input energy temperatures to the regenerator, the solar collector performance and solar energy availability decreases with higher collector temperatures. If the machine would work with a lower regeneration temperature, even at a lower value of the COP, the availability of solar hot water in the storage tank would be greater. This measure can be taken by lowering the concentration of the absorbent solution. Figure 6 shows the performance characteristics of an Yazaki absorption chiller especially developed with variable concentration of the absorbent solution by automatic control to a given hot water temperature to the regenerator.

For example, take the cooling capacity at a 13 °C chilled water temperature for an 85 °C hot water temperature against a 29.5 °C cooling water temperature as the 100 % base case, as can be seen in the lower graph of Fig. 6, when the COP is 0.6 in the upper graph. Now if the hot water temperature from the solar tank becomes lower with less solar radiation, heating the generator at only 75 °C, the cooling capacity is reduced to 45 % and the COP to 0.3 for the same conditions of chilled water and cooling water temperatures. In that case, the chilled water temperature was controlled to 15 °C instead of 13 °C as less cooling energy maybe required, the machine can achieve a cooling output of 65 % with a COP of 0.45. It is very important, therefore, to regulate the chilled water temperature to be as high as possible to meet the cooling requirement for effective use of solar energy.

5. Single-Double Effect Convertible Absorption Chiller

As technical developments of absorption chillers allowed for lower regenerating temperatures, as low as 73 °C, the percentage of the solar contribution to air-conditioning became higher. If it worked at a lower COP with the same regeneration temperature when conventional fuel was used as the auxiliary, it would not be considered sensible. Tanaka [4] proposed some years ago an absorption chiller that works in double effect by using fuel at a higher COP and in single effect using solar energy so as to achieve an overall high coefficient of performance.

If during the morning hours the chiller operates under the double effect cycle with fuel at 1.1 COP against the cooling load of 20 kW: for example, the fuel consumption from 9 to 11 a.m. may amount to 36 kWh. Then, if the solar tank temperature became high enough after 11 a.m., the machine can operate under the single effect cycle with solar energy at a 0.6 COP against the cooling load of 30 kW, for example, for 6 hours. In this case, the fuel consumption is 36 kW against the cooling load of 20 x 2 + 30 x 6 = 220 kWh and the overall COP becomes 220 ÷ 36 = 6.1 for the day. The machine, originally suggested by Tanaka, is being manufactured. It was first installed at a new building of the Energy Engineering Department of Oita University, Kyushu. [5] After successful operation for a two year test period, the solar air-conditioning system was operated with steam as the auxiliary fuel. The identical single-double effect absorption chiller was also operated at the Kabe Office Building of the Chugoku Electric Company in Hiroshima. Figure 7 shows the diagram of this type of absorption chiller where the high pressure regenerator for steam is independently located from the low pressure regenerator for solar and hot water vapour from the high pressure regenerator before being condensed. A high pressure generator gives a primary effect and a low pressure one a secondary effect, thus being called a double effect.

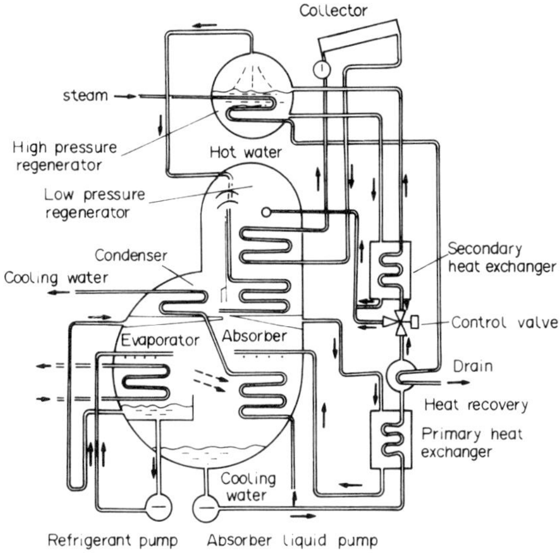

Fig. 7. Single/double effect convertible absorption chiller of water-LiBr type.

Double effect absorption chillers ranging from 2RT (tons of refrigeration) to 10RT in modular designs are also available. If the customer needs a larger capacity than 10 ton chillers, multiple 10 ton units may be installed so that some of them work under the conditions of lower solar intensity and less cooling load by automatic control. The performance of the small sized machines is almost the same as larger ones. They use the bubble pumping effect in transporting the LiBr solution from the regenerator absorber, while larger ones use a circulating pump for that purpose.

6. Solar Air-Conditioning System Design and Control

Different from the conventional air conditioning system, the solar absorption chiller operates at a lower regenerating temperature to make greater use of solar energy than the conventional absorption chiller. It is stressed that every effort must be paid to realize an energy efficient performance for driving the solar absorption chiller as much as possible.

A solar air conditioning system consists of a collector, storage tank, absorption chiller, air-conditioner, piping and duct works.

In order to make a solar air-conditioning system work effectivel it is necessary to design the system correctly both in theory and practice. There are several points quite different from the way in which conventional air-conditioning system design is practised. For example, oversizing the equipment often leads to an unexpectedly lower solar percentage because the available heat source is limited in the case of solar energy systems. Similarly, pipework insulation is particularly important because heat losses through the entire system are equivalent to a reduction in the availability of the heat source. It must be stressed, however, that the pipework insulation for conventional systems is even more important than that for solar systems, because the fossil fuel accumulated in the past is more precious than solar energy available at present. Control strategies are also important to make the system work with greatest efficiency. Care must be taken in designing control systems, because it has been found that over-complicated or sophisticated control system often do not work as anticipated.

7. Optimum Collector Area and Storage Tank Capacity

For typical office buildings, the optimum collector area against 1RT of refrigeration may be 25-30 m² and the optimum water storage tank capacity should be 30-40 litre per square meter of collector area, according to the studies made by long term simulation under Japanese climatic conditions in summer. Note that the size of the solar storage tank is often expressed in terms of litre of water per square meter of collector. This may be roughly interpreted as follows. Suppose that solar radiation in the daytime on a fine day in summer amounts to 600 W/m². A collector of 25 m² area with overall collector efficiency of 0.4 is capable of supplying 600 x 0.4 x 25 = 6000 W of solar energy to the regenerator of an absorption chiller to produce a cooling effect of 6000 x 0.6 = 3600 W. This is nearly equivalent to 1RT of refrigeration.

As solar intensity is lower in the morning, this arrangement would allow the absorption refrigeration machine to start working around 10:30a.m. or 11:00a.m. If excess energy is left over in the solar storage tank, the solar operation may be started earlier. Theoretically a storage tank is not needed for continuous operation if the solar intensity is constant. It is better, however, to provide a small tank of 30 litre of water per square meter of collector area to store solar energy collected during a couple of hours in the morning. This lets the temperature reach the minimum level required for operation of the absorption refrigeration machine. Assuming that the temperature difference between inlet and outlet of the regenerator is 5 °C and the temperature drop during the night 3 °C, an optimum capacity of water to raise the temperature by 8 °C in total would be around 30 l/m² of collector in which solar energy if 280 Wh/m² collected in the morning could be stored to be ready for operation as early as possible.

8. Combination of Collector and Storage Tank for Solar Heating, Cooling and Hot Water Supply System

Optimum tank capacity may be different for cases where space heating and hot water supply are needed in addition to space cooling. Since the temperature required for different purposes are different, it is necessary to provide a proper arrangement of piping junctions on both sides of the storage tank, i.e. the collector side and the load side. Figure 8 shows an arrangement for a vertical type of water storage tank to allow for efficient operation proposed by the author [2].

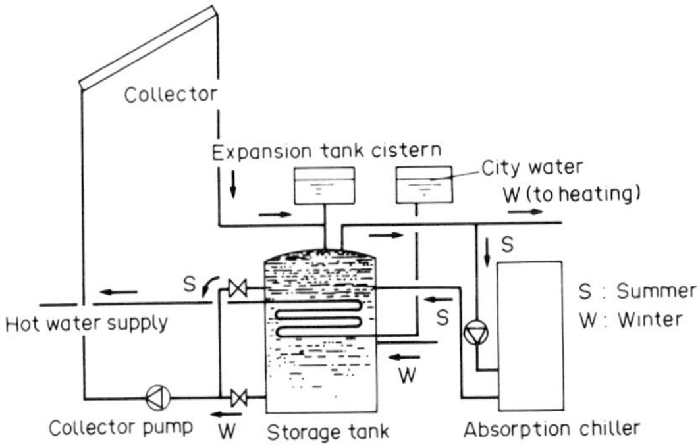

Fig. 8. Piping junctions for solar heating, cooling and hot water supply system.

The upper portion of the tank stores high temperature water to be used for driving the absorption chiller. Both the inlet and outlet piping junctions for the collector and chiller circuits must be located to this higher position so as not to disturb a stable temperature distribution. If the water was taken out of the lowest part of the tank, the return temperature from collector would possibly be lower than the upper part, resulting in complete mixing of the whole tank. The second high temperature is required for the hot water supply and the heat exchanger piping must be located below the upper part lest is should withdraw the higher temperature water for driving the absorption chiller. The water at the bottom should be forwarded to the collector only in winter and the return pipe connection may remain the same as in summer. If is necessary, therefore, to change the outlet piping junctions to the collector twice a year so that more effective solar collection may be realized for the purposes of heating, cooling and hot water supply than with fixed junctions.

9. Combination of Absorption Chiller with Auxiliary Cooling

Provision of auxiliary cooling is necessary for solar absorption air-conditioning especially for commercial buildings as the weather is not always fine. The simplest idea is to install an oil fired or gas fired boiler as shown in Fig. 9 and hot water from the boiler may be fed to the regenerator when solar heat is not available from the storage tank [6]. In this case, precious fuel must be used for driving the absorption chiller at this low value of COP if the chiller is designed only for solar with single effect operation. Compared with a centrifugal chiller in terms of primary energy requirements, the solar absorption system as shown in Fig. 9 would not be as energy saving as a conventional cooling system if less than 72% of total cooling load was supplied by solar energy as illustrated in Fig. 10 [4], [7]. Line A in Fig. 10 represents the primary energy consumption per ton of refrigeration (RT) including parasitic power against percentage of solar for a combination of single effect solar absorption chiller and auxiliary boiler. Line B shows that the primary energy consumption by a combination of single effect absorption chiller and centrifugal chiller driven by electricity is always lower than energy required by a centrifugal chiller. In Figure 10, parasitic power means the electric power required for pumps and fans. Percentage of solar means the percentage supplied by solar energy to the total energy required.

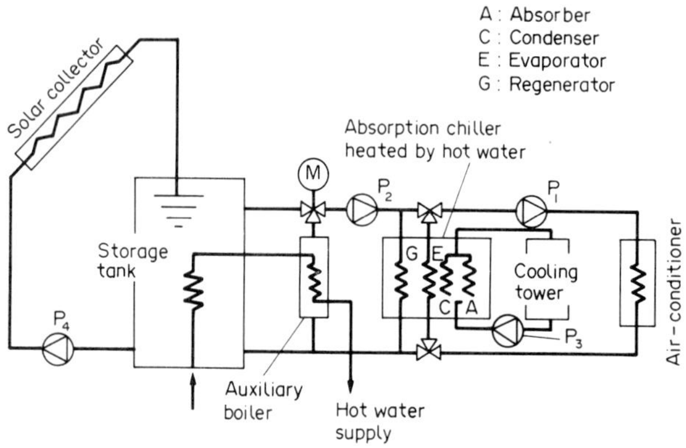

Fig. 9. Standard solar air-conditioning system with single effect absorption chiller heated by hot water.

Otherwise the double effect type of absorption chiller may be used as explained earlier. This machine operates in double effect at a COP around 1.0 when steam was fed to the regenerator, and works in single effect at a COP of 0.6 with solar energy. According to Line C of Fig. 10, the break even point on energy efficiency between solar and conventional is approximately 40%.

10. Parallel Installation of a Solar Absorption Chiller and a Heat Pump

It would be considered more practical, therefore, to install a solar absorption chiller coupled with collector and storage independently of a conventional heat sink source. This arrangement seems to bring higher initial cost but it would be rational to prepare both chillers if partial load should be supplied by solar cooling, because this always achieves a higher energy efficiency than conventional systems for any value of percentage solar. Figure 11 shows a diagram of the Soka Solar House [8]

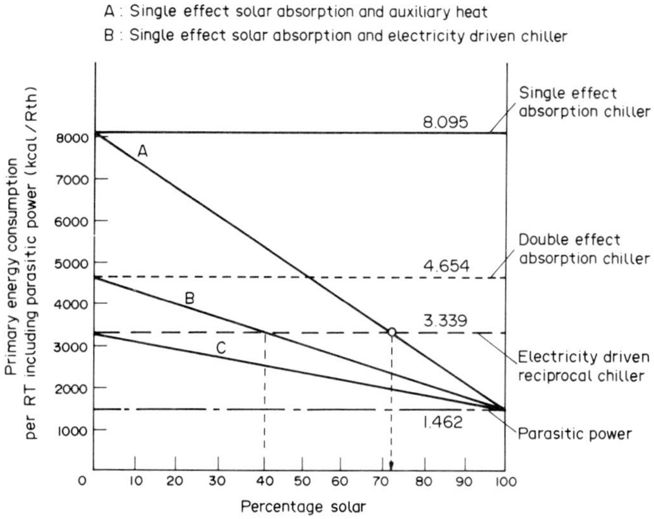

Fig. 10. Relationship between percentage solar and primary energy consumption for cooling.

where a solar absorption chiller and a heat-pump were installed in parallel to produce cold water to be stored in the low temperature thermal storage tank. Figure 12 shows several different solar cooling system diagrams with LiBr absorption chillers installed in Japan [9].

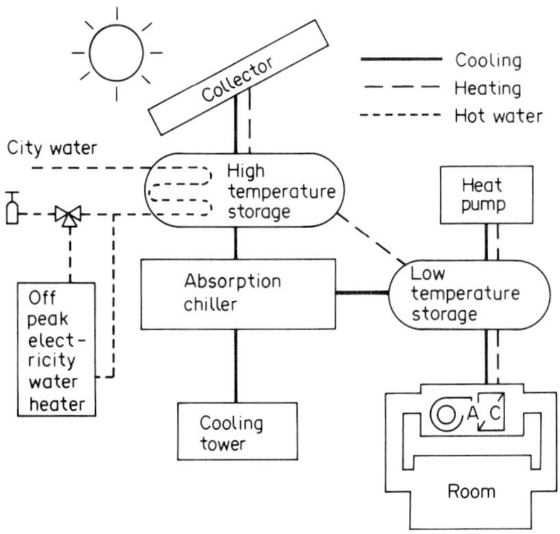

Fig. 11. System diagram of Soka Solar House.

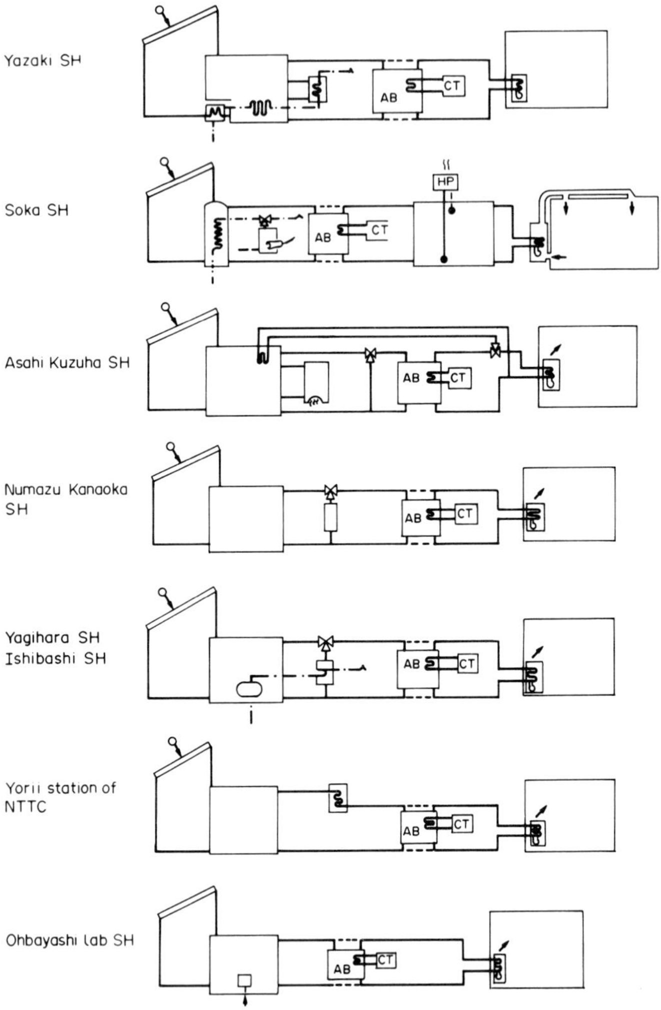

Fig. 12. Solar cooling systems using the absorption refrigeration machine.

11. Efficient Use of Solar Air Conditioning

Because auxiliary heating for the regeneration of the absorbent solution is not energy efficient, a solar absorption air-conditioning system without auxiliary heating may be most sound and economical. This means that the occupants must put up with an overheated environment when the solar system does not work, but the period when the solar system is idle is the period when the cooling load may be less because it is likely to be cloudy.

Natural ventilation may suffice for reasonable comfort during such periods.

If solar cooling is considered as the auxiliary in the total air-conditioning system to share the partial load, it then may be wiser to have the absorption chiller used to precool the outside air intake. Since an absorption chiller works more efficiently at higher chilled water temperatures, the outside air precooling by solar absorption is considered suitable from the viewpoint of the overall energy efficiency of the system. The control strategy in this arrangement of solar and auxiliary would be as described before, requiring an automatic changeover between the two operations.

12. Performance Evaluation of the Total System

The Coefficient of Performance (COP) is used for evaluating the energy performance of refrigeration machines, the value of COP being around 0.6 for solar absorption chillers of the water-LiBr type. For evaluating an overall solar air-conditioning system including collector, storage, chiller, air-conditioner, piping and ducting, the System Coefficient of Performance, SCOP, must be used. The definition of SCOP may be expressed as follows, taking account of all forms of energy on a periodical total basis, such as daily, weekly, monthly or annually;

$$\text{SCOP} = \frac{\text{cooling output (+heating output + hot water supply output)}}{\text{auxiliary energy input + parasitic power input}}$$

For example, if the total cooling load for one month is 10 MWh and 80% of it was shared by solar energy with a parasitic power consumption of 7 MWh, the SCOP may turn out to be:

$$\frac{10}{2+7} = 1.1,$$

whereas a conventional chiller would only give a SCOP of 0.59. When all these items are converted into the values equivalent to primary energy, the ratio may be called a Total System Coefficient of Performance, TSCOP. For example, a parasitic power in kWh must be converted into the primary values using 2.45 Mcal/kWh or 10 MJ/kWh.

Various arrangements of the solar air-conditioning system using an absorption chiller of the water-LiBr type can be provided with specific features. As solar intensity is variable and not predictable, care must be taken to design the system so that the overall performance ratio in terms of cooling output compared to total consumption by auxiliary fuel and parasitic power is always lower than the conventional air-conditioning system. Control strategies are particularly important for the selection of optimum values of set point temperature for all locations in the system. This is as important as the location of piping junctions from an understanding of the overall energy flows.

References

1. Ishibashi, T., Absorption Refrigeration Machine, Solar Energy Application Handbook, Japan Solar Energy Society (1985).
2. Kimura, K., Introduction to Solar House, Ohm-sha, (1980).
3. Yazaki Solar Air Conditioning System Equipment, Yasaki Corporation.
4. Tanaka, S., Solar Heating and Cooling Systems, Ohm-sha (1977).
5. Ishibashi, T., Solar Air-Conditioning System, Fundamentals and Applications of Solar Energy, edited by Japan Solar Energy Society, Ohm-sha (1976).
6. Ishibashi, T., The Operation Results of the Yazaki Experimental Solar House, Silver Jubilee Congress of ISES, Atlanta, (May 1979).
7. Kimura, K., Utilization of Solar Energy, The Japanese Experience, Lim. Ed., Solar Energy Applications in Tropics, D. Reidel Pub. Co., (1983) 187-216.
8. Kimura, K., Solar House Development, International Symposium on Solar Energy for Development, Tokyo (Feb. 1979).
9. Kimura, K., Solar Houses in Japan, Sayigh Ed., Solar Energy Application in Buildings, Academic Press (1979) 287-315.
10. Ishibashi, T., Estimation of Improved Solar Energy Cooling and Heating System, Proceedings of Solar World Congress, Perth 1983, Pergamon Press.
11. Absorption Air-Conditioning and Refrigeration Equipment, ASHRAE HANDBOOK EQUIPMENT Chapter 14, (1983).

CHAPTER 4

DESICCANT COOLING

Harry I. Robison

Professor Emeritus
CCC/USC Energy Laboratory
University of South Carolina
Conway, South Carolina, USA

INTRODUCTION

The term "air conditioning" has come to imply the maintenance of pleasant indoor conditions in buildings even when outdoor conditions are too hot, too humid, or both. In the highly industrialized nations, air conditioning is usually accomplished by electrically driven vapour-compression systems that depend upon non-renewable energy sources.

Many research organizations around the world have developed heating and ventilating systems that produce comfortable indoor conditions in winter yet use only renewable energy sources, chiefly solar energy. A much smaller number had developed comfort cooling systems that are partially powered by thermal energy from solar radiation. These systems are appealing because the building cooling load, unlike the heating load, is nearly in phase with the available solar energy; consequently, energy storage requirements are greatly reduced.

The solar-powered closed cycle cooling systems that have been investigated are 1) solar-assisted vapour compression heat pumps utilizing solar collectors operating at sub-ambient temperature as evaporators, 2) freon Rankine expanders driving vapour compressors, 3) freon-jet compressors, and 4) cooling systems using absorption refrigeration. Although the solar-assisted heat pumps save energy, they still require an electrically-driven compressor. The other three systems all operate at source temperatures well above 80 °C and therefore require expensive concentrating collectors or very large flat-plate collection arrays.

In addition to energy from direct solar radiation, indirect solar energy is available worldwide in the form of the earth's unsaturated atmosphere. Only a few researchers are studying the use of this energy source. Open cycle cooling systems, using water as a refrigerant, can operate solely from the energy of the unsaturated atmosphere. These evaporative cooling systems (direct, indirect, indirect/direct, and regenerative) require no solar energy thermal input and can operate in any region where the atmosphere is only slightly saturated. However, if dehumidification is necessary in addition to cooling, the open cycle air conditioning systems must make use of desiccants to lower the moisture content of the process air before evaporative cooling takes place. Desiccants have a high affinity for water vapour. When heated, desiccants store potential energy in the form of chemical absorption potential or physical adsorption potential. The sorbents lose this potential energy during the dehumidification process as they absorb water vapour. Solar energy may be used to desorb water from the desiccants in a process known as regeneration. For some desiccants, regeneration can be accomplished at temperatures as low as 45 °C. Recently-developed cycles suggest the possibility of cooling and dehumidification by using only the energy of the unsaturated atmosphere; such cycles require no solar-thermal input.

BACKGROUND

Energy Sources and Sinks

Air conditioning systems that use refrigeration are heat pumps that move energy from the conditioned spaces to the outdoors. Heat pumps require energy sources and sinks.

At the present time most air conditioners use electricity from fossil fuel as an energy source to drive a freon-vapour compressor. Some commercial air conditioners use absorption refrigeration systems using waste heat from steam plants; some are fired by natural gas. A few commercial systems use steam jet compressors utilizing waste process steam.

Freon vaporized by solar thermal energy has been used to drive both Rankine expanders, which in turn operate vapour compressors, and also to drive jet compressors directly. Both commercial-size and household-size absorption-refrigeration air conditioners are available powered by solar thermal energy.

Most of the systems described use ambient air as a heat sink to absorb excess energy in the summer. When heating is necessary, vapour compression heat pumps use solar energy collected and stored in the ambient air as a heat source. When cooling is necessary in any climatic region, air source heat pumps are quite effective. However, in regions where the temperature drops below the freezing point, the efficiency of these heat pumps used for heating becomes very low.

Because the earth's temperature at about 10 to 15 meters below the surface is equal to the average annual temperature at the surface of the earth, the earth can be substituted for the atmosphere as a heat source and sink. Therefore, heat pumps employing shallow well-water, where available, are superior in operational costs to other energy sources. Although shallow-well water is a renewable resource, it is not available in many locations. In those regions where ample well water is not available, an earth-coupled heat exchanger can be fabricated by using buried polybutylene tubing. In the past five years over 20,000 residential and commercial closed-loop, earth-coupled heat-pump systems have been installed in the United States, Canada, Sweden and Germany. These systems produce savings in excess of 50% for heating and greater than 25% for cooling in comparison to other heating systems.

Open Cycle Cooling Systems

Direct Evaporative Coolers

Direct evaporative coolers became quite popular in the southwestern United States during the 1930's. Cooling of the process air is accomplished by using water as a refrigerant in an open cycle. The refrigerant (water) has been considered by the atmosphere, so no compressor is required. When the water is brought into direct contact with the atmosphere, evaporation and the resulting adiabatic cooling of the process air take place.

Closed cycle, Rankine, jet, and absorption systems, in which the compressed refrigerant is cooled in a condenser and therefore can not be absorbed by direct contact with the atmosphere, can not take direct advantage of the readily available energy of the unsaturated atmosphere. Many commercial systems use the earth's unsaturated atmosphere indirectly by discharging heat to a stream of water which in turn is cooled in an open cycle by direct contact with the atmosphere in a wet cooling tower.

Direct evaporative coolers lower the dry-bulb temperature of the process air, but unless the wet-bulb temperature of the process air is unusually low, the final high moisture content of the cool air results in conditions that are only marginally improved. Because direct evaporative coolers lower only the temperature (remove sensible heat) and can not lower the moisture content (remove latent heat), they can not be used when the atmosphere is near saturation and therefore are limited to use in arid or semi-arid regions. The primary drawback of this method of cooling in the arid regions where it operates best is the requirement for water, which is usually precious in arid climates. Each ton (3.5 kW) of cooling requires six litres of water.

Two and Three Stage Evaporative Coolers

During the late 1930's several new cooling cycles were devised. Indirect evaporative coolers evaporate water into an airstream, and the resulting cool but wet air is used to sensibly cool the process air. These coolers operate well in overheated and temperate non-arid regions.

If the cool, dry air of an indirect cooler is, in turn, evaporatively cooled by the addition of water to the airstream, the process air will have an even lower temperature. This is an indirect/direct evaporative cooling cycle. Like the direct cooler, it produces a very humid airstream, and it requires an even greater amount of water.

Regenerative Evaporative Coolers

Regenerative evaporative cooling equlipment can be used even in slightly humid overheated or temperate regions. In this application building air is exhausted through the evapoartive cooler. The resulting wet, cool exhaust airstream is used to sensibly cool the incoming air. The process airstream is cool and dry.

While direct evaporative coolers could not obtain temperatures below the wet-bulb temperature of the process airstream, the advanced systems could achieve temperatures very close to the dew-point temperature, thus resulting in a lowering of the enthalpy of the process airstream without a corresponding expenditure of thermal energy. The thermal COP of these machines is quite high and can be infinite. Even when the small amount of energy needed to operate fans and pumps is considered, the COP is extremely high. However, the more complex cycles were never able to compete economically with simple evaporative coolers or even with conventional vapour compressors. Volume production of vapour compressors reduced the initial cost; moreover, the reliability of these machines combined with the availability of cheap electricity produced from oil- and coal-fired power plants made it impossible for the indirect evaporative coolers, which required huge plate air-to-air heat exchangers, to compete with the very reliable vapour compression system. Even though the evaporative coolers could actually reduce the enthalpy of the process air, they could not dehumidify, and some dehumidification is necessary in most areas of the world in order to overcome excess internal moisture and produce comfortable conditions.

Dehumidification

Excess internal moisture can result from the following sources: from occupants; from combustion and cooking; from washing and drying clothes; from showers, wet surfaces, and standing water; from infiltration air; and from water vapour permeating walls, floors, and ceilings. Additionally, ventilation and make-up air add to the internal moisture load. Control of excess moisture is absolutely essential, not only for human comfort but for preventing condensation on or in materials within the building. Moisture can react with many inorganic materials, resulting in inorganic corrosion. The moisture absorbed or adsorbed by hygroscopic organic materials may cause organic corrosion, severe deterioration, rotting, and mold or mildew formation, as well as rapid bacterial growth. By removing excess water from the indoor air, dehumidification prevents surface condensation or moisture retention within materials.

Whenever warm and humid air contacts any surface-including water-that has been cooled below the dew-point temperature of the air/water vapour mixture, dehumidification takes place. Water will condense on the cooled surface, thus drying the air/water vapour mix.

Often those people who could best benefit from dehumidification techniques are unaware of the existence of these useful techniques: dehumidification by use of surface water, by refrigeration, and by desiccants.

Dehumidification by Use of Surface Water

An often overlooked renewable resource is surface or shallow-well water. This totally renewable resource should not be confused with water from deep wells that are fed by non-renewable aquifers. Those regions most in need of dehumidification usually have abundant surface water, for the very abundance of surface water is the cause of the saturation of the local atmosphere.

When shallow-well water is available and the temperature of the water is below 13 °C, both cooling and dehumidification can be accomplished by using this renewable resource. Coils are not necessary. Highly efficient direct contact with well-water sprays, fountains, or wetted packing can be used to cool and dehumidify the process air.

Dehumidification by Refrigeration

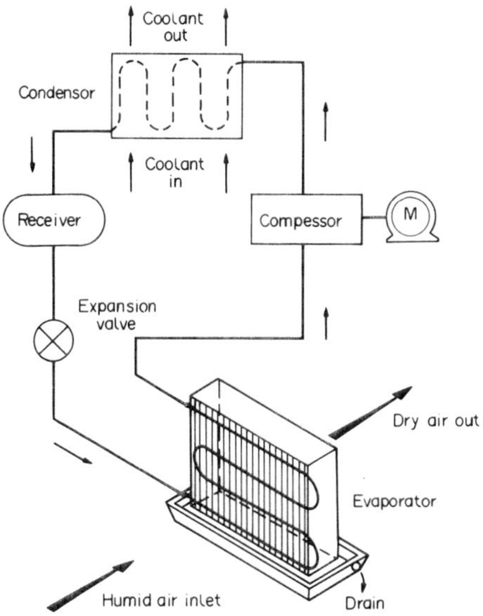

Fig. 1. Vapour-compression dehumidifier.

A common method of dehumidification in highly industrialized nations utilizes a vapour compression system as shown in Fig. 1. This process is illustrated as steps 1-2-3 in Fig. 2. Freon vapour is first compressed by a compressor driven by electricity. The compressor raises both the pressure and the temperature of the freon vapour, which must then be cooled by a fluid (air or

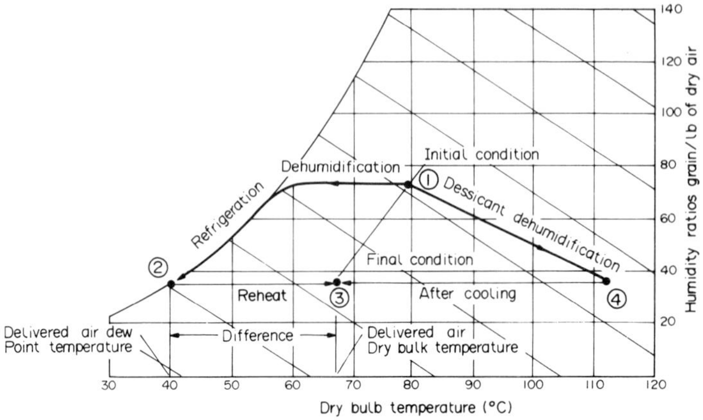

Fig. 2. Dehumidification by refrigeration and by desiccants.

water) as the freon passes through the condenser. The compressed vapour discharges energy to the cooling fluid and condenses to a liquid still under high pressure. The liquid freon is then allowed to expand, and the refrigerant, cooled by expansion, flows through a finned evaporation coil. Humid air passing over the cold evaporator fins is cooled below the dew-point temperature of the airstream.

Moisture condenses on the surface of the evaporator fins. The condensing water vapour gives up energy which continues to evaporate the freon liquid that is under a low pressure. Although the process airstream is saturated at the low temperature at which it leaves the wet evaporator coil (relative humidity is 100%), the process air is much drier, as indicated by the reduced humidity ratio or specific humidity. As the process air is reheated (often by using the process airstream as the condenser cooling fluid), the relative humidity falls, but the amount of moisture as indicated by the humidity ratio does not change with the change in temperature.

It is important to understand that as long as the air is not completely saturated, raising or lowering the dry-bulb temperature does not affect the moisture content of the air. The often-used parameter called ralative humidity, which depends upon both moisture content and temperature of the air, can not be used alone to indicate the moisture content (humidity ratio or specific humidity) of an air sample. A change in relative humidity can be used to judge a change in the moisture content of an air sample only in those rare cases in which the temperature of the air sample does not change. Also, relative humidity is not an indicator of human comfort. The body's temperature is regulated by conduction, convection, and evaporation. Conduction and convection depend upon air temperature and airflow rate. Evaporation depends upon the partial pressure of the water vapour in the air and upon the airflow rate. The partial pressure of the water vapour in an air sample depends upon the amount of moisture in the sample and can be expressed as the humidity ratio, the partial pressure of the water vapour in the sample, or the dew-point temperature of the mixture of air and water vapour. All of these measurements are independent of the temperature at which the measurements are made and are, therefore, to be preferred to relative humidity for expressing the moisture content of an air/water sample. However, relative humidity is useful in determining the relative saturation of any air sample at the temperature at which the relative humidity is measured.

Although dehumidification by refrigeration is quite effective at high temperatures, it becomes more difficult at lower temperatures. A drop of one degree in the temperature of saturated air at 7 °C removes less than one-third of the amount of water vapour that a corresponding drop removes from 29 °C saturated air.

Dehumidification by Desiccants

Another, but lesser known, method of dehumidification is that of exposing the moist process airstream to a desiccant, a material which has a high affinity for water vapour. Moisture is extracted from the process air and collected by the desiccant. Later the desiccant is heated, driving the collected moisture into a scavenger airstream, which is then exhausted. The dry desiccant can be reused.

Some solid desiccants owe their water-retention properties to surface adsorption and capillary condensation. Absorbents undergo no physical or chemical changes when picking up moisture. Silica gel, alumina, and molecular sieve are examples of such solid adsorbents.

Some solid materials such as the anhydrous salts of metal halides combine with water by absorption to form hydrates. If even more water is absorbed, the hydrates form liquid desiccants.

Another group of desiccants is composed of liquids such as sulfuric acid, sodium hydroxide, and triethylene glycol, all of which are strong absorbents of water vapour.

For use in dehumidifying, sorbents should be stable, non-toxic, oderless, non-corrosive, non-flammable, and economical. They should possess good heat transfer characteristics. Additionally, if the desiccants are hygroscopic solutions, they should have a low vapour pressure at the temperature at which they contact the process airstream, and they should not crystallize in the operating range of temperatures. If solar energy is to be used to remove the absorbed or adsorbed water, the desiccant should be capable of being regenerated at temperatures well below 85°C.

Figures 3, 4 and 5 illustrate three types of dehumidification equipment available commercially. The use of solid desiccants is shown as steps 1-4-3 in Fig. 2. Both the dual-bed packed-

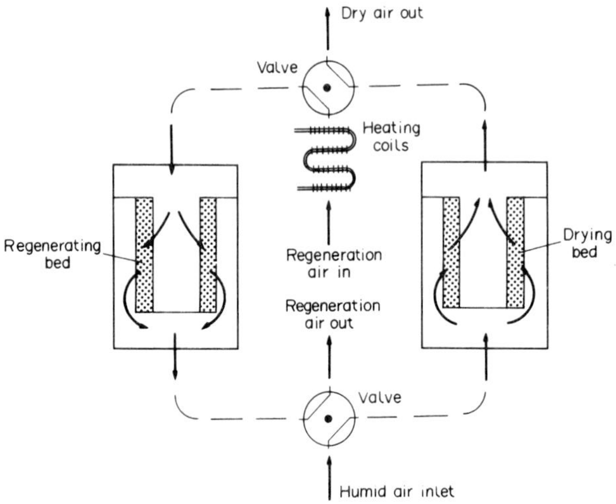

Fig. 3. Dual-bed packed-tower dehumidifier.

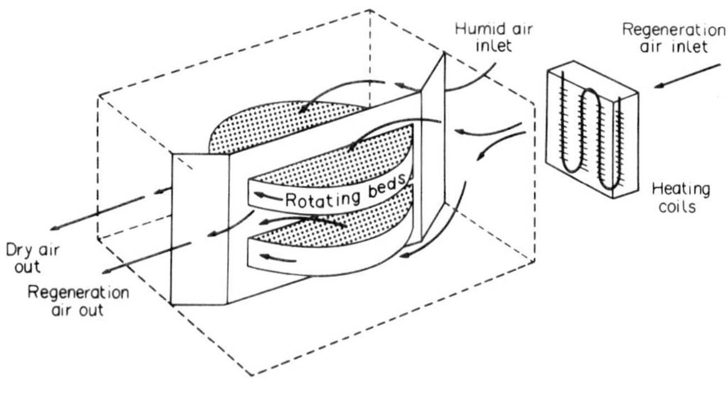

Fig. 4. Rotary-bed dehumidifier.

tower system (Fig. 3) and the rotary-bed system (Fig. 4) employ silica gel or activated-alumina adsorbent beds. A system such as that in Fig. 5 is available with the honeycomb drying wheel impregnated with lithium chloride as an absorbent, or with molecular sieve as an adsorbent.

The difference in the vapour pressure between the water vapour in the airstream and that of the trapped water on or in the solid desiccant creates the driving force moving the water vapour. If the desiccant is cooled, the resulting vapour pressure of the moisture in the process air stream is greater than the vapour pressure of the water held by the desiccant, so water will be driven from the air into or onto the desiccant; dehumidification of the air will result. Conversely, when the desiccant is heated, the vapour pressure of the water held by

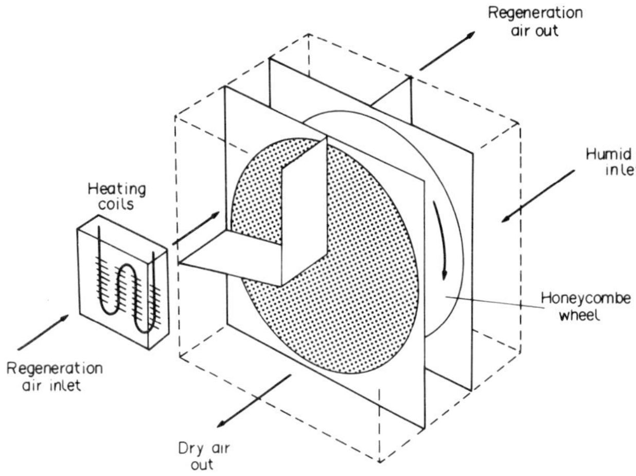

Fig. 5. Desiccant-wheel dehumidifier.

the desiccant exceeds the vapour pressure of the moisture in the air and water is driven from the desiccant into the scavenger airstream; the desiccant is desorbed or regenerated. Unless the process airstream can be used for some purpose which can utilize hot as well as dry air, the process air must be cooled before it can be used. Additionally, before the desiccant can be reused, it must be cooled in order to establish a pressure difference between the vapour pressure of the moisture in the process airstream and the vapour pressure of the remaining water trapped in the desiccant.

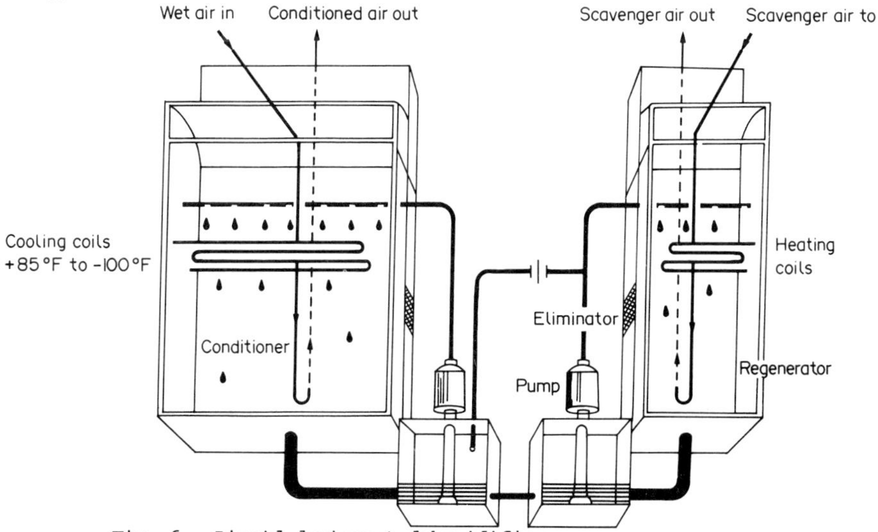

Fig. 6. Liquid-desiccant dehumidifier.

Figure 6 illustrates a liquid absorbent dehumidifier. Triethylene glycol and solutions of calcium and lithium chloride are commonly used as the desiccant for these systems. The operating cycle of a liquid dehumidifier is shown as steps 1-3 in Fig. 2. The liquid systems have the following advantages: continuous cooling during absorption by removal of not only the heat of sorption but sensible heat as well, thus resulting in very high mass transfer rates; heat and mass transfer from a liquid surface, resulting in low fan-power requirements; direct heating of the desiccant for reconcentration, resulting in high reconcentration efficiency at very low temperatures; utilization of liquid-to-liquid heat exchangers, resulting in a partial recover of energy used for reconcentration of the liquid sorbent; microbiological decontamination, resulting in removal of airbourne viruses and bacteria; and, probably the most important advantag of all, if solar energy is utilized, the ability of the liquid desiccant to store chemical potential energy in an uninsulated tank for an indefinite period.

SOLAR-POWERED DESICCANT AIR CONDITIONING

Solar-powered desiccant air conditioning systems use a desiccant dehumidifier, a heat sink, and an evaporative cooler in an open cycle to both dry and cool the process airstream. Solar energy is used to regenerate the solid desiccants or to reconcentrate the liquid desiccants that are used in the dehumidifier.

As early as 1955, Pennington in the U.S.A. patented a cooling system using a rotary desiccant wheel and a heat exchanger wheel coupled with a double regenerative evaporative cooler that removes sensible heat as well as the heat of sorption. In the same year, also in the U.S.A., Löf proposed a solar space-cooling system in which room air is dehumidified by using triethylene glycol in an absorption chamber, which is cooled by cooling tower water; then the dehumidified air is evaporatively cooled. Solar-heated air is used to reconcentrate the dilute glycol in a stripping tower.

Another early attempt to cool a building in Liberia by using solar energy was described by Dannies in 1959. The "drying house" contained desiccant-blocks of wood on the east and west walls. In the morning, the east wall was exposed to the sun so that the heated sorbent wooden blocks could be dried by the room exhaust air while the incoming air was being dehumidified as it passed over the blocks on the west wall. During the afternoon, the role of each of the walls was reversed. A water bath was installed at the base of each wall to evaporatively cool the incoming dry airstream. This system lacked a vital component, a method of adequately removing the heat produced by the drying process (heat of sorption) before the dry air was evaporatively cooled. If this is not done, the airstream merely returns to its previous condition before the dehumidification took place. Dehumidification removed the moisture but heated the air. Humidification cooled the air back to the original temperature but also replaced the moisture that had been removed.

This result demonstrates the least understood fact about open-cycle systems designed for human comfort, the fact that humidification or dehumidification alone is usually quite ineffective in producing comfortable conditions. Both direct-evaporative cooling by humidification and moisture removal by desiccant dehumidification are adiabatic processes. Although the temperature and moisture content of the air change in both cases, no enthalpy is added or removed from the process air in either case. Desiccants alone can not produce comfortable conditions. Additionally, some means must be devised to remove the heat of sorption produced by the dehumidification process.

Solid Desiccant Systems

In Australia in 1965 Dunkle proposed a method of air conditioning that used solar air collectors and a rock-bed energy storage system. The process airstream was dehumidified by passing it through a slowly turning silica-gel desiccant-bed. Two rotary heat recovery wheels, in conjunction with evaporative coolers, cooled the process airstream by discharging heat to an outside airstream. The preheated outside airstream then passed through a solar air heater. The hot air from the solar collectors became the scavenger airstream, which regenerated the rotary desiccant bed and was exhausted outside the building.

In 1968 Munters in Sweden greatly improved the performance of the Pennington desiccant wheel by introducing a desiccant matrix with parallel passages. Using this improved desiccant wheel, the Institute of Gas Technology (IGT) in the U.S.A. built a number of units that used natural gas to regenerate the desiccant wheel. After a rapid increase in energy proces in the early 1970's, IGT redesigned the desiccant air conditioner to utilize solar energy. The two systems at IGT developed by Rush and his associates are shown in Figs. 7 and 8.

Several complete experimental operating systems utilizing solid desiccants were built by IGT, Molecular sieve was used as the desiccant in the dehumidification wheels for these systems. Low-temperature solar energy was used to do most of the regeneration of the sorbent wheels. A natural gas burner was installed downstream from the solar hot water coil, and the high-temperature exhaust was used directly to efficiently complete the regeneration process. The sulfer-containing exhaust from burning natural gas damages a lithium chloride desiccant wheel, thus the choice of a molecular sieve, which does not absorb sulfur compounds. Molecular sieve can achieve a process airstream with a much lower vapour pressure than can a wheel impregnated with lithium chloride. However, the lower vapour pressure can be achieved only by using higher ption temperature than can be achieved from flat plate solar collectors alone.

In the early 1980's both Exxon in the U.S.A. and Sharp in Japan conducted research with similar cycles. Also in Japan, Toyobo developed latent heat wheels using carbon and fiberglass fibers, while in the U.S.A. AiResearch experimented with a thin rotating drum packed with silica gel. A more recent method for continuously removing the heat of sorption from a desiccant bed is shown in Fig. 9. This cross-flow dehumidifier/heat exchanger

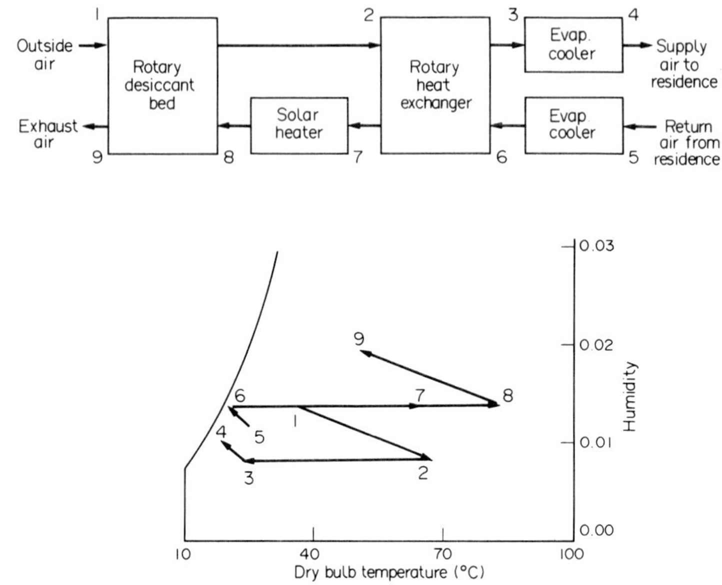

Fig. 7. The ventilation cycle desiccant cooling system.

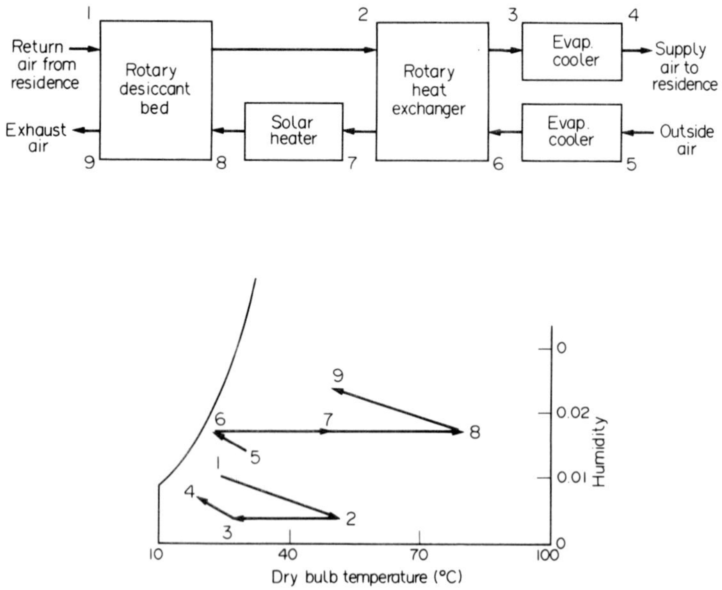

Fig. 8. The recirculation cycle desiccant cooling system.

uses a fluffy layer of Teflon impregnated with silica gel or molecular sieve to remove moisture. The heat produced by the drying process is removed by a cross-flow scavenger airstream. If silica gel is used as the desiccant, the dehumidifier designed by Lavan and co-workers in the U.S.A. is regenerated intermimittently by solar energy.

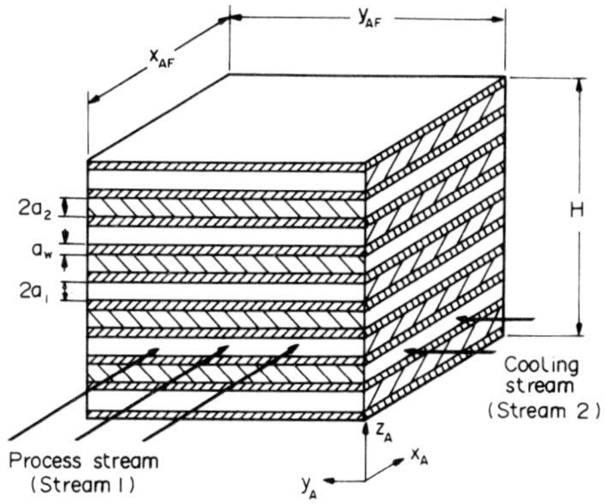

Fig. 9. Schematic of cross-cooled dehumidifier.

As the success of the solid desiccant systems depends upon efficient recovery of the heat of sorption in the heat recovery wheel, many investigators have devoted a good deal of time to improving the effectivity of the heat exchange wheels. Most of these wheels used in the various systems have an aluminium matrix with parallel passages. An improved heat exhange wheel has been developed in Australia. It consists of a polyester ribbon spirally wound to form parallel-walled laminar-flow passages.

Component and Systems Analysis

Early efforts to improve the COP of regenerative dehumidifier cycles were hampered by the lack of satisfactory analytical methods of predicting their performance. Maclaine-cross, Banks, and their associates in Australia have introduced combined potentials, analogous to specific heat ratios, in order to predict combined heat and mass transfer in solid desiccant dehumidifiers. These quantities can be calculated from the vapour pressure isotherms, the specific heat of the matrix, and the thermodynamic properties of moist air. Maclaine-cross has added the effect of a Lewis number other than unity in the analogy theory, and his program, MOSHMIX, has been widely used by subsequent investigators. Mitchell's research group at the Solar Energy Laboratory at the Univ. of Wisconsin has continued to improve the understanding of the solid desiccant cycles. Jurinak, after simplifying the analogy theory for rapid calculation, has studied the effect of isotherm shape, maximum water content, heat of adsorption, regenerative matrix thermal capacitance, matrix moisture diffusivity, adsorption hysteresis, and purge rate on dehumidifier performance. Nelson used the analogy theory to predict performance of complete cycles in several different locations and, more importantly, proposed using a humidifier matrix consisting of silica gel separated by uniform air passages.

Both UCLA and the Solar Research Institute (SERI) investigators have measured effective Lewis numbers of silicon gel packed-beds between 3 and 30, indicating expressive mass transfer resistance. In order to reduce this resistance, researchers at UCLA designed and tested a fixed-bed dehumidifier constructed of lexan sheets coated with fine silica gel particles.

Schlepp and associates at SERI subsequently developed a laminar-flow, parallel passage dehumidifer using a spirally-wound polyester ribbon wheel (similar to the Australian-designed heat exchange wheel) coated with fine silica gel particles (similar to those used by the UCLA research group). The small particles greatly reduced the mass transfer resistance (effective Lewis number is 1.2), and the performance of the desiccant wheel now approximates that of the matrix proposed by Nelson.

Exxon has also developed a parallel passage dehumidifier wheel. In order to produce a higher COP, scientists at Exxon have taken a different approach from that taken at SERI. Instead of attempting to lower the mass transfer resistance, they have chosen to add heat capacity to the desiccant wheel by adding an inert material to the matrix along with the silica gel. This allows a higher COP but greatly increases the cost of the machine by substantially reducing the capacity of the desiccant wheel.

avan and associates conducted a study of the open cycle solid desiccant cooling systems and performed a second-law analysis of the operation of these machines. The following conclusions were reached:

1) The COP increases with decreasing outdoor humidity ratio, with improved heat exchangers, and with increased evaporative cooler effectiveness.
2) The performance in the ventilation mode is better than the performance in the recirculation mode in dry climates.
3) In wet climates, the recirculation mode gives better performance than does the ventilation mode.
4) At ARI conditions, the performance in the recirculation mode is slightly better than the performance in the ventilation mode.
5) The performance of iso-thermal or cooled-bed desiccant systems is higher than that of adiabatic dehumidifiers.
6) The maximum possible reversible COP for open-cycle desiccant systems at ARI conditions is 4.66 as compared to 1.86 for closed cycle systems at the same conditions.
7) The open-cycle COP approaches infinity at low outside humidity ratios.

SERI computer simulations predict that a system with a 1-m-diameter laminar-flow parallel passage silica gel wheel and a 95% effective heat exchanger wheel would have a capacity of 4.7 kW with a thermal COP of 1.07 and an electric COP of 7.9 at ARI conditions.

Computer simulations show the Exxon system to have a potential to reach a COP of 1.3. Actual systems have attained a thermal COP of 1.05 with a 0.76-m-diameter desiccant wheel and a 93% effective heat exchange wheel. However, the cooling capacity at this COP was only 2.4kW.

The third generation desiccant air conditioner from IGT is operating with nonuniform regeneration temperature distribution and unbalanced process and regeneration flows. At ARI conditions the thermal COP is only 0.5 and the electrical COP is 8.1. However, its capacity is over 9.5 kW. IGT predicts that the design thermal COP can be increased to 0.95 by increasing the heat exchanger effectiveness from 0.9 to 0.95. The predicted performance of the laminar-flow, parallel-passage dehumidifier designs of SERI and Exxon is nearly double that of current state-of-the-art systems. A gas-fired system with this performance could be competitive with electric vapour compression systems. However, if solid desiccant systems are to be solar-powered, either a higher COP, reduced capital costs, or both, will be necessary in order for these systems to become competitive with conventional systems.

Licensed by Exxon in 1982, American Solar King is field-testing at eight sites, mostly in the southern part of the U.S.A., a residential solar regenerated desiccant cooling system, which is packaged with eight flat-plate collectors and a storage tank. The company thinks that it can meet the target cost of $10,000 for the system, said to be about the size of a conventional 10.2 kW unit. No performance details have been made available from laboratory or field tests.

In 1977 Lunde suggested the use of a hybrid system consisting of an absorption chiller and a desiccant dehumidifier as the best way to utilize solar energy for cooling. Closed cycle air conditioning systems lower the temperature (remove sensible heat) directly, but are forced to overcool the process airstream in order to indirectly remove moisture (remove latent heat) by condensation on the evaporator coils. Desiccant air conditioners remove moisture directly, but are forced to overdry the process airstream in order to indirectly cool the air by evaporation of water back into the airstream. The ideal air conditioning system would be a hybrid unit consisting of a closed cycle refrigeration system to remove sensible heat and an open cycle desiccant system to remove layent heat. Each system would be controlled independently of the other, and the waste heat from the closed cycle system could be used to regenerate or reconcentrate the desiccant. The open cycle desiccant systems can be powered very efficiently by 60-80°C solar thermal energy. However, the freon jet, freon Rankine, and lithium bromide absorption closed cycle systems all require higher temperature sources, and consequently attempts to build a practical hybrid system that can be totally powered by solar energy have so far been unsuccessful. In 1978 Robinson suggested the use of a hybrid system consisting of an electrically powered vapour compressor and a desiccant dehumidifier. The waste heat from the vapour compressor would be used to regenerate a solid desiccant wheel. Since that time, both Northrup and Griffiths have received patents describing the details of such hybrid systems, and Griffiths and Meckler have described a number of commercial systems.

The Gas Research Institute (GRI) is conducting one of the larger gas-fired cooling research efforts in the United States. All hardware development projects currently involve HVAC manufacturers on a cost-shared basis. GRI feels that integrated dehumidification/chiller systems that separate latent and sensible load control show good near-term potential. GRI is also looking into the technical and economic feasibility of combining desiccant dehumidification with passive and other sensible cooling techniques for residential buildings. Two parallel development projects for prototype units are in place, each involving a major HVAC manufacturer.

Fairey at Florida Solar Energy Center is analyzing a proposal to use night-sky radiation as the heat sink for a solid desiccant dehumidification system. While radiation to the night sky works well in the southwestern United States, where daytime roof and building temperatures are high and where the night sky is very clear, in places such as Florida, the humid air acts as a blanket between roof surfaces and the night sky. To overcome the effects of high humidity and the small temperature difference between the roof and the night sky, FSEC proposes the use of heat of sorption given up as process air is dehumidified in a shallow silica gel bed to raise the temperature of the metal roof radiator.

The building would have well-insulated concrete block walls with internal cores as cooling ducts. See Fig. 10. The thin silica gel bed would be contained on the floor on a plenum with the metal roof as the upper surface of the plenum. During the day solar energy would be used to regenerate the desiccant. Outside air would carry away the evaporated moisture. At night, room air would be recirculated through the plenum. The desiccant would dehumidify the process airstream, and the heat of sorption would be radiated to the night sky. For this plan to be successful, the well-insulated massive walls must be capable of storing not only the daytime heat load but must also be capable of absorbing the excess water vapour from the indoor air during the daytime. Although the building material and the furnishings have a high moisture capacity, if they are allowed to absorb water vapour during the daytime, the resulting heat of sorption will still lead to uncomfortable daytime conditions.

Maclaine-cross, in 1985, proposed a new ideal adiabatic regenerative open cycle that could require no thermal energy input at ARI conditions. The maximum possible reversible COP would be infinite. This cycle depends upon several very effective evaporators and heat exchangers. Actual machines will certainly require some thermal input.

Liquid Desiccant Systems

In 1969 Kakabaev and co-workers in Russia developed a solar-powered refrigeration unit (chiller) using a mixture of lithium chloride and water as the absorbent solution. The desiccant solution was reconcentrated by solar energy as the dilute liquid

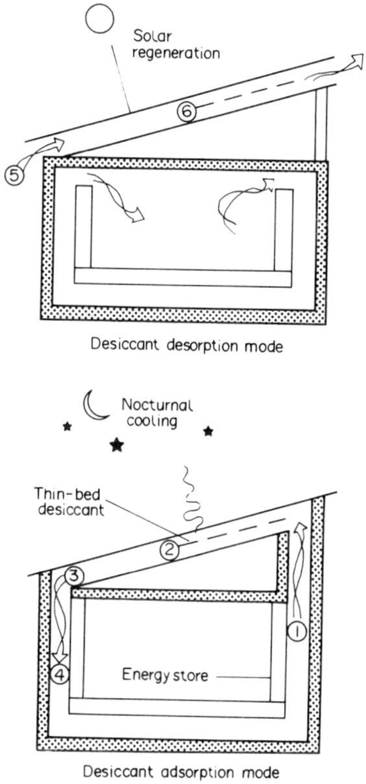

Fig. 10. Schematic diagram of operating modes of the desiccant enhanced nocturnal radiator system.

was allowed to flow down the roof in a thin film. See Fig. 11. No glazing was used. However, only the concentrator/collector operated in an open cycle. The refrigerant (water) was vaporized at a low pressure in the evaporator and was absorbed by the concentrated desiccant solution in the absorber in a closed system. A cooling tower removed the heat produced in the absorption process. In order to remove the air picked up by the sorbent solution in the open reconcentration cycle, it was necessary to continuously evacuate the system.

In 1974 Gupta and associates in India devised a solar collector/concentrator for a calcium chloride/water sorbent solution that could be used in a liquid desiccant cooler. The dilute solution, heated by solar energy, flowed down a black roof under a single sheet of glass while ambient air entered at the roof eave and thermosyphoned upward. After removing moisture from the dilute desiccant solution, the scavenger airstream exhausted at the roof crest.

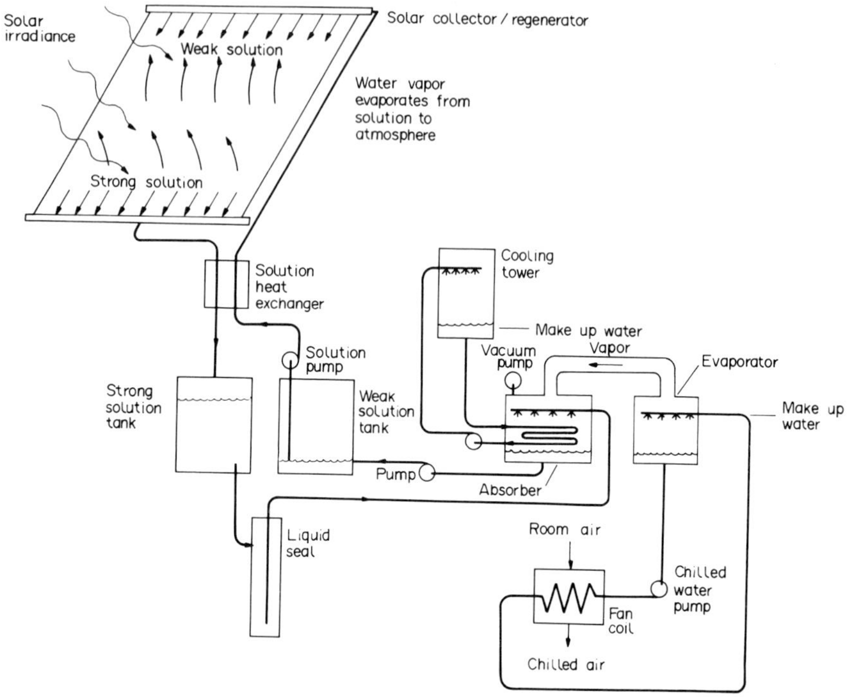

Fig. 11. Schematic of open-cycle absorption refrigeration system.

In 1977 Robison in the U.S.A. built and tested several collector/concentrators similar to that designed by Gupta. Triethylene glycol was chosen for the desiccant because it is non-corrosive and non-toxic. Although the vapour pressure of glycol is very low, it is not zero, and because of carry-over while being re-concentrated, this desiccant was unsatisfactory for use in the open cycle concentrator. Calcium chloride was substituted for the glycol and proved to be highly satisfactory for operation in a desiccant cooler.

Some advantages of the open cycle collector/concentrator are as follows:

1) The rooftop acts as both collector and concentrator, resulting in low collector and no concentrator costs.
2) Solar energy directly heats the dilute desiccant solution, resulting in concentration at low temperature.
3) Water is evaporated to the unsaturated atmosphere where vapour pressures are much lower than those in the condenser of a closed cycle machine, resulting in efficient concentration.
4) Air is moved through the concentrator by natural convection, and 3.4 kW of cooling requires only one litre per minute of dilute sorbent flow, resulting in very low parasitic

power requirements.

Liquid desiccants have a unique property: the ability to store chemical potential energy at high energy densities at ambient temperature. The energy density of concentrated alkali-metal salt solutions is greater than that of heat storage systems employing water, rocks, or ice and other liquid/solid phase-change materials, and is even somewhat greater than the energy density of lead-acid batteries. As conventional solar heating systems store energy as sensible heat, and therefore require a large energy store, a solar heating system to be used in conjunction with a liquid desiccant cooling system should also take full advantage of the desiccant's unique ability to store energy as chemical potential energy. Therefore, in 1978 Robison designed an open cycle chemical heat pump (liquid desiccant heat pump).

This open cycle desiccant heat pump can heat, cool, and humidify or dehumidify air, and it can also heat domestic water. It is a heating, ventilating, and air conditioning system that uses no freon, no compressor, no gas-fired generator, no condenser or evaporator coils, and no vacuum or pressure systems. The machine uses water as the refrigerant, a mixture of calcium and lithium chloride as the absorber, and the process airstream as the heat transfer medium. The heat pump is driven by solar energy stored as chemical potential energy in a hygroscopic liquid. During the cooling cycle, the heat pump uses a desiccant solution to cool the process air to a temperature lower than that of a shallow-well heat sink. During the heating cycle, the concentrated desiccant is used to achieve a process air temperature higher than that of the shall-well heat source. The final design of the heat pump is described in the seciton on demonstration systems.

In 1980, twenty-five years after his original work, Löf began the study of an open cycle refrigeration system consisting of four major sections: the evaporator, the absorber, the concentrator, and the solar collector. Water, the refrigerant in this cycle, is sprayed into an evaporator (evacuated to 5mm Hg pressure ure), where it flashes to a vapour. This phase change produces the cooling effect. Chilled water is pumped from the bottom of the evaporator to a fan-coil in the building. No compressor or condenser is needed because the water vapour from the evaporator is absorbed by a concentrated absorbent spray. Continuous absorption of water vapour maintains a low pressure in the system and permits continuous vaporization of water in the evaporator. A vacuum pump is necessary to remove noncondensible gases from the absorber. Cooling tower water is used to carry away the heat of absorption. To concentrate the dilute sorbent, the desiccant solution is allowed to trickle down through a packed column concentrator in the opposite direction to a forced stream of solar-heated air from the flat-plate solar collectors.

This sytem is advantageous for those homes that already have solar air-heaters in place. However, the rooftop collector/concentrator used by Kakabaev, by Gupta, and by Robison allows the black rooftop, with or without a single sheet of glass, to replace the expensive flat-plate collectors and the packed-column concentrator. The building roof need only be modified by the addition of a weak-solution distribution-header at the top and a gutter at the bottom to collect the concentrated desiccant.

Component and Systems Analysis

The performance of a liquid desiccant cooling system can be quickly determined by measuring the quantity of water absorbed from the dilute sorbent in the collector/concentrator. When a quantity of water is driven from the desiccant in the concentrator, a like quantity will be absorbed from the process airstream by the reconcentrated sorbent in the absorber, and an equal quantity of water will be evaporated into the process airstream in the evaporative cooler stage, thus determining the cooling effect of the machine.

In Australia in 1963, Hollands first experimentally and theoretically analyzed the reconcentration of lithium chloride in a solar still. Although solar stills are quite adequate for concentrating very dilute sea water, they can not be used for achieving high desiccant concentrations. A temperature difference of 24 °C between the solution and the glass cover is necessary for useful evaporation to take palce from a lithium chloride solution with a 40% concentration. This condition only exists during a few hours on very clear days.

The collector/concentrator of a solar liquid desiccant system has a function unlike that of any component of conventional solar systems. It is driven by solar thermal energy, yet it does not collect and transfer thermal energy as do conventional collectors. The collector/concentrator uses solar energy to directly produce chemical potential energy that can be stored at high energy densities, that can be stored indefinitely, that can be stored at ambient temperature, that can be stored at atmospheric pressure, that can easily be transported, that can be used to heat, cool, humidify and dehumidify air, and that can be used to drive an exhaust-free heat expander to generate electricity or to drive pumps and fans for HVAC systems.

Because of the unique function of the collector/concentrator, most of the liquid desiccant analysis has been directed toward improvement of this component. Over twenty analytical computer models of collector/concentrators have been developed by various research groups. In 1985 Wood at Arizona State University described these physical and analytical models, beginning with the work of Kakabaev in 1969 and including work by Gupta, Ganhidasan, Collier, Grodzka, Howell, Löf, and others. By using the data from a number of these physical models and the predictions of the corresponding analytical models, and with a prototype 11x11-m collector/concentrator, Wood was able to design a new collector/concentrator model based on derived correlations, a model which accurately predicts actual experimental temperature and concentration data. This model can be used to test the effect of various climatic, operational, and collector design variables on the evaporation rate.

Griffiths at Somerset Technologies has developed analytical models for both packed-column and for sprayed-coil absorbers and concentrators. These models, which accurately predict the performance of these components, have been in use for several years.

Demonstration Systems

In 1976 Kakabaev's research group reported on their experience with two open cycle absorption cooling systems operating in the southwestern Soviet Union. The smaller of the two systems produced 7 kW of cooling. The collector/concentrator consisted of the 60-m² unglazed black roof of the two-room laboratory building. See Fig. 11. The second system cooled a three-story, nine-unit apratment building, providing 176 kW of cooling and used the building roof to provide a 180-m² unglazed collector / concentrator. Both systems used lithium chloride and water as the desiccant solution.

In 1976 Gandhidasan and Gupta built a complete calcium-chloride desiccant solar air conditioning system for a small building in India, and in 1979 Johannsen described the design and operation of a complete solar-powered glycol-desiccant system for air conditioning a large laboratory in South Africa.

Even though the first desiccant solar air conditioning system was a liquid system devised in the U.S.A. in 1955, U.S. investigators showed no further interest in liquid desiccant systems, and for many years all research in the U.S.A. was limited to solid desiccant systems.

In 1978 through 1981, Griffiths at Somerset Technologies, Collier at SERI, Gandhidasan at the University of West Indies, and Robison at the University of South Carolina designed, built, and operated a solar-powered liquid desiccant heat pump. The absorber and packed-column concentrator were built by Griffiths and coworkers at Somerset Technologies.

During the first year of operation, the heat pump heated, cooled, and provided humidity control for a 3-story, 260-m² home. No backup equipment was installed. None was needed. Calcium chloride brine was reconcentrated by direct solar radiation in a 45-m², very inexpensive, simple, collector/concentrator and stored for continuous operation as chemical potential energy in a small uninsulated tank. The cooling output of the machine is 12 kW. The average solar energy COP (refrigeration divided by insolation) over a 24-hr period is 0.6 when the system is operating with 22 °C well water. This is the COP of the combined collector and heat pump and should not be confused with the usually reported thermal COP of only the heat pump.

During the next year and a half, triethylene glycol, calcium chloride, and mixtures of calcium and lithium chloride were tested as absorbents in the machine. A small 40-cm-square packed column was also tested as a concentrator. The dilute desiccant solution was heated by plastic flat-plate solar collectors, by off-peak electricity, and by waste heat from a vapour compressor. Figure 12 shows the inherent simplicity of this system. The absorber, evaporator, and saturator are Munters cross-fluted Humi-Kool cooling-tower packing.

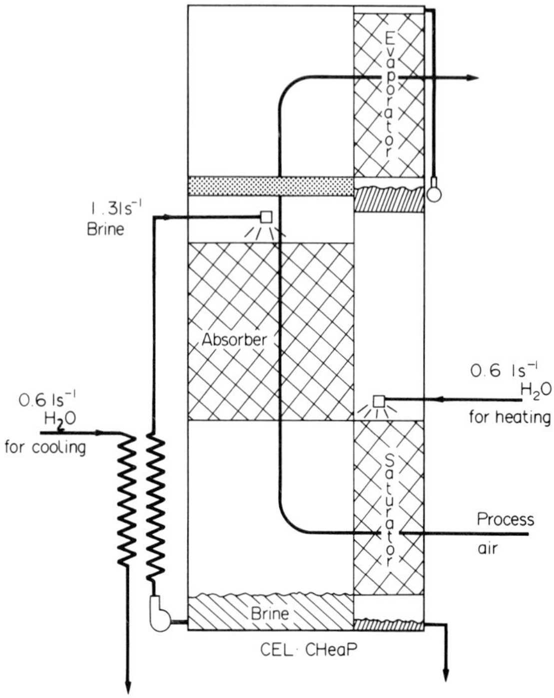

Fig. 12. Liquid desiccant heat pump.

During summer (cooling) operation, return air from the building is first passed through the absorber and is dehumidified by contact with a sorbent brine. The latent heat of absorption, and possibly some sensible heat, is rejected to a non-refrigerated heat sink, such as cooling tower water, river water, or well water. The air, at reduced enthalpy and humidity, is then passed through an evaporative cooler, and is evaporatively cooled. After being cooled, the air is delivered to the building.

A portion of the brine in the absorber, having been slightly diluted by the absorption of water from the air, is transported to the brine storage reservoir, and is replaced by more concentrated brine from the reservoir. The water absorbed from the air is thus transported to the brine storage reservoir.

During periods of favorable insolation, brine is withdrawn from the brine storage reservoir and passed through the glazed solar collector/concentrator. The brine is heated within the collector/concentrator, and scavenger air thermosyphons through the collector/concentrator, carrying away the water absorbed in the absorber and is transported to the brine storage reservoir. The concentrated brine is returned to the brine storage reservoir.

During the hot summer weather, the system typically would have the following performance:

Building-maintained conditions	26°C, 50% RH
Air leaving absorber	27°C, 23% RH
Air leaving evaporative cooler	15°C, 90% RH
Air entering concentrator	35°C, 24°C DP
Air leaving concentrator	60°C, 31°C DP

This performance can be attained with a heat sink temperature of 21°C and a solar collector/concentrator brine temperature of 68 °C. Ground water temperatures of 21°C or less are available in nearly all of the continental United States, and cooling tower water temperatures of 21°C are attainable throughout most of the western United States. Solar collector/concentrator fluid temperatures of 68°C are easily obtainable and substantially improve the efficiency of the collector when compared to the 85°C discharge fluid temperatures required by other heat-activated air-conditioning devices.

During winter operation, return air from the building is first passed through a first-stage saturator and is humidified by contact with water, and the enthalpy is raised by heat extracted from a natural heat source, typically ground water. The first-stage saturator consists of packing through which ground water flows. The air next passes through a second-stage saturator and is further humidified by contact with water, and its enthalpy is further raised by heat extracted from the brine storage reservoir. The second-stage saturator consists of an extended-surface coil with recirculated water contacting the air and the airside surface, and brine recirculated from the brine storage reservoir passing through the tubes.

The air next passes through the absorber and is dehumidified by contact with the sorbent brine. As the air is dehumidified, its dry-bulb temperature is raised by an amount equal to its enthalpy increase through the first- and second-stage saturators, plus the chemical heat of mixing associated with the absorption of the water vapour into the sorbent brine. The air thus heated is delivered to the building.

A portion of the brine from the absorber, having been slightly diluted by the absorption of water from the air, is transported to the brine storage reservoir, and is replaced with more-concentrated brine from the reservoir. The water absorbed from the air is thus transported to the reservoir.

During winter the system would typically have the following performance:

Building-maintained conditions	20°C, 40% RH
Air leaving first-stage saturator	18°C, 90% RH
First-stage saturator source temp.	21°C
Air leaving second-stage saturator	22°C, 90% RH
Second-stage saturator brine temp.	24°C, minimum
Air leaving absorber	37°C, 23% RH

The heat supplied by thermochemical heat-pumping (brine concentration) is 48% of the total. The minimum brine storage reservoir temperature is only 24°C as compared to 49°C necessary for the storage of sensible heat.

Using operational data from research conducted by Griffiths and Robison, a second-generation liquid-desiccant air conditioner and energy sotrage system has been designed and constructed. Unlike the original liquid desiccant heat pump which uses shallow-well water for a heat sink and source and thus is limited to regions where surface water is available, the new machine has no such restrictions; it can discharge heat to the outside air. An all-plastic crossflow heat exchanger designed by Griffiths acts as absorber, heat exchanger, and regenerative evaporative cooler.

Operation in the laboratory at conditions such as those on the Louisiana State University campus where the unit has now been installed resulted in these conditions:

Building-maintained conditions	26°C, 50% RH
Outside air entering absorber	36°C, 46% RH
Air leaving absorber	26°C, 25% RH
Air entering building from evap.	17°C, 70% RH
Exhaust air leaving regenerative evaporative cooler	30°C, 90% RH
Lithium chloride concentration	40%

The capacity of the machine is approximately 10 kW. Operational results should be available after operation during the summer of 1986.

Another method of utilizing the chemical potential energy in a concentrated liquid desiccant has been demonstrated by Isshiki in Japan. Isshiki and his associates used the difference in vapour pressure between that of a concentrated salt solution and that of pure water to drive a steam expander. This expander could in turn generate electricity or drive pumps and blowers for desiccant cooling systems. The concentration difference energy (CDE) engine is powered by a pure-water boiler, which is totally immersed in a solution of lithium and calcium chloride rather than in a firebox. The discharged vapour from the expander is injected through nozzles into the aqueous solution surrounding the pure-water boiler. Energy to provide steam in the boiler is generated by absorption of the exhaust vapour into the hygroscopic aqueous solution.

A unique feature of this engine is that the exhaust steam is totally absorbed in the brine; therefore, all of the heat rejected by the engine is recovered. This feature reduces the input energy for a given output, producing efficiencies in excess of limits predicted by the customary definition of Carnot efficiency. Several CDE's have been operated successfully. An automobile and a motorcycle have been operated for several years, both powered by salt solutions that have been concentrated by solar energy.

To operate in conjunction with a desiccant heat pump, the input brine for a CDE engine would have a concentration of about 50%. This engine could furnish blower and pump power (and electricity if needed) and would discharge brine to a desiccant heat pump at a concentration of about 45%. The brine, further diluted to a concentration of about 40% in the heat pump, would be reconcentrated in a simple rooftop collector/concentrator.

Future Research

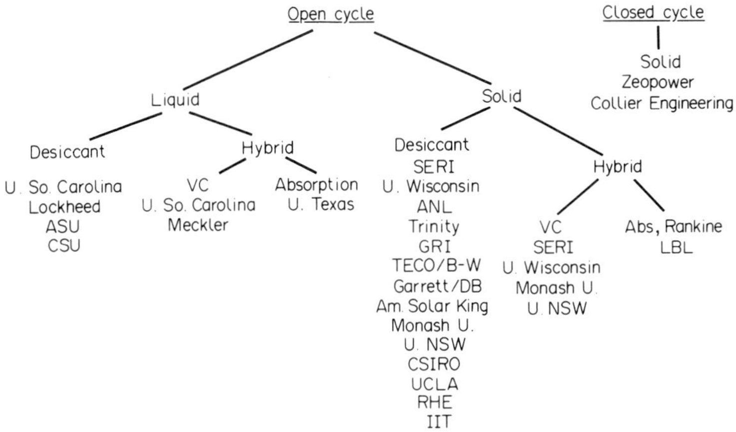

Fig. 13. Organizations contributing to desiccant cooling research.

Figure 13 shows some of the organizations contributing to desiccant research. Other groups conducting research are Mayer, Universidad Simon Bolivar, Venezuela; Lazzarin, Instituto di Fisica Technica, Padova, Italy; Grossman, Technion-Israel Institute of Technology, Haifa, Israel; Kimura, Waseda University, Tokyo, Japan; Givoni, Ben-Gurion University, Sede Boger, Israel; Kamoshida, Shibaura Institute of Technology, Oomiya, Japan; Clark, Trinity University, San Antonio, Texas, and Saito, Osaka Institute of Technology, Osaka, Japan.

Some suggested research ideas follow.

Desiccant Heat Pump Powered by Solar Pond

Griffiths has suggested the use of either a salt-gradient or a shallow solar pond to provide summer cooling and winter heating by utilizing the energy collected and stored in a solar pond. Energy could be stored in the form of a concentrated solution or as the heat of fusion of the salt in the pond. This sytem could eliminate the need for cooling towers or indirect coolers to reject heat extracted from the building. The use of the sorbent brine directly in the pond could substantially reduce the operating temperature, the size, and the cost of the solar pond. A 6-m-diameter pond could heat and cool a small home.

Improved Operation of Desiccant Heat Pump by Radiation to Night Sky

Robison has suggested the use of radiation to the night sky to enhance the performance of the second generation liquid desiccant heat pump, which uses a regenerative evaporative cooler to discharge the heat of sorption to the exhausted building airstream.

In addition to the brine-distribution-header installed at the crest of the roof under the glazing, a water-distribution-header would be installed above the glass. At night, water would be pumped over the outside of the glass and would be collected in a gutter at the bottom of the glazing. The water, cooled by radiation to deep space, would be circulated through a polybutylene coil immersed in the brine storage tank. This tank typically operates at about 32 °C. The cooling water, warmed by the heat of sotprion, would then be returned to the roof. At night, the reduced temperature of the brine entering the absorber would compensate for the dilution due to absorbed water; during the daytime, the increased brine concentration would compensate for the increase in the brine temperature.

An unglazed system would require only the brine-tank cooling coil. At night, water would be distributed and collected in the same collector/concentrator circulation system that is used for brine in the daytime.

Proposed Passive Liquid-Desiccant Heat-Pumping System

A proposed desiccant heat-pumping system would passively collect, concentrate, and store solar energy in the form of a concentrated solution in order to heat, cool, humidify, and dehumidify process air.

In the summer, the air would be dehumidified as it is exposed to a liquid desiccant in a packed-column absorber, and subsequent evaporative cooling would produce cool, comfortable air with a temperature lower than ambient. In winter, air would be humidified in a saturator, and subsequent dehumidification in the packed-column absorber would use the heat of sorption to produce warm, humid air with a temperature higher than ambient. A ground coupled closed-loop heat exchanger would act as the heat sink to remove the heat of sorption during the cooling cycle and as the heat source to increase the enthalpy of the air in the saturator for the heating cycle. See Fig. 14. When adequate insolation is available, the desiccant would be reconcentrated as solar energy evaporates water from the dilute solution as it flows in a thin film across the roof under a single glazing. As warm air from an attached sunspace, Trombe wall, or greenhouse exhausts between the black roof and the glazing by natural convection, makeup air will be drawn into the building through the absorber without benefit of blowers.

A small pump would circulate the brine through the polybutylene closed-loop heat exchanger and the absorber during the cooling cycle. The same pump would circulate water through the closed-loop heat exchanger and the saturator during the heating cycle.

In summer, the hot, wet scavenger airstream is discarded. See Fig. 15. However, during the heating season, the scavenger airstream flows through an all-plastic air-to-air heat exchanger. During moderate weather, sensible heat is recovered from the

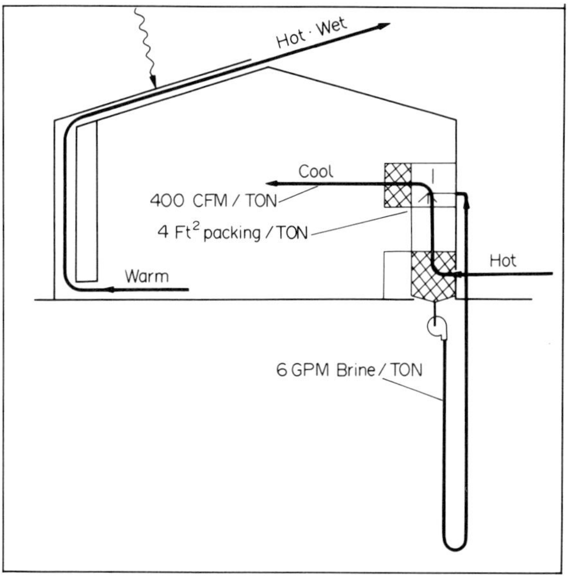

Fig. 14. Daytime cooling.

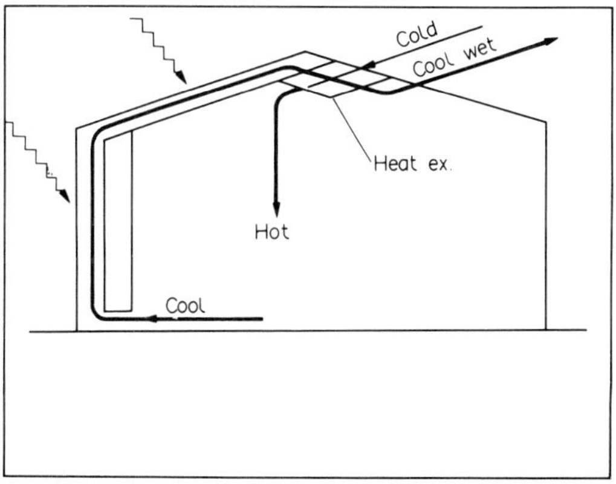

Fig. 15. Daytime heating.

scavenger airstream and used to heat the incoming makeup airstream to the building. When the outside air is cold, both latent heat and sensible heat are recovered, as the cold incoming air lowers the temperature of the scavenger stream below the dewpoint, thus causing condensation of some of the water vapour in the scavenger airstream. The dry outside makeup air may be humidified as needed by returning a portion of the scavenger airstream to the building.

The passive collector/concentrator would use large rooftop areas that would not normally be used for passive solar conversion.

Liquid-Desiccant Energy-Storage Improvements

Stratified storage (by concentration/density gradient) techniques to separate concentrated and dilute desiccant solutions should be investigated. Isothermal phase-change material could be placed in the stored desiccant to limit temperature swings from solution dilution and reconcentration. It may be possible to use the desiccant as an isothermal phase-change heat source. (Calcium hexahydrate is one example). Seasonal energy storage is also a possibility.

Conclusion

Solar desiccant systems have now reached a point in their development such that further research advances will bring about a rapid transfer of technology from the research laboratories to industry.

SUGGESTED READING

Bjurström, H., and W. Raldow (1981). The absorption process for heating, cooling, and energy storage - an historical survey. Energy Research, Vol. 5, 43-59.

Isshiki, N. (1980). Storage and generation of power and heat by aqueous solution of salts. In G. Wettermark (Ed.) International Seminar on Thermo-chemical Energy Storage. Royal Swedish Academy of Engineering Sciences, Stockholm.

Jurinak, J.J. and J.W. Mitchell (1984). Effect of matrix properties on the performance of a counterflow rotary dehumidifier. Journal of Heat Transfer, 106, pp. 638-645.

Lenz, T., G.O.G. Löf, R. Iyer, and J. Wenger (1983). Open Cycle Lithium Chloride Cooling System, Solar Energy Applications Laboratory, Colorado State University, Fort Collins, Colorado.

Lenz, T. and co-workers (1985). Open Cycle Lithium Chloride Cooling System, Solar Energy Applications Laboratory, Colorado State University, Fort Collins, Colorado.

Novak, K.S., and B.D. Wood (1985). Combined Heat and Mass Transfer on an Open Flow Liquid Absorbent Solar Collector/Regenerator ERC-R-85012, Arizona State University, Tempe, Arizona.

Penny, T.R., I. Maclaine-cross (1985). Promising Advances in Desiccant Cooling, SERI/TR-252-2464, Solar Energy Research Institute, Golden, Colorado.

Robison, H.I. (1984). Passive and low-energy research and practice dehumidification. In A. Bowen (Ed.), Energy and Ecology in Houses - PLEA '84, Pergamon Press Ltd., Oxford, England.

Robison, H., W. Griffiths, and O. Arnas (1984). Liquid desiccant vapour compressor hybrid air conditioner and energy storage system for load management. In D.P. Colvin (Ed.), Solar Energy in the Southeast, N.C.S.E.A., Raleigh, North Carolina.

Schlepp, D., and K.J. Schultz (1984). High-Performance Dehumidifier for Solar Desiccant Cooling Systems, SERI/TR-252-2497,

Solar Energy Research Institute, Golden, Colorado.
Schlepp, D., K. Schultz, and F. Zangrando (1984). Facility Design for Cyclic Testing of Advanced Solid Desiccant Dehumidifiers, SERI/TR-252-2464, Solar Energy Research Institute, Golden, Colorado.

CHAPTER 5

DESICCANT SYSTEMS

George O.G. Lof,
Solaron-Colo 80110, USA

INTRODUCTION

The maintenance of comfortable conditions in a living space requires control of air temperature and humidity. Refrigeration produces a reduction in air temperature, and if decreased below the dew point, also removes humidity by moisture condensation. Air may be dried, however, without cooling, by use of hygroscopic materials which absorb or adsorb part of the water vapour present in humid air. These hygroscopic materials are known as desiccants, and may be liquid or solid.

If used for comfort air conditioning, the dehumidification or desiccation process must be associated with another process for air temperature reduction. Not only is it necessary to cool the air below its original temperature, but the moisture removal process results in the liberation of heat and an increase in air temperature. One method for reducing the temperature of the dried air is cooling by means of a vapour compression or an absorption type refrigeration machine. Another method is direct or indirect evaporative cooling of the air by means of water. Although the desiccant process is separate from the subsequent cooling step, the two may be closely interconnected and mutually dependent, particularly if the final step is evaporative cooling.

The relationship of the desiccant process to a solar heat supply results from the requirement for regeneration of the desiccant material after it has absorbed or adsorbed moisture from the air being dried. By heating the desiccant material to temperatures of 50 to 150 °C, part of the moisture it contains is evaporated into the atmosphere so that the material can then be reused for air dehumidification. The temperature to which the desiccant must be heated depend on its properties, particularly its equilibrium moisture contents at various temperatures. Regeneration temperatures are usually low enough for the heat to be supplied from typical flat plate solar collectors.

SOLID DESICCANT SYSTEMS

Three types of materials are used in drying air for industrial purposes, silica gel, "molecular sieve", and the chloride salts of lithium and calcium, Silica gel or zeolite (molecular sieve) in granular form may be packed into a flat cylinder through the broad faces of which air can be circulated. The stream of air to be dehumidified is supplied to one of the circular faces of the cylinder and withdrawn from the other face after passing through the desiccant. Moisture is absorbed in the silica gel, the heat of absorption producing a temperature rise in the air being dried.

In order that the process can be continuous, the cylinder slowly rotates (one to three revolutions per minute), so that the silica gel which has become moisture-laden moves into another stream of air which has previously been heated. As the hot air stream passes through the silica gel, moisture is evaporated from the desiccant into the air which is discharged to the atmosphere. The dried silica gel slowly rotates again into the stream of air being dehumidified, and the process continues.

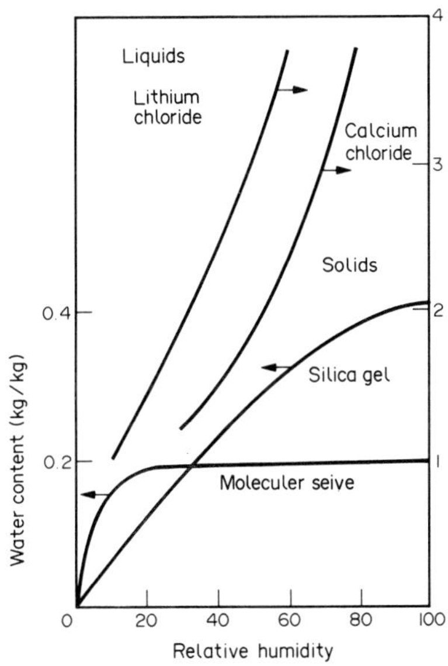

Fig. 1. Equilibrium isotherms for solid and liquid desiccants.

Figure 1 shows the relationship between air relative humidity and equilibrium moisture content in typical desiccant materials. At a given percent relative humidity, the equilibrium moisture content of the desiccant is not appreciably dependent on temperature. It is evident that at low humidity, silica gel has a low equilibrium moisture content and that it has high moisture at high humidity. In cycling between these conditions, the material can therefore pick up moisture and subsequently discharge it. The path of humidity and temperature followed by the air being dried is shown on a psychrometric chart from point (1) to point (2), Figure 2.

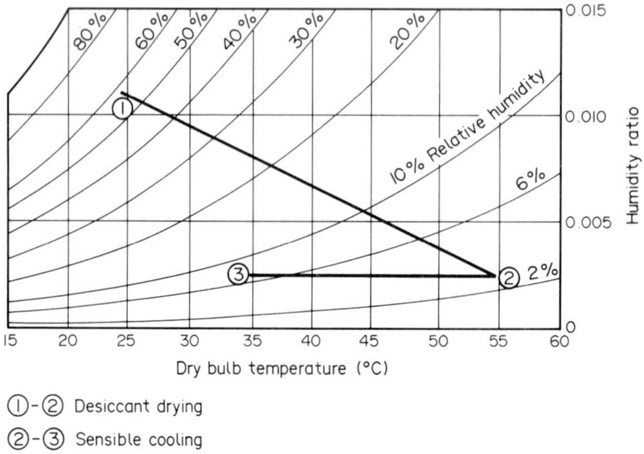

Fig. 2. Desiccant cooling process.

Another commercial system using a solid desiccant is based on the use of fibrous material impregnated with lithium chloride or calcium chloride. The hydrogroscopic character of these salts results in a process similar to that described above. A third type of desiccant, known as a molecular sieve, is a naturally occurring silicate mineral called a zeolite.

This hygroscopic compound has characteristics similar to those of silica gel. Of all these materials, silica gel has the greatest absorption and desorption capacity over practical ranges of temperatures for dehumidification and regeneration.

There has been no practical use of solid desiccants in comfort air conditioning systems. The principal reason is the unavailability of a fully developed system in which air cooling is integrated with the desiccant unit. Research on such a system comprising a desiccant wheel, evaporative coolers, and air-to-air heat exchangers has led to some theoretical and experimental combinations which may prove practical. Although operating in a temperature range compatible with solar heat supply for regeneration, a full scale system of this type has not been operated with solar energy as its principal heat source.

Two combinations of desiccant humidification and evaporative cooling are known as the ventilation type and the recirculation type. Figures 3 and 5 show these systems diagramatically, and Figures 4 and 6 indicate the humidity--temperature paths of conditioned air and regeneration air in both systems. The primary components in each system are the desiccant wheel, a heat exchanger, two evaporative coolers, and a heater in the regenerating air stream. A fan is also needed in each of the two air

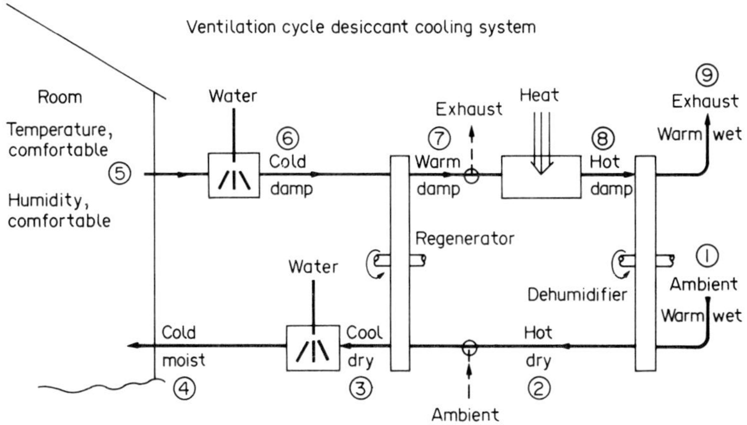

Fig. 3. Ventilation cycle desiccant cooling system - schematic.

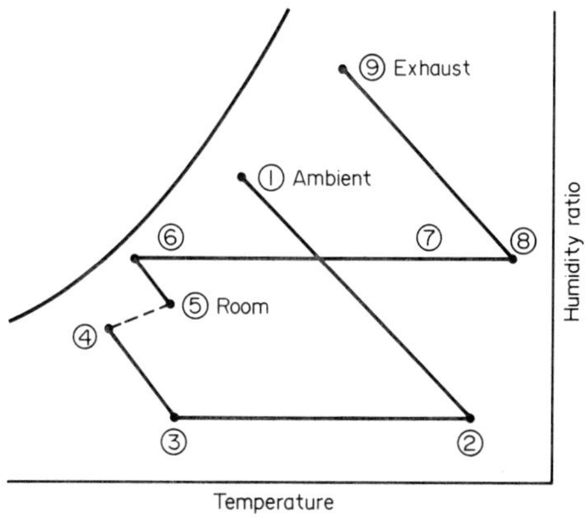

Fig. 4. Ventilation cycle desiccant cooling system - psychrometric chart.

flow channels. The numbers on the diagrams and psychrometric charts show the conditions of the two streams of air at each point in the systems.

In the ventilation cycle, Figures 3 and 4, outdoor air at condition (1) is first dehumidified in the desiccant wheel, a process which also results in an increase in air temperature. Hot dry air at (2) is then ccoled by heat exchange in a rotary type pregenerator from which it leaves at (3). Air temperature is further reduced by evaporative cooling to condition (4) at which it enters the rooms of the building. The temperature and humidity of the air increase in the rooms, and this air is supplied to the regeneration cycle at condition (5).

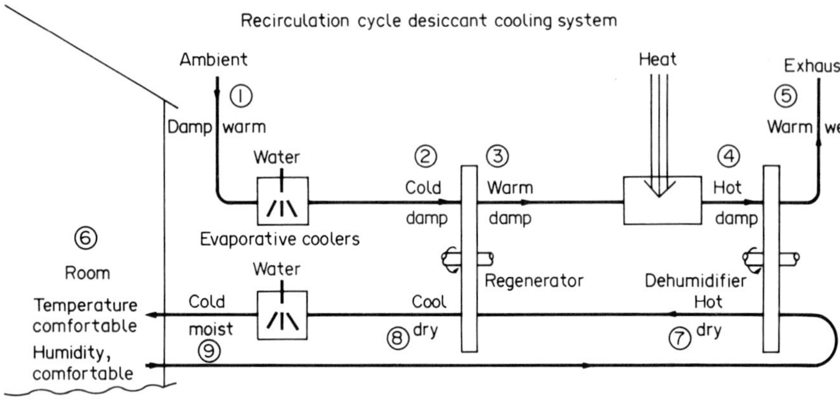

Fig. 5. Recirculation cycle desiccant cooling system - schematic.

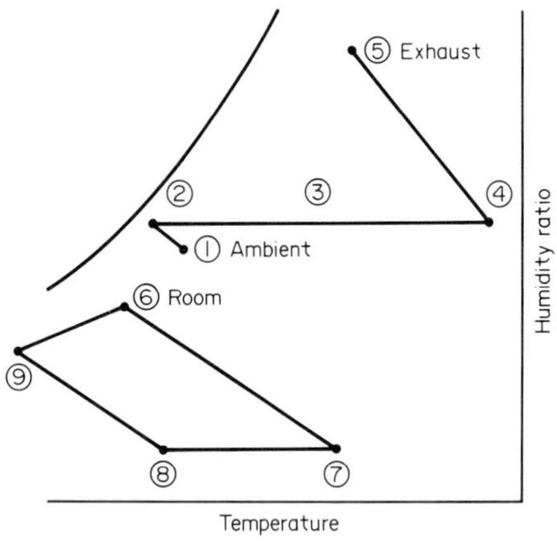

Fig. 6. Recirculation cycle desiccant cooling system - psychrometric chart.

Air for regenerating the desiccant is first cooled by evaporative cooling to condition (6) so that it can be used in the rotary heat exchanger to cool air for the building. After it's temperature rises in the heat exchanger to condition (7), the air is further heated by solar energy or by fuel or by a combination of the two sources, to condition (8). This hot air then passes through the moisture-laden desiccant, evaporates most of the absorbed water, and is discharged to the atmosphere at (9).

In the recirculation cycle, room air is recirculated for dehumidifying and cooling rather than supplying it from outdoors. Figures 5 and 6 show the same operation and conditions as in the ventilation cycle for the air being dried and cooled for use in the building. Room air at condition (6) is first dehumidified to condition (7) then cooled by heat exchange to condition (8), and further cooled to condition (9) by evaporative cooling. Air entering the regeneration side of the equipment is at outdoor condition (1) rather than at room condition. The steps in the regeneration process are the same as those in the ventilation mode, but the conditions are usually different because of a more favorable air source for regeneration.

Theoretical analysis of the systems have been extensive, but only limited experimental data have been procured. No systems have been operated with solar energy as the primary supply. Effective regeneration of silica gel requires air temperatures near 100 °C, so well-designed and operated solar flat plate collectors or evacuated tube collectors must be used if the major heat supply is from solar. Alternatively, solar energy can be used to increase air temperature to an intermediate level, fuel then being used for final heating. Solar heat may be supplied from liquid collectors to the air by means of a heat exchanger, or air collectors may be used directly in the regeneration air circuit.

In the experimental systems which have been tested, power consumption for fan operation has been excessive. Pressure loss through the several components on each side of the system has required almost as much electric energy as a conventional vapour compression cooling system. One refrigeration ton (about 3.5 KW) of cooling has required at least one-half kw of electric input, and is some designs, nearly one kw. Efforts to reduce power requirements have involved decreasing the resistance of the various units to air flow. The desiccant wheel has usually been the largest resistance, and new designs have led to substantial reductions in friction loss through this unit. In one design, granular silica gel material has been replaced with spirally wound strips of paper or fabric impregnated with very small silica gel particles. Air passes through the narrow spaces (one milimeter or less) between layers of material, absorption and desorption of water occurring in the desiccant particles. It has been estimated that power requirements can be reduced to 300 watts per refrigeration ton, equivalent to 12 KW of cooling per KW of electric energy supply, or an electric COP of 12. Theoretical performance coefficients (thermal COP) for this type of equipment range from 1.1 to 1.3, based on heat actually supplied to the cycle.

If the heat of absorption of moisture in the desiccant bed can be partially removed during the absorption process, temperature rise can be minimized and the effectiveness of the bed increased. The equilibrium moisture content of the desiccant is substantially higher at the resulting lower temperatures. Stationary beds, for example, may be internally cooled by circulating water through coils or other heat exchange surfaces around which the desiccant material is packed. The higher equilibrium moisture content of the desiccant at the lower temperature results in a greater portion of the regeneration heat supply being used for evaporation of water rather than for heating the solid desiccant material, with the result of a higher thermal COP.

Internal cooling of rotating desiccant beds is difficult, but an approach to this objective has been developed by increasing the thermal capacity of the desiccant material. A mixture of desiccant and intert solids, such as metal particles or mineral grains, has a higher heat absorption capacity per degree temperature rise than does the pure desiccant. Use of the mixture for dehumidification results in a lower rise in air and solid temperatures than if pure desiccant is used. Thermal COP's above 1.0 have been achieved with this concept. Figure shows a diagram of such a system which has been experimentally marketed in the United States.

A substantially different concept, known as a hybrid cooling system, is also based on the use of a desiccant. Air is dehumidified in a desiccant bed, as described above, and then partially cooled by heat exchange with the incoming regeneration air stream. This dehumidified air is then cooled in a conventional vapour compression refrigeration cycle to the desired temperature. The rationale in this design is based on the usual requirement to remove both the sensible heat and latent heat in air conditioning processes. Humidity reduction in typical residential and commercial applications usually requires one third to two thirds of the energy supply, temperature reduction requiring the balance. Thus, if humidity is reduced in a desiccant system, the load on the refrigeration unit is decreased and the capacity of that component can be smaller. Less electric energy will be required than if both latent and sensible loads have to be removed in the vapour compression machine.

An additional gain in energy saving may be achieved in those situations where the latent load is high and where vapour compression is normally used. Under those conditions air temperatures must be reduced below the comfort level in order that adequate condensation of moisture can occur. The air then has to be reheated to the desired supply temperature, employing additional energy.

PERFORMANCE OF SOLID DESICCANT SYSTEMS

Performance models for the components and for the ventilation and recirculation desiccant cooling systems have been developed, and values of COP and capacity have been predicted. Upper COP limits for the ventilation and recirculation cycles have been projected at 2.0 and 1.5 respectively.

A number of experiments have been conducted with desiccant systems employing molecular sieve materials, and it was found that a cooling capacity of 2.7 tons (9.5 KW) could be obtained at a COP of 0.5 with a regeneration temperature of 180 °F (82 °C). The solar heat input was simulated by use of fuel. A high heat exchanger effectiveness (95 percent) was found to be important in obtaining good performance. Experiments with a system based on a silica gel desiccant bed showed a capacity of 1.35 tons (4.75 KW) and a COP of 0.6 in a recirculation mode, with regeneration at 200 °F (93 °C). Experiments with a cooled desiccant bed have also been conducted, a COP of 0.63 being obtained with a regeneration temperature of 145 °F (63 °C), one ton of cooling capacity (3.5 KW) being obtained in the recirculation mode.

The only published results of operating a solid desiccant cooling system primarily with solar energy are those obtained with a cooled crossflow desiccant dehumidifier prototype. Solar heated air was used directly in the regeneration of a switched bed in which fabric sheets coated with very small silica particles were used. In the dehumidification cycle, one bed was cooled with atmospheric air while the second was being regenerated with solar heated air. Capacities of tons (KW) and COP's of were obtained at a regeneration temperature of °C.

LIQUID DESICCANT SYSTEMS

Air can be dehumidified by absorbing water vapour in a hygroscobic liquid. The important requirements of the liquid are its miscibility with water and a water vapour pressure over the solution substantially lower than the partial pressure of water in the moist air. Under those conditions, the liquid will absorb water from moist air with which it is brought into contact.

In order that the absorbing liquid can be recycled and reused, it must also have a reasonably high water vapour pressure at higher temperatures. By heating the liquid, its absorbed water can then be vapourized into an air stream and rejected to the atmosphere. The ideal combination of properties is a low water vapour pressure over a solution with a high concentration of water at room temperature, and a relatively high vapour pressure over a solution with considerably lower water content at elevated temperature. Other desirable properties are low toxicity, corrosiveness, and volatility. Low odor, flammability, and cost are also beneficial.

Liquids which have been commercially used as desiccants in air dehumidification systems, with suitable combinations of the foregoing properties, are aqueous solutions of calcium chloride, lithium chloride, and triethylene glycol. Equilibrium data (water content of solutions in equilibrium with air at various relative humidities) at conventional operating temperatures are shown in Figure 1.

The graph shows that if humid air and a 40 percent solution of lithium chloride, for example, (1.5 kg water per kg LiC1) are allowed to come to equilibrium at room temeprature, the air will be dehumidified to a relative humidity of 20 percent. Another air stream can be used to regenerate the solution (evaporate the absorbed water) if its relative humidity is less than 20 percent. So if atmospheric air at 70 percent relative humidity is heated from 25 °C, thereby reducing its relative humidity to slightly below 20 percent, and contacted with the solution, some of its water can be evaporated into the air stream. Practical operation of dehumidification and regeneration equipment requires vapour pressure differences between liquid and gas phases so that the rates of vapour transfer are sufficient for the operation of equipment of practical size.

SYSTEM DESCRIPTION

A liquid desiccant system for air dehumidification employs a dehumidification unit in which moisture is removed from the air and absorbed in the liquid, and a regeneration unit in which water is evaporated from the heated desiccant solution into another stream of air which is discharged to the atmosphere. A heat exchanger for energy recovery from the hot, concentrated absorbent solution, one or more circulating pumps, two fans, and suitable piping, valves, and controls are also used. Figure 7 shows a system of this type in which liquid and air are brought into contact in spray chambers in which air passes through a spray of fine droplets of either the strong or the weak solution

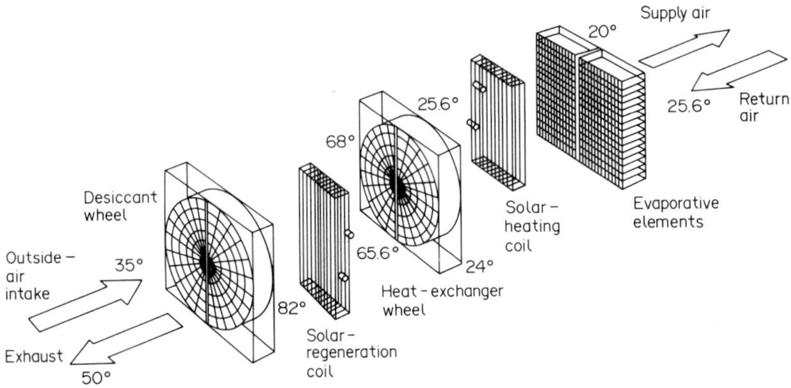

Fig. 7. Solar heating and solid desiccant cooling system - schematic.

Absorption of water vapour in the concentrated solution is accompanied by the liberation of heat and a rise in liquid and air temperatures. Cooling coils may therefore be used in the dehumidifier chamber unless warm air is required. Water flowing through the cooling coils may be circulated through an evaporati cooler for reuse.

Dilute solution from the dehumidifier chamber is pumped through a heat exchanger in which its temperature is first increased by exchange with the hot solution leaving the regenerator. Further temperature increase is achieved by passing the dilute solution through a heater supplied with some type of fuel or, in the application discussed here, with solar energy. Although the solution could be pumped directly through a solar collector, practical considerations would usually require heating the solution in an exchanger by means of solar heated water. Anothe method for supplying heat for evaporation is through a heat transfer surface in the regenerator chamber itself. A third method, which has been experimentally used with a solar air heating system and a triethylene glycol desiccant, involves supplying solar heated air to the evaporator chamber. Experiments with lithium chloride and lithium bromide solutions have also been carried out in a regenerator comprising a packed bed or column of loose solids down through which the solution trickles counter current to a rising stream of solar heated air. Finally, solar regeneration can be accomplished by trickling the solution across a gently sloping roof so that direct exposure to the sun results in the evaporation of moisture to the atmosphere. Several experimental studies in the U.S.A. and Japan have provided data on the regeneration of lithium chloride and calcium chloride solutions in open flow, truckle-type solar collectors.

The psychrometric chart, Figure 8, shows the conditions of the two air streams as they pass through the two contacting chambers; the dehumidifier is internally cooled and the regenerator is internally heated. Number on the chart corresponds with number on the system diagram.

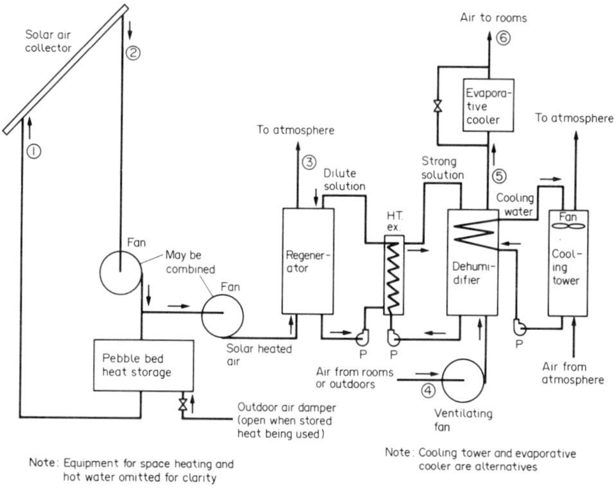

Fig. 8. Solar heating and liquid desiccant cooling system - schematic.

If spray chambers are used for absorbing and evaporating water in the two units, droplets of mist must be removed. The demisters may be some type of mesh or a clotted obstruction on which droplets impinge and collect.

An important component in the system is the heat exchanger in which the solution passing from the dehumidifier to the regenerator is heated by the concentrated solution being pumped from the bottom of the regenerator to the dehumidification unit. If the effectiveness of the exchanger were 100 percent, the dilute desiccant solution would enter the regenerator at a temperature equal to that of the concentrated solution leaving the regenerator, and the only additional heat supply would be that required for water evaporation, approximately 1,000 Btu per pound (2333

KJ per kg), plus a small amount of heat for increasing the temperature of the regeneration air stream from its inlet to its exit condition. Disregarding the air heating requirements for the moment, the thermal COP of such an operation would be 1.0, that is, one unit of "cooling" (actually the heat of condensation of moisture from the air being dehumidified) would be obtained per unit of heat supply to the regenerator. With a typical rise in regeneration air temperature of 50 degrees C, and the use of a liquid to air mass flow ratio of 2.0, the heat capacity of the air is about 1/8 of the heat capacity of the liquid stream Assuming the solution concentration increased from 50 percent to 55 percent lithium chloride, 9 percent of the solution is evaporated. Per 100 kg. of solution, evaporation of 9 kg. of water requires 21,000 KJ of energy, and 50 kg. of air require about 2,320 KJ for increasing its temperature. Under these conditions, about 21,000 out of 23,320 KJ are used for evaporation, corresponding to a COP of 0.9. If the liquid-to-liquid heat exchanger has an effectiveness of 0.85, the overall COP would be approximately 0.85 x 0.9, which is 0.77. This result can be considered a practical level for the performance of this type of system.

If solar heat is the primary source of energy for this operation, an overall system efficiency may be obtained by multiplying the above estimate by the solar collection efficiency. A well designe and operated solar liquid system can achieve an average summer efficiency of about 40 percent, that is 40 percent of the solar energy incident on the collector during a month can be collected and used. The product of 40 percent and a desiccant unit efficiency of 0.77 is approximately 0.3. Thus, the solar cooling performance factor (SCPF) of 0.3, or 30 percent, may be considered a practical maximum for this system.

COMBINATION DESICCANT AND COOLING SYSTEM

As outlined in the discussion of solid desiccant cooling systems, a liquid dehumidification unit may be combined with an air-to-air heat exchanger and one or two evaporative coolers. Although the capital cost of these combinations is considerably greater than that of the dehumidifier alone, they offer a higher overall system COP, better control of air temperature, and elimination of a cooling tower. One such system is shown in Figure 9. The dehumidifier is operated without internal cooling. Hot dry air from that unit is then cooled in a heat exchanger, such as the rotary type shown, and finally evaporatively cooled with water to the desired final condition. Air for regeneration may first be evaporatively cooled, then preheated by exchange with the dehumidified air stream, and then supplied to the regenerator for evaporating water from the dilute desiccant solution. Solar energy may be supplied to the regenerator either by further heating the air in an air collector or by solar heating the solution entering the regenerator. Solar operation may thus be achieved either by heating the air or heating the solution supplied to the regenerator.

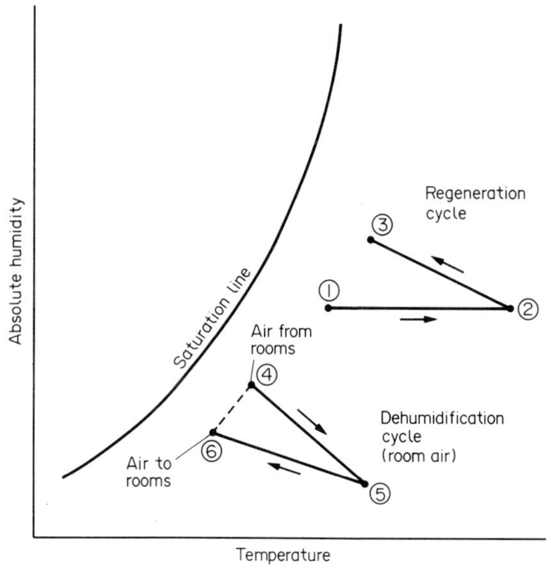

Fig. 9. Air cooling and dehumidification with liquid desiccant system - psychrometric chart.

CHAPTER 6

RADIATIVE COOLING I: THE SKY RADIATION

Xavier BERGER and Bernard CUBIZOLLES

Centre National de la Recherche Scientifique,
Laboratoire d'Ecothermique
Sophia - Antipolis, B.P. 21
06561 VALBONNE CEDEX FRANCE

INTRODUCTION

Cooling by radiative exchange with the sky is a natural proces commonly used in hot countries, and generally avoided in tempe ate climates. The emission from atmospheric constituents of t atmosphere is increased during daytime by solar direct radiati and scattering of the sun radiation (sunlit clouds and clear a (Fig. 1). Therefore, the clear sky, which may be considered a the coldest environmental heat sink during night-time, is discredited during daytime when the powerful sun changes the sky background radiation from a cold body emission to a hot one. Radiative cooling systems consequently require solar protectio selective coatings, association with or against the other heat exchanges (convection, conduction, evapo-condensation) accordi to the moment in time and the working of the system itself. B this is the purpose of the next chapter.

In the present chapter we try to describe the atmosphere in it infrared radiative interest, the sky radiation and its measurement (models and instruments). We also discuss some convenien formulae to express its intensity in a simple form, and we poi out the uncertainties or limits still attached to this resourc knowledge.

The atmosphere can be considered to consist of horizontal supe posed slabs of gas, each one emitting, absorbing and transmitt radiation according to its concentration, temperature and spectral characteristics. Oxygen and nitrogen which compose about 99% of the atmosphere, are transparent to infrared radiation (beyond 3 microns). Water vapour, carbon dioxide, ozone and other asymmetrical molecules are the main constituents which affect the atmospherical radiation. Of these, water vapour is the most important.

In a simple approach, atmospheric absorption may be considered as resulting from the theory of molecular band absorption: fo a single Lorentz spectral line (Fig. 2) the emissivity of a

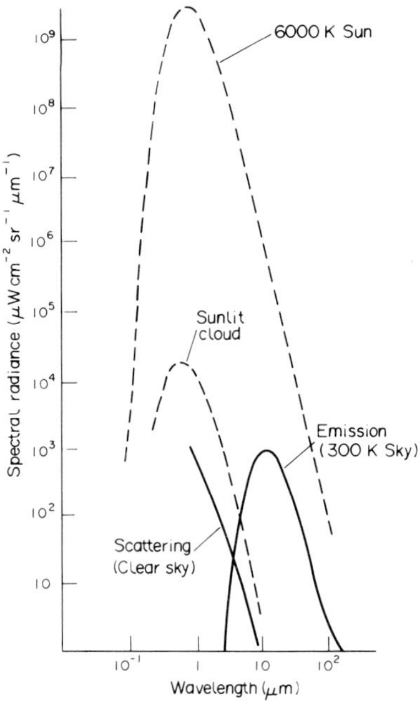

Fig. 1. Background radiation (1).

column of radiating gas may be given by an expression of the form:

$$\varepsilon_\lambda = 1 - \exp(-k_\lambda m_g)$$

with: k_λ: absorption coefficient

m_g: number of grams of the radiating gas contained in the column of a cross section unity.

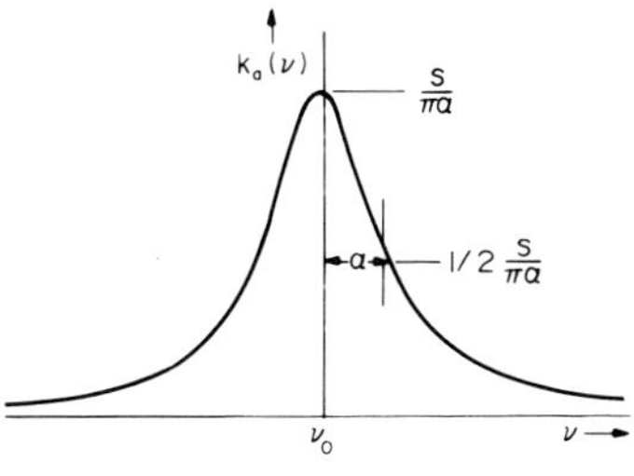

Fig. 2. A Lorentz line (2).

k_λ not only depends on the line strength S, but also on the width α which, according to kinetic theory, depends on the pressure and absolute temperature (2). The number of bands, their strengths, spacing, widths, the Döppler or scattering effects, the directional aspects and the variable constitution of each atmospheric slab lead to more or less complicated models. Therefore, besides the absorbing characteristics of each constituent the physical properties of the atmosphere affect the computation of the sky radiation.

I - DESCRIPTION OF THE ATMOSPHERE

a) Density, temperature and pressure

From ground level up to 100 km in altitude, density and pressure decrease exponentially (the decrease rate halves every 5 km) with, however, small variations. Thus, the troposphere influence is the most important, mainly the turbulent layer (between ground level and the peplopause) where concentrations in water vapour and in scattering constituents, and where the quotient $(P/P_O)^a$ (between the pressure P at the altitude z and a standard pressure P_O) are the largest. Temperature descreases in the troposphere by about 6-7 C/km and increases in the stratosphere because of U.V. absorption by ozone. Figure 3 shows the normal vertical variations of these three parameters (3) for tropical climatic conditions.

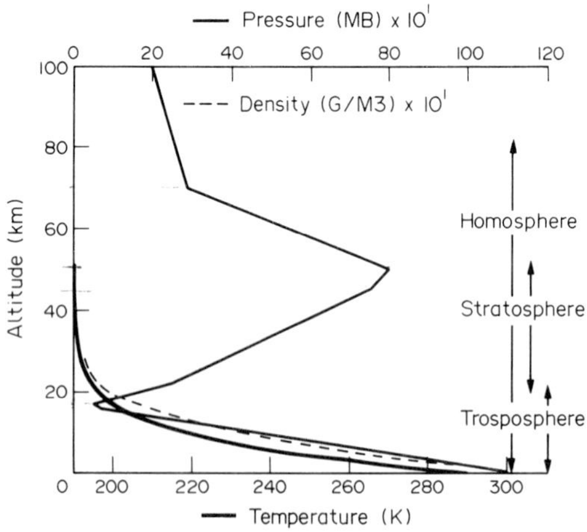

Fig. 3. Model of vertical variation of air temperature, pressure and density (3). Tropical conditions.

The temperatures of the atmospherical layers being around 300 K, the spectral radiance of a blackbody is a maximum at a wavelength about 10 μ (Wien's law). Less than 1% of the total amount of energy is emitted below 5 μ, 35% in the band 8-13.5 μ, and 28% beyond 20 μ. This schematic representation points out the importance of the spectral characteristics of each layer in the determination of the infrared sky radiance. A difference of 1 °C corresponds to a variation of about 6 w/m² in the emitted energy. This shows the difficulty of sky temperature determination with a high precision and the lack of interest in such an accuracy for practical use.

b) Water vapour

Water vapour distribution is ordered by the air movements and the phenomenon of constitutional phase changes (ice, liquid, vapour). This explains the large variability of water content in the troposphere, and its low percentage at higher levels (Fig. 4). The infrared action of this constituent is characterized by a strong absorption (3) below 8 μ (vibrational transitions) and above 13 μ (rotational transitions). Beyond 20 μ the absorption becomes so much stronger that the spectral emission at ground level is that of a black body at the corresponding temperature. In the sophisticated models, the water vapour influence is generally computed in two steps:

* a main influence (water vapour) due to lines whose centers are in the region of study

* a weak influence (H_2O continuum) due to H_2O, H_2O dimers and wings of many lines whose centers are outside of the continuum region 7-30 μ (2).

Water vapour is responsible for about 90% of the sky radiation (Fig. 5) and the first four kilometres are sufficient to compute the emitted energy with an accuracy of about 90%. This explains the search for a simple formula only based on the water vapour influence, and on the air temperature at ground level.

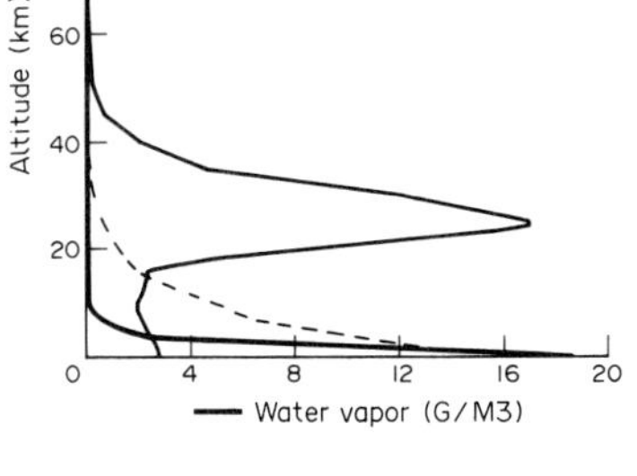

Fig. 4. Vertical variation of the densities of water vapour, carbon dioxide and ozone (2,4).

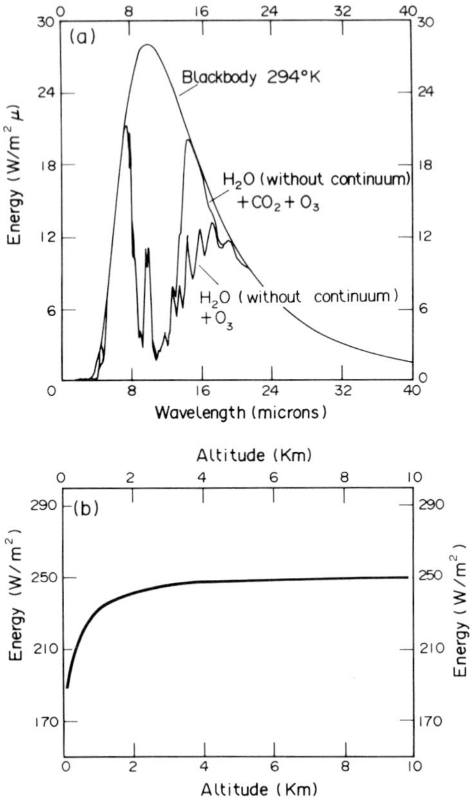

Fig. 5. Contribution to sky radiation
A: of the main constituents of the atmosphere (Low tran code)
B: of the atmospheric layers in altitude (radio-prospecting)

c) Carbon dioxide

After O_2, N_2, Ar which have no significant radiative action, and H_2O previously presented, CO_2 is the important constituent at ground level: 5% of the total mass of the dry air.

Oceans represent a pond fifty times larger than the atmosphere and thus have a stabilizing action on the CO_2 rate in the homosphere (5). As for most of the other constituents, its density decreases with altitude (Fig. 4).

The main radiative action of CO_2 is a strong band at 15 μ (half-width 2 μ), which increases the spectral radiance at this wavelength almost up to the spectral radiance of a black body at ground level. A weaker band at 4.3 μ also exists, but its action is small on account of the weak emission.

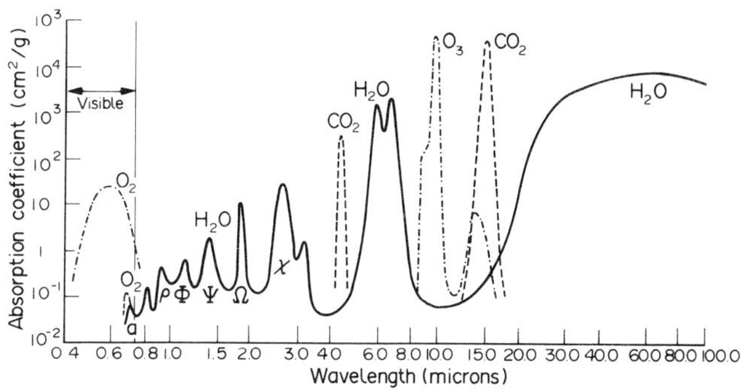

Fig. 6. Absorption spectrum of water vapour, carbon dioxide and ozone (5).

d) Ozone

The concentration peak of ozone at 30 km (Fig. 4) is a characteristic of this constituent as well as the high variability (from 2.4 to 4.5 mm) of the reduced thickness with seasons and latitude.

A strong absorption band at 9.6 μ and a weaker one at 9 μ (Fig. 6) increase the radiance by about 5 w/m² in the centre of the "atmospheric window" where the emitted energy is at a minimum.

The weak band at 14 μ is of low interest because it is blended with the strong CO_2 band.

e) Other constituents: Aerosols

The radiative action of CO, N_2O, CH_4, O_2, HNO_3, N_2 is small enough to be neglected if an accuracy of half a degree centigrade is required for T_{sky}.

The effects of aerosols are both absorption and scattering, which increase the radiation, but only in the atmospherical window. The size and the number of particles per cm³ strongly decreases with altitude: aerosols models (2) give a decrease of about 60% per kilometre for the particle density for the first five kilometres, both for hazy skies and for clear ones, the density is five times larger for a visibility of 5 km than for a visibility of 23 km; but it is the same beyond 5 km in altitude.

Many varieties of aerosols exist, e.g. urban, maritime, dust (industrial, sand), smokes, volvanic eruptions, forest fires, spores or pollens, and gas molecules. Winds which carry them have their own characteristics e.g. the harmattan. The day-night vertical motion of the boundary layer is also a reason for peculiarities. Rather than referring to works concerning aerosols in temperate climates (2,6), it seemed interesting to present the results obtained under two hot climatic conditions:

* In Kuwait (7), during dust storms, the visibility may fall below 2 km. Weather parameters have been correlated with infrared radiation (pyrgeometer measurements - 133 cloudless days over a period of more than one year). The correction on the total sky emissivity (and not to the 8-13.5 μ band emissivity considered by the authors) has been found to be:

$$\Delta\varepsilon = 0.072 \ \ln(10900/\text{vis})$$

where vis is the visibility in metres. This result is restricted to the following limits:

$$2000 \text{ m} < \text{vis} < 10{,}000 \text{ m}$$

$$-7°\text{C} < \text{dew point} < 22°\text{C}$$

*In the Ivory Coast, observations (8) were made in the south (Abidjan), the centre (Bouafle) and the north (Korhogo) of the country both with an Eppley pyrgeometer and a Buriot pyrradiometer with a good agreement. The weather was characterized by a light harmattan in the north and the centre and a very light harmattan in association with maritime aerosols in the south. The Linke coefficient corresponds to the number of pure atmospheres, without water vapour, which gives the same extinction of direct solar radiation as the real atmosphere (it is about 3.8 in France, in non polluted zones and in blue sky conditions). This coefficient, obtained from solar radiation measurements, was about 6.5, and the average visibility estimated to 2-3 km (such a visability is given by ASECNA, the official local meteorological group, in the same conditions of weather, in the same places and for the same months of the previous years). The observed correction to the total sky emissivity appeared larger for low emissivities than for high ones (Fig. 7). It is an anticipated result according to the dust effect increasing the radiation up to the radiation of a black body at ground level. Thus, the obtained expression contains two terms, one constant, and one relating to the dew point as follows:

$$\Delta\varepsilon = 0.0221 \quad 0.0016 \ t_{dew} \qquad (t \text{ in C})$$

This correction is valid for a 2-3 km visability and is in agreement with the formula established from Kuwait observations (Fig. 7). The lower effect of maritime aerosols (in comparison with the effect of dry mists) may be interpreted as consequent to a higher sky emissivity under humid climates.

f) Clouds

The presence of clouds increases the total sky emissivity. This increase depends on the cover n (O<n<1: fractional area of sky covered by clouds) and on the cloud base height. The radiation from the overcast sky may be given by:

$$S = (1-f_W)\ \sigma\ T_a^{\,4} + f_W\ \varepsilon_B\ \sigma\ T_a^{\,4} + n\ f_W(1-\varepsilon_B)\ \sigma\ T_c^{\,4}$$

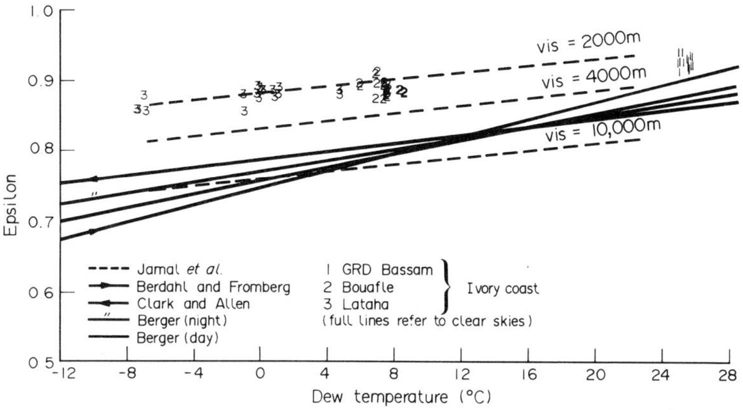

Fig. 7. Observations of clear sky emissivity in Ivory Coast. Comparisons with corrections proposed in reference 7.

$$S = \sigma(T_{cs} + \Delta T_{cs})^4$$

where: f_w : fraction of blackbody radiation in the atmospheric window
T_a : absolute air temperature (k)
ε_B : hemispherical band emissivity
T_c : absolute temperature of the cloud base
T_{cs} : absolute clear sky temperature

The factor $f_w(1-\varepsilon_B)$ in the last term correspondence the fraction of radiation emitted by the cloud in the atmospheric window which can only be received at ground level.

$$\varepsilon_n = \frac{\varepsilon_0 - (1-f_w)}{f_w}$$

where ε_0 is the total sky emissivity

$$S = \sigma\, T_{cs}^4 + n\, f_w(1-\varepsilon_B)\, \sigma\, T_c^4$$

By linearisation we get:

$$T_{cs} = n\,\frac{1-\varepsilon_0}{4T_{cs}^3}\,T_c^4$$

$$= n\,\frac{1-\varepsilon_0}{4\varepsilon_0^{0.75}T_a^3}\,T_c^4$$

Berdahl and Martin (9-10) give results adapted to cloudy conditions, The relations adopted, using Lowtran models and questionable hypothesis were:

$$T_a = T_{sky} = f_w(T_a - T_c) \qquad \text{overcast sky}$$

with f_w = 0.23, 0.17, 0.14 for low (1km), middle (4km) and high (10km) level clouds, and:

$$\varepsilon = \varepsilon_0 + (1-\varepsilon_0)\, n \exp(h/8.2)$$

where h is the cloud base height in km. Table 1 gives their results for tropical model of atmosphere in comparison with ours.

Table 1. Δ_{cs} for 3 cloud base heights (tropical conditions) and for 3 expressions. $n=1$, $\varepsilon_0=0.8$; $T_a=300k$.

	1km ($T_c=294$)	4km ($T_c=277$)	10km ($T_c=237$)
Berger	16.3	12.8	6.9
ref. 9	14.9	12.4	8.4
ref. 10	14.5	10.2	5.1

II ATMOSPHERIC MODELS OF CLEAR SKY IRRADIANCE

Atmospheric models may be classified in three main categories:

i) initial models based on the expression of the emissivity of a column of a radiating gas, and considering monochromatic radiations. Without denying the difficulty in getting an absolutely accurate analysis of the problem, they however propose a convenient solution.

ii) more recent models which are compilations of models solving the spectral transmittance of absorbing gas with peculiar considerations concerning the lines and their distribution in some delimited wavelength ranges. They use numerical integrations, theoretical and empirial formulae, and observed spectra.

iii) simple relations based on parameters at ground level.

A brief overview of these models is given which may prove to be sufficient to evaluate the adopted procedure.

a) Bliss - Kondrat'yev model

The spectral emissivity of water vapour strongly varies in narrow ranges of wavelength (11). The expression valid for a monochromatic radiation: has been considered as a convenient approximation applying to a finite wavelength interval $\Delta\lambda$;

$$\varepsilon_{\Delta\lambda} = 1 \quad \exp(k_{\Delta\lambda}\ m_g)$$

($k_{\Delta\lambda}$, m_g previously defined), then we get for the total emissivity

$$\varepsilon = (1/q) \sum_{\lambda} \varepsilon_{\Delta\lambda}\ I_\lambda\ \Delta\lambda$$

where q is the intensity of blackbody radiation at temperature T, I_λ the blackbody spectral intensity per unit wavelength (Planck's equation). Knodrat'yev gives a set of 32 values for $k_{\Delta\lambda}$ (wavelength intervals from 5 to 35 μ). Outside the tabulated spectrum, $\varepsilon_{\Delta\lambda}$ is assumed to be 1.

Measurement of the emissivity of water vapour is obtained from a radiometer observing the variations in spectral intensity of a narrow beam crossing a column of gas of known physical conditions (composition, length, temperature, pressure). The emissivities of a column (length L) and of a hemisphere (radius L) of gas are equal. But for a slab (thickness L) which contains a supplementary radiating matter, the emissivity is greater, as if the density length product m_g were multiplied by 1.66.

The radiant energy contribution from carbon dioxide lies from 13 to 17 microns where 18.5% of the total blackbody radiation at ambient temperature is concerned. From measurements of the emissivities of water vapour and carbon dioxide in this wavelength range, similar approximated expressions have been established which allow an amount m_c of carbon dioxide to be considered as equivalent to a supplementary amount of water vapour. The adopted correction to ε is (11):

$$\Delta\varepsilon = 0.185\ (\exp(-1.7\ m_g) \quad \exp(-6.7\ m_g))$$

Figure 8 shows the total slab emissivity as a function of the precipitable water at 1 atmosphere pressure. When considering not only a slab but all the layers which compose the atmosphere the computation is done with the following procedure: the atmosphere is considered as a series of layers of slabs. The emissivity of a slab at total pressure P is assumed equal to the emissivity at P_0 obtained with the reduced amount of water m*, with:

$$m^* = (P/P_0)m$$

The emissivity is assumed not to vary with temperature. The adopted variation of pressure with altitude z is:

$$P = P_0 \exp(-1.2\ 10^{4}\ z) \qquad \text{(z in meters)}$$

The adopted variation of water vapour density ρ with height is:

$$\rho = \rho_0 \exp(-4.5\ 10^{4}\ z)$$

where ρ_0 is a function of the surface dew point only. So we get:

$$m^* = {}_0^z\ dm^* = \frac{\rho_0}{5.7\ 10^{-4}}\ (\ -\ \exp(-5.7\ 10^{-4}\ z)).$$

The first step in the determination of the emissivity is to reduce the water vapour content of each slab. The contribution ε_1 of the lowest slab alone is directly obtained from Figure 8. The increase in emissivity when considering the first and second slab together (at T_1 and T_2 temperatures) is associated to an equivalent emissivity contribution of the second slab calculated as a fraction of T_1 blackbody radiation;

$$\Delta_z\varepsilon = (\varepsilon_2 - \varepsilon_1)(T_2/T_1)^4$$

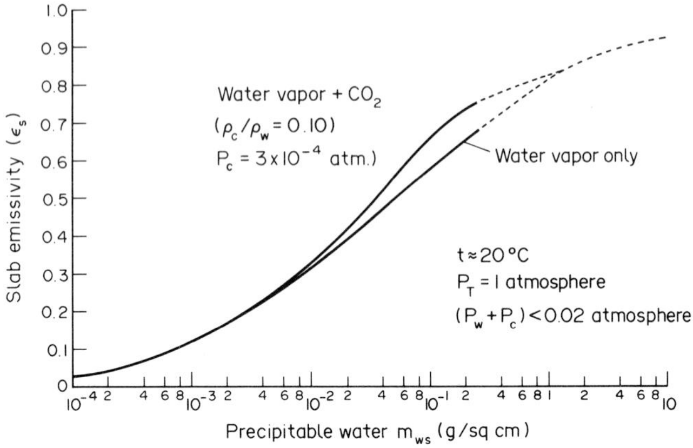

Fig. 8. The slab emissivity as a function of the reduced amount of water (11).

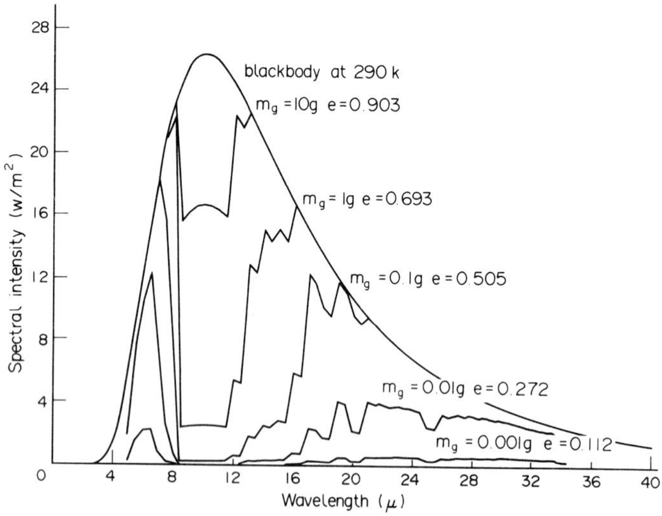

Fig. 9. Spectral intensity of radiation for different amounts of water vapour. Corresponding emissivities (13). Case of a column of gas.

where ε_2 is read in Figure 8 from the total reduced amount of water vapour of the first two slabs. A similar procedure is followed with the other slabs:

$$\Delta_3\varepsilon = (\varepsilon_3 - \varepsilon_2)(T_3/T_1)^4$$

$$T_{sky}^4 = \varepsilon\, T_1^4$$

with : $\varepsilon = \varepsilon_1 + \Delta_2\varepsilon + \Delta_3\varepsilon + \ldots$

The spectral emissivity may be obtained in a similar way. Figure 9 shows the spectral radiation obtained when considering only the water vapour influence and a hemisphere or column of gas.

b) Goody and Lowtran models.

Computers have enabled the development of more sophisticated models of atmospheric absorption. The Aggregate method is a compilation of various models, each one applied to a specific wavelength range. Lowtran method is simpler and requires graphs and empirical formulae.

The extinction of ... iation through an atmospherical layer is proportional ... ensity of the dimmer medium and to the covered dista...

$$dL_\lambda(\lambda,s) \ldots s)\ L(\lambda,s)\ \rho\ ds$$

where: L ... al radiance at the point s

ρ

$k(\lambda,s)$: ... ient of extenuation (absorption and ... ng)

The numerical for ... olution to this equation is (2):

$$L_\lambda(z_0) - \varepsilon_{tar} \ldots \tau_n(\lambda) + \sum_\lambda L_\lambda(T_i)(\tau_{i-1}(\lambda)\tau_i(\lambda)/\tau^*_{si}(\lambda))$$

with: $L_\lambda(tar)$: sp ... ckbody radiance corresponding to the target temperature. ... ver (at altitude z_0) is looking vertically at a targ ... tude z_n)

$\varepsilon_{tar}(\lambda)$: spectr ... vity of the target

$L_\lambda(T_i)$: spectra ... blackbody radiance corresponding to the average temperature T_i of the ith layer of the atmosphere

$\tau_i(\lambda)$: spectral transmittance of the atmosphere from the top of the ith layer to the observer (Fig. 10).

$\tau^*_{si}(\lambda)$: spectral transmittance of the ith layer accounting only for scattering

The last factor on the right side of this equation comes from the following considerations: the expression of τ_i can be written from three terms:

1. the transmittance from the top of the (i-1)th layer to the observer

2. the non-scattered part of the energy in the ith layer

3. the non-absorbed part of the energy in the ith layer:

$$\tau_i = \tau_{i-1} \, \tau^*_{si} \, (1-\sigma_i)$$

According to Kirchoff law ($\varepsilon_i = \sigma_i$ for monochromatic radiations), the second term of the right side of the equation is:

$$\Sigma \, L_\lambda(T_i) \, \varepsilon_i \, \tau_{i-1} = \Sigma \, L_\lambda(T_i)(1 - \tau_i/(\tau^*_{si}\tau_{i-1}))\tau_{i-1}$$

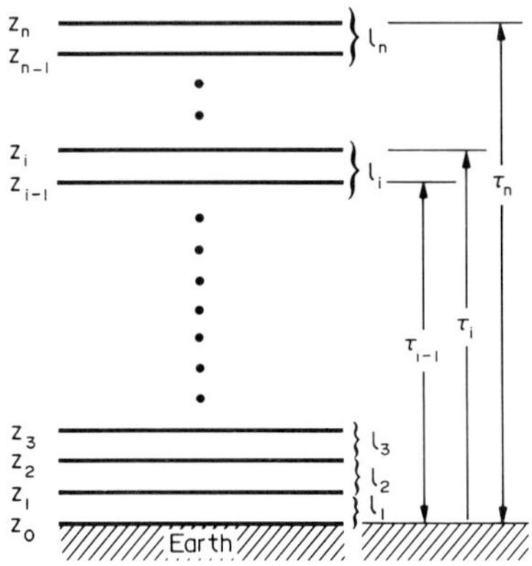

Fig. 10. Schematic and layered atmosphere for transmission and radiance calculations (2).

Slant paths are considered by multiplying the optical path by sec θ (θ:zenithal angle of the line of sight). Integration can be achieved from θ = 0 to 80 degrees, and by adding the contribution of a blackbody radiation at ground temperature for the last ten degrees. As suggested by Bliss, it may be done by introducing the factor 1.66 in the density length product, then by considering a 53 degrees average value of θ.

The problem consists on determining the values of $\tau_i(\lambda)$, or $\tau_i(\Delta\lambda)$ for small wavelength intervals (0.1 μ in our computed Lowtran method, 0.5 μ in our Goody one).

The absorption of radiation can be got by a combination of the Lorentz or Doppler profiles (or mixed Lorentz-Doppler lines) according to the pressure of the surrounding medium. Several models have been elaborated for a practical approach to compute the atmospheric absorption; they use assumptions on the distribution, spacing and widths of the lines over a band. (Regular models such as the Elsasser model, random models such as the Goody model, line-by-line calculations (12) ...).

The Aggregate model uses for H_2O the Strong Line Goody Model in the band 1.0 to 2.0 μ and 4.3 to 15.0 μ, and the Goody Model in the band 2.0 to 4.3 and 15.0 to 30.0 μ. Other models are used depending on the constituents and on the bands. The Lowtran method is empirical and depends on a single adjustable parameter: the equivalent absorber amount. It is simpler, and thus more attractive.

b) - 1) Strong Line Goody Model, and Goody Model

The Strong Line Goody Model is a statistical band model which supposes that in a spectral interval, n lines exist with equal intensity, large strength line S and large amount w. Then the transmittance is calculated with the following form of expression:

$$\tau(\lambda) = \exp(-(w^* \ k(\lambda))^{0.5})$$

where w* is the equivalent absorber amount given by:

$$w^* = \rho_0 \int_{path} M(z) \ (P(z)/P_0)^2 \ (T_0/T(z))^{1.5} dz$$

$k(\lambda)$: spectral coefficient

ρ_0 : air density (kg/cm³) at standard temperature and pressure

$M(z)$: mixing ratio at altitude z (g H_2O/kg air)

$P(z)$: pressure at z (mm Hg)

P_0 : standard pressure (760.0 mm Hg)

$T(z)$: temperature at z (kelvin)

T_0 : standard temperature (273.16 kelvin)

All these quantities can be either measured, or determined from tables, and by considering the relation:

$$\rho_0 \ M(z) \ (P(z)/P_0)(T_0/T(z)) = \rho(H_2O)$$

$$= H_2O \text{ vapour density}$$

The Goody Model considers an exponential distribution for the probability of the strength of the lines. Then the transmittance becomes:

$$\tau(\lambda) = \exp(-x \ w/(1+2/P^* \ y)^{0.5})$$

with: $w = \rho_0 \ M(z) \ (P(z)/P_0)(T_0/T(z)) \ dz$

$P^* = P_0 \ w^*/w$

x,y given in tables

b) - 2) Lowtran Method

In the Lowtran Method the spectral transmittance is directly given from the equivalent amount of absorber w* by the way of curves. For H_2O absorption w* is obtained from the relation:

$$w^* = \rho(z) \ (P(z)/P_0)^{0.9} 0.1 \qquad g \ cm^{-2} km^{-1}$$

with ρ in g/m³. Figure 11 shows as a comparison the application of these two Goody models, and of Lowtran model for H_2O molecular absorption, vertical path through an hypothetical layer, and for various absorber amounts. The more detailed curve of Figure 11b is due to the lower stepsize of the wavelength interval used in the Lowtran method.

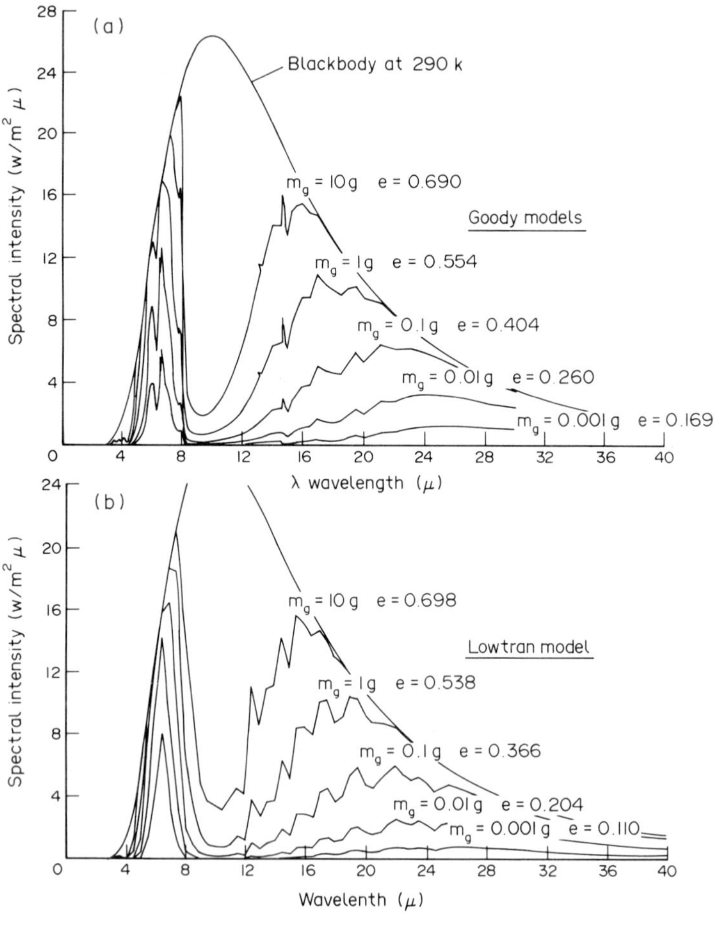

Fig. 11. Comparisons between Goody and Lowtran models for H_2O (without continuum) influence (13). The effect of H_2O continuum can be observed by comparison with Figure 9.

b) - 3) Remarks

The computation of these methods requires complementary informations: * the transmittance for the other atmospherical constituents is computed by similar ways as for H_2O molecular absorption. The total transmittance is their product:

$$\tau = \tau_{H_2O}\ \tau_{CO_2}\ \tau_{O_3}\ \ldots$$

The integration uses numerical models of atmosphere (2), or/and measurements from radio-prospectings. It is done over the altitude, the zenithal distance (solid angle), and the wavelength. Concerning this last variable, a special care must be carried to the upper limit: up to 40 μ only 93.5% of the emitted energy of a blackbody at 8 °C is obtained. 16.1 w/m² are missing, that is 3.1 °C less for the corresponding temperature. The F function gives the part of the energy of a blackbody which is in a wavelength range:

$$F_{40\mu\rightarrow\infty}(T) = 1.1030\ 10^{-9}\ T^3 - 1.4878\ 10^{-7}\ T^2 + 7.1356\ 10^{-6}\ T$$

for: $248 < T < 318$ (kelvin)

$0 < F < 1$ and $\Delta F\ \sigma\ T^n < 0.2$ w

The accuracy of the computation depends on the radiative influence of each constituent. The relative importance has been found by computation of the downward flux and by considering successively the only effect of each constituent (the total effect is not the sum on account of the product of the transmittances). For mid-latitude summer atmosphere these fluxes are (13):

H_2O (lines only)	:	263 w/m²
H_2O (lines + continuum)	:	328 w/m²
CO_2	:	76 w/m²
O_3	:	5 w/m²
H_2O (lines) + CO_2 + O_3 + CH_4 + N_2O	:	301 w/m²
H_2O (lines + continuum) + CO_2 + O_3 + CH_4 + N_2O	:	341 w/m²
$\sigma(294)^4$	:	423 w/m²

c. Simple expressions for the clear sky emissivity

Measurements at ground level taken in the last decades have lead to formulations in which air temperature and humidity were the only two parameters. The small amount of data, and the accuracy of the pyrradiometers did not allow these studies to go further. Either the dew temperature T_d (kelvin), the resulting depth of condensed water w (cm), or the water pressure vapour c (mb) were the variables in the expression of the emissivity ($\varepsilon = T_{sky}^4/T_{air}^4$), but one has to bear in mind the relations existing between these parameters:

$w = 0.17\ c$ $\qquad \rho_0 = 0.623\ (353/T)\ (c/1013)$ $\qquad$ (kg/m³)

$$c = (1013/760) \exp(20.519 - 5179.25/T_d)$$

It is not useful to remember all these formulations because of the large deviations. Observations with very high quality instruments allow the following relations to be considered to be the most reliable ones in spite of a non negligible deviation from one to another.

Recent work by Clark and Allen (14), summarizes the results of more than 800 clear sky night measurements in several centres with modern equipment (Eppley and Funk type infrared radiometers) and lead to the expression:

$$\varepsilon_{sky} = 0.787 + 0.0028\, t_{dew} \qquad (t_{dew} \text{ in Celsius})$$

With a few thousand measurements with pyrgeometers and angular spectral radiometers in some cities of the U.S.A., Berdahl and Fromberg pointed out a day-night effect (15):

$$\varepsilon_{night} = 0.741 + 0.0062\, t_{dew}$$

$$\varepsilon_{day} = 0.727 + 0.0060\, t_{dew}$$

More than 1600 data obtained with a compensated pyrradiometer of the French National Meteorological Institute situated in its Carpentras station allowed us a determination of the emissivity. The influence of the ground thermal inertia, of the height of the atmospheric boundary layer, and of the sunrise (16) was added to the day-night effect which has already been pointed out. The air temperature became a corrected air temperature:

$$T_{air} \rightarrow T_{air} + K(H) + L(H)\,(T_{dew} - T_{air}) = T_{air,\ corrected}$$

with k and l: parameters which depend on the trihourly index (4 for sunrise, 5 for sunrise + 3 hours, ...). cf. Table 2.

Table 2.

hourly index	1	2	3	4	5	6	7	8
K	6.45	7.18	7.65	3.86	0.18	2.90	4.85	5.65
L	0.435	0.620	0.616	0.399	0.276	0.290	0.311	0.364

$$\varepsilon_{night} = 0.770 + 0.0038\, t_{dew}$$

$$\varepsilon_{day} = 0.752 + 0.0048\, t_{dew}$$

Physically, the zone in the atmosphere which is between the

ground and the boundary layer contains the quasi-totality of the humidity, and thus, strongly influences the T_{sky} determination. This boundary layer, from a lower altitude at sunrise, lifts up with a delay of about 1 hour due to the ground heating. From the middle of the afternoon, it comes downwards slowly and regularly, without any specific effect at sunset (19). k refers to the ground inertia, l($T_{dew}-T_{air}$) to the altitude h of the atmospheric boundary layer, according to the simple relation for its evaluation:

$$h = (T_{dew}-T_{air})/8$$

These expressions are in good agreement in the case of temperate climate conditions where the most part of the observations were obtained. Local climatic effects, frost or high humidity may perhaps explain the observed differences. The spreading of the observed results may also be responsible for these differences.

Sophisticated models and simple relations were compared in order to propose a convenient spectral description of the sky irradiance. The value of the emissivity only differs from unity in the atmospherical window (spectral range: 8 - 14 μ). In this range, ε_λ is given by the relation:

$$\varepsilon_\lambda = 1-\exp(-k_\lambda\ w)$$

where k_λ is a coefficient given in Table 3, and w the total equivalent absorber amount. The study of radio-prospectings (20) lead to an evaluation of these quantities:

$$w_{night} = 0.9996 + 0.060\ t_{dew}$$

$$w_{day} = 0.8974 + 0.044\ t_{dew}$$

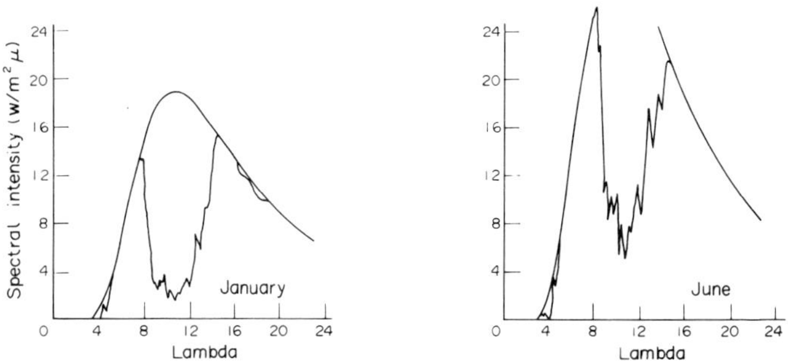

Fig. 12. Spectral intensity of radiation in Nimes. (average monthly radio-prospectings for clear sky conditions).

Figure 12 shows an example of spectral irradiance obtained from radio prospectings. Figure 13 shows 3 examples of data obtained

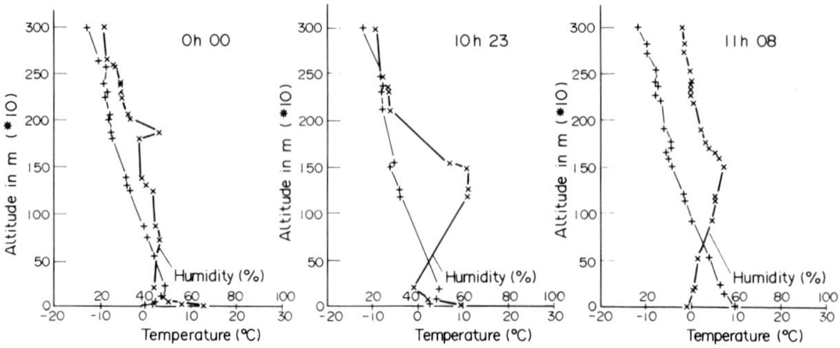

Fig. 13. 3 atmospheric profiles from radio-prospectings in Ajaccio (15-1-1979) (National Meteorological Institute).

from radio prospectings to point out the observed irregularities near ground level. According to the high influence of the climatic conditions in the first hundred metrês of altitude on the sky irradiance, it seems impossible to propose a determination in any place, and at any moment, with a universal relation or model. Thus, local measurements still remain a necessity. Too sophisticated models also appear as only useful for theoretical studies, and not as convenient for practical uses in radiative cooling.

Table 3.

λ in μm	k_λ
8.0 - 8.5	1.000
8.5 - 9.0	0.404
9.0 - 9.5	0.251
9.5 - 10.0	0.245
10.0 - 10.5	0.178
10.5 - 11.0	0.164
11.0 - 11.5	0.222
11.5 - 12.0	0.298
12.0 - 12.5	0.406
12.5 - 13.0	0.691
13.0 - 13.5	0.873
13.5 - 14.0	1.231

III. INSTRUMENTATION

Sky radiation knowledge is associated with a variety of measurements: downward infrared flux (spectral or total, hemispherical or angular), total flux (solar + infrared) and solar flux, net flux (downcoming celestial - outgoing terrestrial), state of the atmosphere (temperature, pressure, composition, hygrometry, nebulosity, visibility...). We may consider several types of instruments among those specifically intended to radiation itself:

* usual instruments situated at ground level for global measurements of atmospheric thermal flux.

* sophisticated radiometers for angular and spectral measurements at ground level. Normally only a few examples are built.

* balloons (or satellites) for measurements of the atmospherical parameters in altitude.

Some of these are described below not in great detail but in order to show the complexity with which they have been conceived.

a) The Eppley pyrgeometer

The Eppley pyrgeometer is a widespread instrument used for global measurement of thermal sky radiance. It is constituted by a thermopile covered with a silicon KRS5 hemisphere. A weather protective coating is deposited on the outer surface of the dome. On the inner surface an interference filter makes it opaque at wavelength lower than 3 μ. The transmittance is not greater than 0.5 from 4 to 40 μ. Nevertheless, measurements before sunset and after sunrise have shown significant errors (up to 5 °C) in the T_{sky} determination when a complete blackness of the celestial vault is not obtained (18). Possibilities have been studied (15) of evaluating the sun influence (heating of the dome and transmission).

The cooling of the sensitive surface causes the thermopile to produce an output voltage. A thermistor circuit is incorporated for temperature compensation (needing a 1.4 V mercury battery), the downcoming thermal sky radiance being the addition of the net radiation which affects the receiver surface, and of the outgoing radiation flux due to the temperature of the detector.

The sensitivity is about 3 $\mu V/W/m^2$. Calibration is accomplished by referring to blackbody sources or to radiometers. Direct measurements of the receiver temperature by the additional incorporated thermistor, lead to a severe limitation of the error, the output signal being, in a major part, a correction for temperature compensation. In spite of this, an average precision better than 0.5 °C can be obtained in the T_{sky} determination. Figure 14 shows the apparatus.

b) The Buriot pyrradiometer

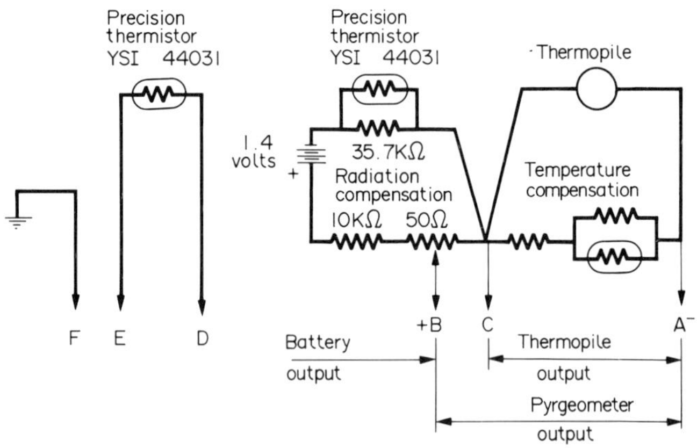

Fig. 14. The Eppley pyrgeometer and the electrical circuit arrangements:

* thermopile and temperature compensation of radiometer response
* radiation compensation for detector temperature
* additional thermistor (from notice of the pyrgeometer).

b) The Buriot pyrradiometer

The evaluation of the sky radiance for the purpose of cooling applications, especially in hot countries, leads to the necessity of the existence of a cheap IR pyrradiometer. Because the weather conditions are frequently bad (wind, dust, sand, humidity...), the considerable expense of any instrument and associated micro-voltmeters, the need for local measurements with a sufficient accuracy but not with a high precision, lead to the conception of the Buriot apparatus (16). The basic principle is the measurement of the temperature of a black copper radiator cooled by radiative exchange with the sky by means of a copper-constant or thermocouple. A transparent polyethylene (IR transmittance $\neq$ 0.6) cover gives a thermal protection against convective exchanges (Fig. 15).

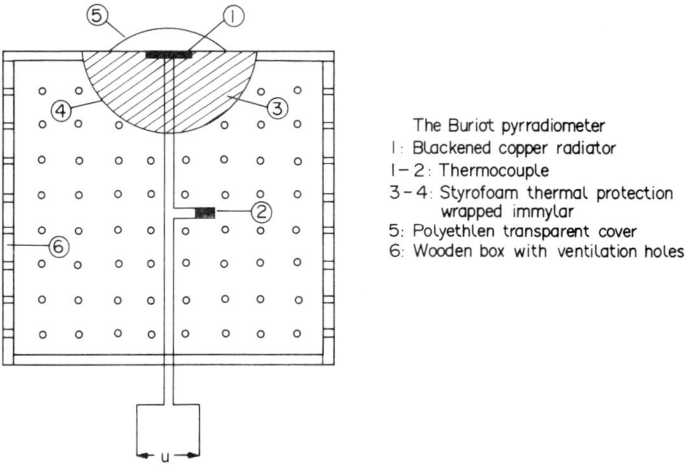

Fig. 15. The Buriot pyrradiometer.
1: blackened copper radiator
1-2: thermocouple
3-4: styrofoam thermal protection wrapped inmylar
5: polyethylene transparent cover
6: wooden box with ventilation holes.

The calibration is done with an Eppley pyrgeometer as a reference. The good agreement between the two instruments is the result of frequent and careful calibration. Radiation compensation is solved by measuring the difference $T_{radiator} = T_{air}$. The observed values fit well with the proposed expression.

$$T_{sky} = T_{air} - C\,U$$

where: C: calibration constant ($\neq$ 0.1 c/μV)
U: voltage.

Later modifications have incorporated an amplifier to allow voltage measurements with an universal tester (19). More recently, two micro-circuits have been substituted to the thermocouple (reference: Thomson EPCIS-TDB 0135 ACM). The new sensitivity of 10 mV/C was sufficient to avoid the previous amplifier.

c) The Gier and Dunkel differential pyrradiometer

This apparatus is made of silver-constantan thermopiles mounted on a black resin plate (20). Two polyethylene hemispheres protect them and forced ventilation is assumed to maintain the same convective losses for the two sides. The temperature gradient corresponding to the radiative fluxes on the two sides is proportional to the net balance. An incorporated thermocouple permits the mean temperature of the thermopile to be recorded (Fig. 16).

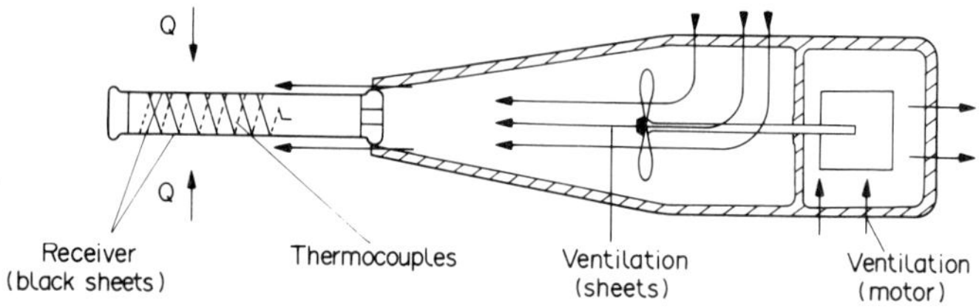

Fig. 16. The Gier and Dunkel differential pyrradiometer.

d) The Funk differential pyrradiometer

This apparatus uses a thermo-electrical battery made of 250 copper constantan thermocouples mounted on two thin aluminium sheets, painted black on their outer side. Two polyethylene hemispheres protecting them are maintained with an excess pressure of nitrogen. Condensation on the hemispheres are avoided by the use of a heated air flux injected through the holes of an external ring girdling the detector (Fig. 17).

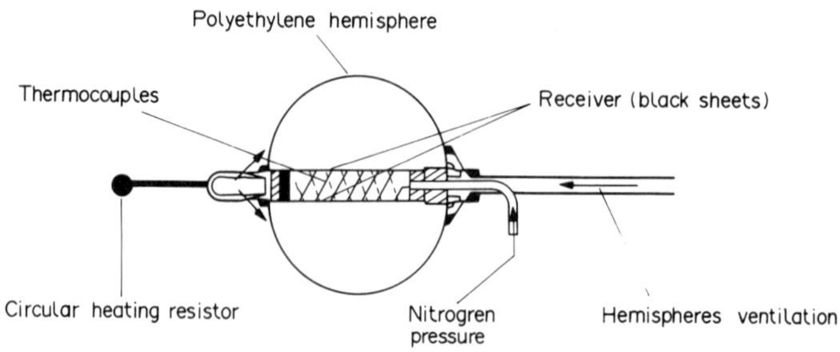

Fig. 17. The Funk differential pyrradiometer.

Other instruments based on the above principles are planned by manufacturers in order to obtain an enhanced performance. Clark used several Funk IR radiometers and Eppley pyrgeometers for his determination of the foreseen relation. But "extremely careful calibration and maintenance of instruments is necessary in order to keep net radiation errors below 10% (14).

e) The compensated pyrradiometer

The compensated pyrradiometer of which the data have been used to determine the above expression of T_{sky} by Berger et al., is an experimental apparatus of the French National Meteorological Institute (21). It is made of a Moll thermopile (14 thermocouples manganin-constantan) painted black and compensated in temperature. A polyethylene (lupolen H) hemisphere protects it. The cold soldering is in excellent thermal contact with the body of the apparatus. The hemisphere is maintained with an excess pressure of dehydrated air, and is thermally insulated from the body of the apparatus.

Wind effects and the infrared emission of the hemisphere are compensated as follows: a second hemisphere, white painted on the external side, and painted black on the internal side, has its temperature measured by a second black painted thermopile. As the temperature of this second hemisphere is not exactly the same as the temperature of the transparent hemisphere, only a part of the measured signal is used for compensation (electrical mounting in opposition with both the measurement circuit and compensation of IR emission of the thermopile). Forced ventilation of the hemispheres is assumed to minimize wind effects and to avoid frost and dew.

Compensation of the IR emission of the thermopile is obtained as follows: a thermistance is maintained in excellent thermal contact with a silver black disk of which the circumference is in excellent thermal contact with the body of the pyrradiometer. The disk is under a third polyethylene hemisphere. The measured signal corresponds to the average adopted temperature of the thermopile. All the parameters are adjusted by calibration. Figure 18 shows the apparatus and the electrical scheme. The extremely careful construction and calibration lead to a high precision and to a possible use, during daytime as well as during night-time.

f) Spectral and angular radiometers

The spectral radiometer used by Berdahl and Fromberg (15) and manufactured by Barnes Engineering Corporation, has six bandpass filters which refer to the atmospherical window (8-14μ), ozone bands (8.8 and 9.6μ), the minimum wavelength for sky emissivity (11μ), the strong band of CO_2 (15μ), and the end of influence of water vapour (17-22μ). The apparatus has a field of view 2 degrees wide, which permits directional measurements. Calibration with a reference blackbody, and comparisons with Lowtran model have shown a random error less than 0.7 °C.

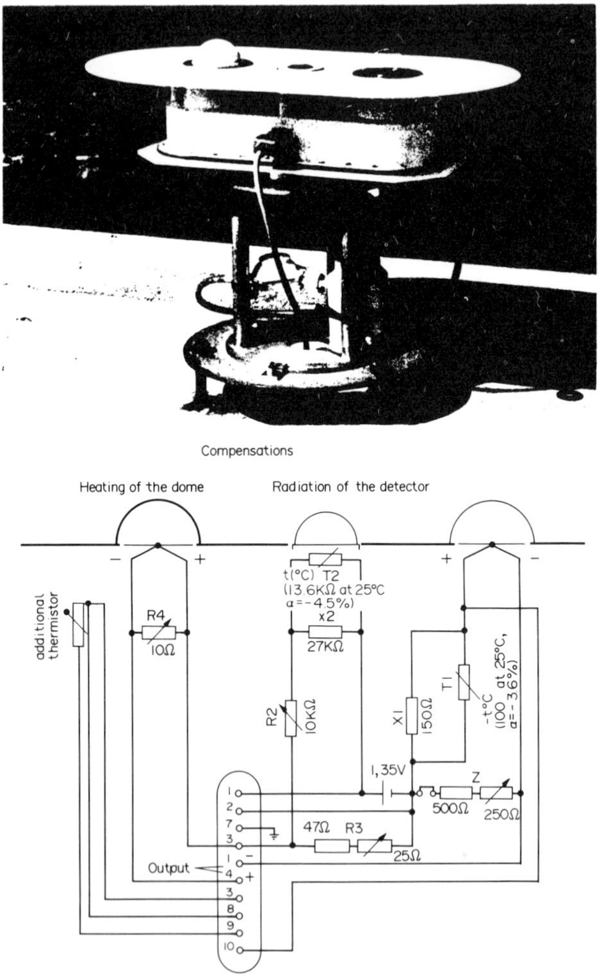

Fig. 18. The compensated pyrradiometer (18).

The pseudo-pyrgeometer is a radiometer with no filter. The spectral sensitivity is high in the atmospheric window. Averaging over the sky dome is accomplished by using the cosine of the zenith angle as a weighting function. Measurements during daytime are possible according to the insensitivity to scattered solar radiation, and if the orientation is opposite to the sun.

Berdahl and Fromberg (15) pointed out a day-night effect with this apparatus.

g) Radio-prospectings

The radio-prospecting technique is now used in order to obtain the values of the main atmospheric parameters at altitude. If balloons launched by Meteorological Services are able to reach altitudes higher than 10 km, this is only achieved by a few of them. Moreover, measurements by the French National Meteorological Institute are only made at 0 H 00 and 12 H 00 UT. Smaller balloons are more frequently launched for pollution control (nuclear or thermal centres of E.D.F.,...) and give the temperature, pressure and humidity up to an altitude of 3 km.

150 profiles have been studied in our Laboratory in association with Lowtran program to compute the sky radiation. The day-night effect was pointed out (Figure 19), and no specific effect of local climate (maritime in Brest, semi-continental in Trappes, Mediterranean in Nimes) could be moticed. Both the accuracy of the computed program, and the possibility of comparing measurements at ground level with measurements of the state of the atmosphere at altitude have thus been confirmed.

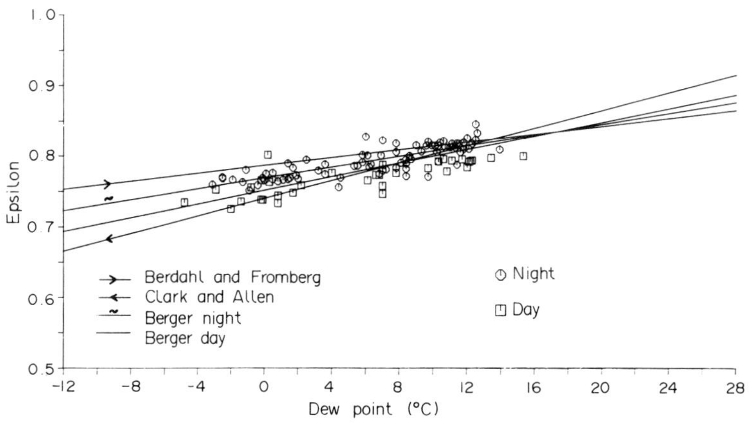

Fig. 19. The sky emissivity, as obtained from radio-prospectings (average monthly radio-prospectings for clear sky conditions. Observing sites: Nimes, Brest, Trappes). Day-night effect. Data of the National Meteorological Institute used in association with Lowtran code.

IV CONCLUSION

Sky radiation knowledge has greatly improved in the last years, either by the computation of codes, or by the analysis of high quality measurements. The predominant influence of water vapour (H_2O continuum effect greater than foreseen), and carbon dioxide concentration near constant (320 ppm) up to the stratopause where it vanishes, confirm the possibility of an accurate determination from measurements of air parameters at ground level. A day-night effect, and even an hourly effect, has been pointed out. However some parameters remain unsufficiently known, or taken into account e.g. visibility, nebulosity, height of the boundary layer, gradients in the first ten metres near ground level, altitude of the site.... Moreover radiation measurements are still affected by a noticeable error. Accounting to the present precision, local climatic conditions do not seem to affect the expressions, as observations in various stations have shown. A correction for the altitude of the observing station has been proposed (22) which, by considering the expression of the pressure with altitude, may be expressed as follows:

$$\Delta\varepsilon = 0.1216\ (\exp(-1.2\ 10^{-4}\ z) - 1) \qquad (z \text{ in metres})$$

Observations under hot climates and countries are needed to check the expressions established from measurements in the north hemisphere. This leads us to undertake measurements either in Africa or in Antilles, as well as at altitude (Fig. 20). But they are actually not numerous enough to extend the knowledge of sky radiation. The interest for very high precision measurements, and for new codes for atmospheric models, is also considerable, but for radiative cooling purpose, local measurements are perhaps more useful than theoretical expressions. Other research goals include a better knowledge of the state of the atmosphere and of the energy balance between the ground and the atmosphere. This research requires a still greater knowledge of sky radiation.

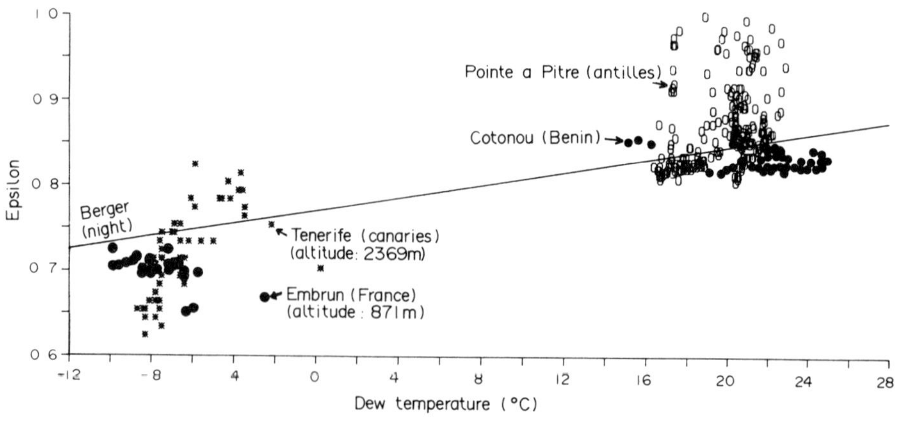

Fig. 20. Some observations in hot climates or in altitude (Buriot pyrradiometer or Eppley pyrgeometer).

Acknowledgements - The authors acknowledge their thanks to Mr Michel Schneider for his contribution in this work.

REFERENCES

1. K.Y. KONDRAT'YEV, Z.F. MIRONOVA, A.N. OTTO; 1964, Pure and applied geophysic, Birkhauser Verlag, Switzerland, vol. 59-3, p. 207-216. Spectral albedo of natural surfaces.

2. W.L. WOLFE, C.J. ZISSIS; 1978, IRIA, Environmental Research Center of Michigan U.S.A., Infrared Handbook.

3. R.A. McCLATCHEY, R.W. FENN, J.E.A. SELBY, F.E. VOLTZ, J.S. GARING; 1972, U.S. Air Force Cambridge Research Laboratory Bedford Massachusetts 72-049, Optical properties of the atmosphere.

4. R.A. CRAIG; 1965, Academic Press New-York, The upper atmosphere. Meteorlogy and physics.

5. P. Queney; 1974, Masson ET Cie Paris, Elements de Meteorologie.

6. C.J. DALRYMPLE, M.H. UNSWORTH; 1978, Quart. J.R. Met. Soc. 104, p. 989-997, Longwave radiation at the ground: comparison of measurement and calculation of radiation from cloudless skies.

7. K. AL JAMAL, M. QUINN, D. JARRAR, N. SHABAN, Y. YAGOUB, J. D'SOUSA; 1984, Solar and wind technology I n 2 p 109-114, Dust effect on infrared sky radiation in Kuwait.

8. X. BERGER, F. GRIVEL, J.C. DEVAL, J. TUBIANA, M.J. TUBIANA; 1985, Rapport Rexcoop A.F.M.E Paris, Le confort thermique vecu en climat chaud.

9. P. BERDAHL, M. MARTIN; 1981, International passive and hybrid cooling conference Miami (AS/ISES). Thermal radiance of skies with low clouds.

10. M. MARTIN, P. BERDAHL; 1984, Solar Energy 33, 3/4, p. 321-336, Characteristics of infrared sky radiation in the United States.

11. R.W. BLISS; 1961, Solar Energy 5, 3 p 103-120, Atmospheric radiation near the surface of the ground: a summary for engineers.

12. F.M. LUTHER; 1984, World climate program, International council of scientific unions and world meteorological organisation (report of a meeting in Italy), The intercomparison of radiation codes in climate models - Longwave clear sky calculations.

13. B. CUBIZOLLES; 1985, Rapport de contrat A.F.M.E. - I.R.B. A.T., Etude du rayonnement atmospherique.

14. G. CLARK; 1981, International passive and hybrid cooling conference Miami, Passive/hybrid comfort cooling by thermal radiation.

15. P. BERDAHL, R. FROMBERG; 1982, Solar Energy 29, 4 p 299-314, The thermal radiance of clear skies.

16. X. BERGER; 1984, Solar Energy 32, 6 p 725-733, About the equivalent radiative temperature for clear skies.

17. H. TENNEKES; 1954, Physics to-day p. 52, The atmospheric boundary layer.

18. K.Y. KONDRAT'YEV; 1965, Pergamon Press, Radiative heat exchange in the atmosphere.

19. C.N. AWANOU, X. BERGER; Colloque Meteorologie et energies renouvelables A.F.M.E. France, Un appareil simple pour la mesure des temperatures de ciel.

20. R. DOGNIAUX; 1984, Colloque Meteorologie et energies renouvelables A.F.M.E. France, Instruments et methodes de mesure en radiometre solaire et terrestre.

21. R. COUDERT, P. GREGOIRE; 1978, Note technique n 1 de la Meteorologie Nationale (France), Le pyrradiometre compense.

22. D. STALEY, G. JURICA; 1972, J. appl. meteor. 11 p. 349, Effective atmosphere emissivity under clear skies.

CHAPTER 7

RADIATIVE COOLING II: APPROPRIATE SYSTEMS

Xavier BERGER and Michel SCHNEIDER

Centre National de la Recherche Scientifique, FRANCE
Laboratoire d'Ecothermique
Sophia - Antipolis, B.P. 21
06561 VALBONNE CEDEX FRANCE

INTRODUCTION

For thousands of years the sky has been recognized as the coldest available resource for passive cooling systems. Today, research is mainly oriented towards higher performance through selective coatings, polyethylene films, movable insulating panels... Instead of reviewing this work, a specific point of view was adopted in this chapter.

For many people in hot countries, radiative cooling is the result of a vernacular understanding attached to the needs of common life; the various aspects of the applications insure comfort, agreement and pleasantness, aesthetics, cheapness, a sufficient efficiency and a traditional continuity.

The present chapter is oriented towards a radiative cooling analysis: In the first two parts we describe the properties of the cooling source and the comfort aspects of man in radiative cooling conditions. The third one points out the limits of radiative cooling and the need for its association with other cooling modes in order to increase performance. The last part describes research which brings together present knowledge and natural inexpensive adaptations to climatic conditions.

I - RADIATIVE PROPERTIES OF THE ATMOSPHERE

a) the equivalent sky radiative temperature

We have shown in the previous chapter that the equivalent black body temperature of the atmosphere may be expressed as follows:

$$T_{sky} = (T_{air}+K(H)+L(H)(T_{dew}-T_{air}))\varepsilon^{0.25}$$

where T is the Kelvin temperature. ε is the emissivity. K and L are parameters which depend on the trihourly index (4 for sunrise, 5 for sunrise + 3 hours, ...) cf. table 1.

Table 1

HOURLY INDEX	1	2	3	4	5	6	7	8
K	6.45	7.18	7.65	3.86	0.18	2.90	4.85	5.65
L	0.435	0.620	0.616	0.399	0.276	0.290	0.311	0.364

The day-night effect pointed out (1) has been confirmed with radio-sounding analysis (2) and lead to the following formulation

$$\varepsilon_{night} = 0.770 + 0.0038\ t_{dew}$$

$$\varepsilon_{day} = 0.752 + 0.0048\ t_{dew}$$

with t_{dew} in °Celsius.

Berdahl et al. (3) proposed a similar formulation.

It must be remembered that the spectral emittance ε of the atmosphere is that of a black body ($\varepsilon = 1$) at a temperature equal to the air temperature near the ground. The value of ε only differs from unity in the atmospherical window (spectral range 8μ - 14μ). In this range a convenient form for the expression ε_λ is according to Kondrat'yev (4):

$$\varepsilon_\lambda = 1 - \exp\ (-k_\lambda . w)$$

with k_λ, a coefficient given in Table 2

w, the total equivalent absorber amount

The study of radio-soundings (5) gives an evaluation of these quantities.

$$w_{night} = 0.9996 + 0.060\ t_{dew}$$

$$w_{day} = 0.8974 + 0.044\ t_{dew}$$

with t_{dew} in ° Celsius.

Figure 1 shows an example of the spectral energy distribution.

b) Selected properties of the sky irradiance

Several graphical representations may be used to set out the radiative properties of the sky:

* on Figure 2, the psychrometric diagram has been used to show the networks of lines

Table 2

λ in μm	k_λ
8.0 - 8.5	1.000
8.5 - 9.0	0.404
9.0 - 9.5	0.251
9.5 - 10.0	0.245
10.0 - 10.5	0.178
10.5 - 11.0	0.164
11.0 - 11.5	0.222
11.5 - 12.0	0.298
12.0 - 12.5	0.406
12.5 - 13.0	0.691
13.0 - 13.5	0.873
13.5 - 14.0	1.231

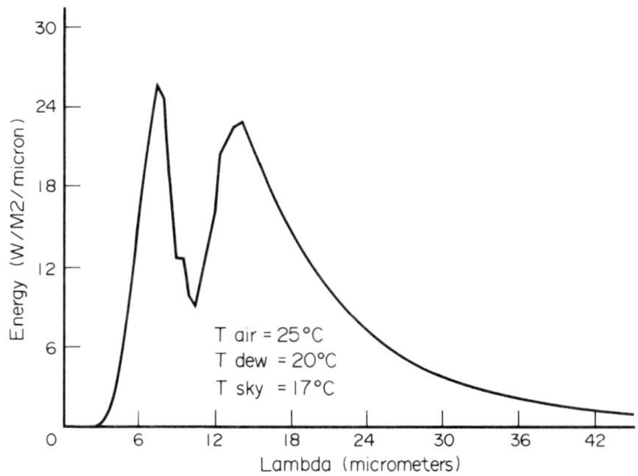

Fig. 1. The spectral energy distribution of the sky radiation as given by the simple proposed model.

$$y = T_{air} - T_{sky}$$

and $z = \sigma \cdot (T_{air}^4 - T_{sky}^4)$

This diagram shows the large cooling possibilities under hot climates. Far from the sea, the water content measured at ground level is almost constant during daytime. Then, with air heating the sky is found to be at its maximum radiative efficiency at mid-day. The conditions to get ice are also pointed out.

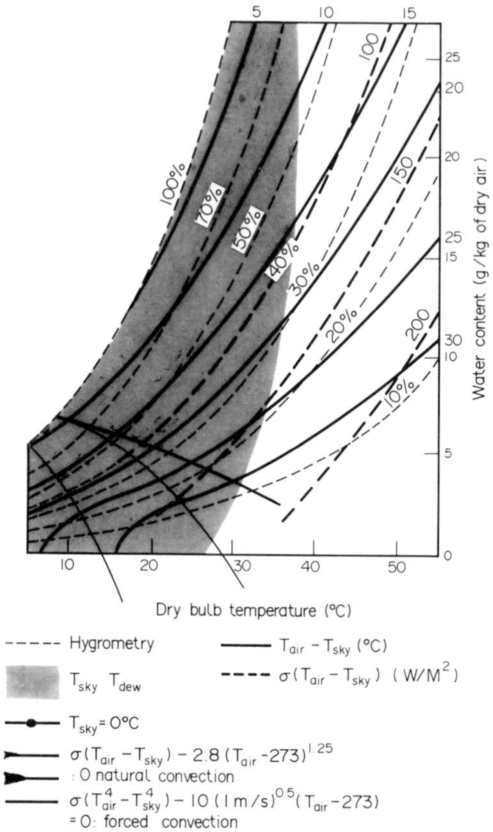

Fig. 2. The sky radiative properties drawn in the psychrometric diagram.

In Mediterranean, and even more in humid countries, the radiati power is reduced, but the dew point may be easily attained. In these cases, the hygrometric degree is almost constant and radiative efficiency is found to be a maximum in the morning (Fig. 3).

Figures 4 and 5 use another graphical representation in T_{air} - T_{sky} versus T_{air} to point out some properties:

for low air temperatures, the level of humidity is insignificant.

for high temperatures, the difference T_{air} - T_{sky} is decreasing for high humidity (more than 50%) when the air tempera ture is increasing. This difference increases with air tempera ture, but only till a maximum for low humidity. Thus, the influence of humidity is large, only for low levels of humidity.

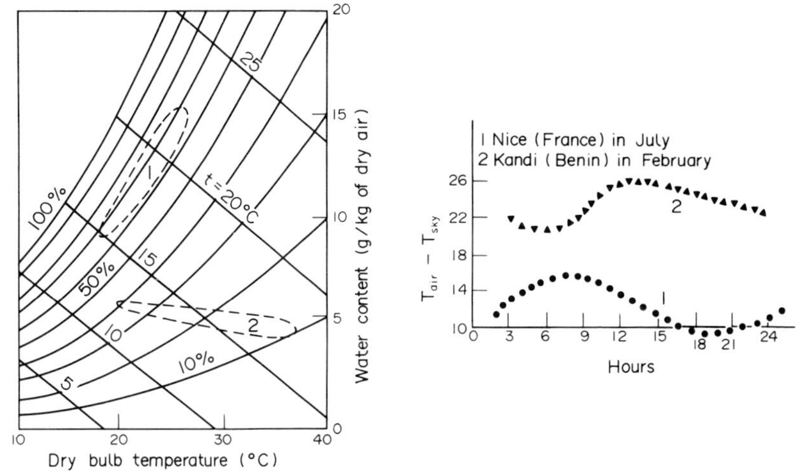

Fig. 3. $T_{air} - T_{sky}$ depending on the maritime or continental aspect of the climate.

c) Pollution, aerosols, clouds.

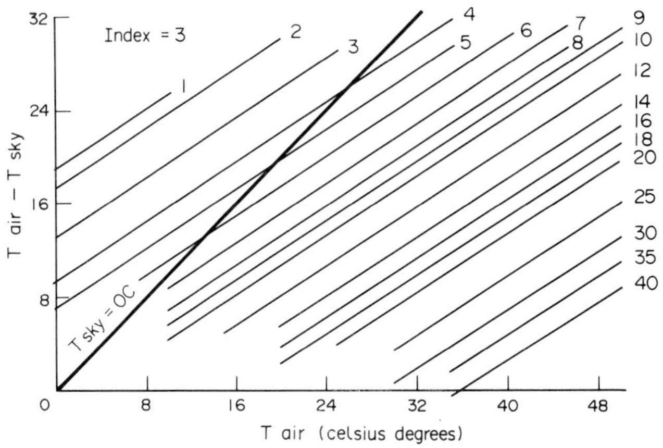

Fig. 4. $T_{air} - T_{sky}$ as a function of air temperature and water content (g/kg of dry air).

Clouds or atmospheric pollution may enhance the atmospheric radiation, and reduce the efficiency of radiative cooling. Two main effects may be quoted:

The effect of clouds has been variously studied to take into account the altitude of clouds in overcast skies. Berdahl and

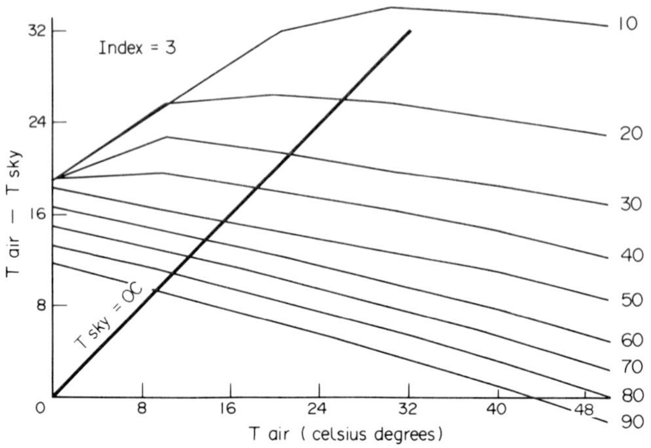

Fig. 5. $T_{air} - T_{sky}$ as a function of air temperature and air Hygrometry (%).

Martin (6) propose the following simple relation:

$$T_{air} - T_{sky} = f_w(T_{air} - T_c)$$

where T_c is the temperature of the cloud base (typical lapse rate 6°C per km). The nominal value f_w equal to 0.32 is well fitted for mid-latitude countries, whereas for tropical countries the f_w values are 0.23, 0.17 or 0.14 respectively for cloud base altitudes of 1, 4, and 10 km. Maritime or urban aerosols may also affect the sky radiation (7), but much more, as very commonly met in sahelian countries, the sand carried by the wind (harmattan) may reduce the radiative cooling power of the sky. Figure 6 shows that the observed emissivity in Ivory Coast for a clear sky is reduced to about a half.

II - THE COLD RADIATIVE COMFORT ZONE

The comfort sensitivity of a man is the result of physiological effects (thermal balance) and psychological perception of his environment.

Among the different thermal zones he is able to meet, the cold radiative comfort zone, well adapted to sky cooling effects is presented here. At the same time, the physiological aspects, and the thermal interest with respect to the other zones are to be considered.

a) Physiological aspects

Homeothermy is the result of a null thermal balance between internal heat production (metabolism) M and various external

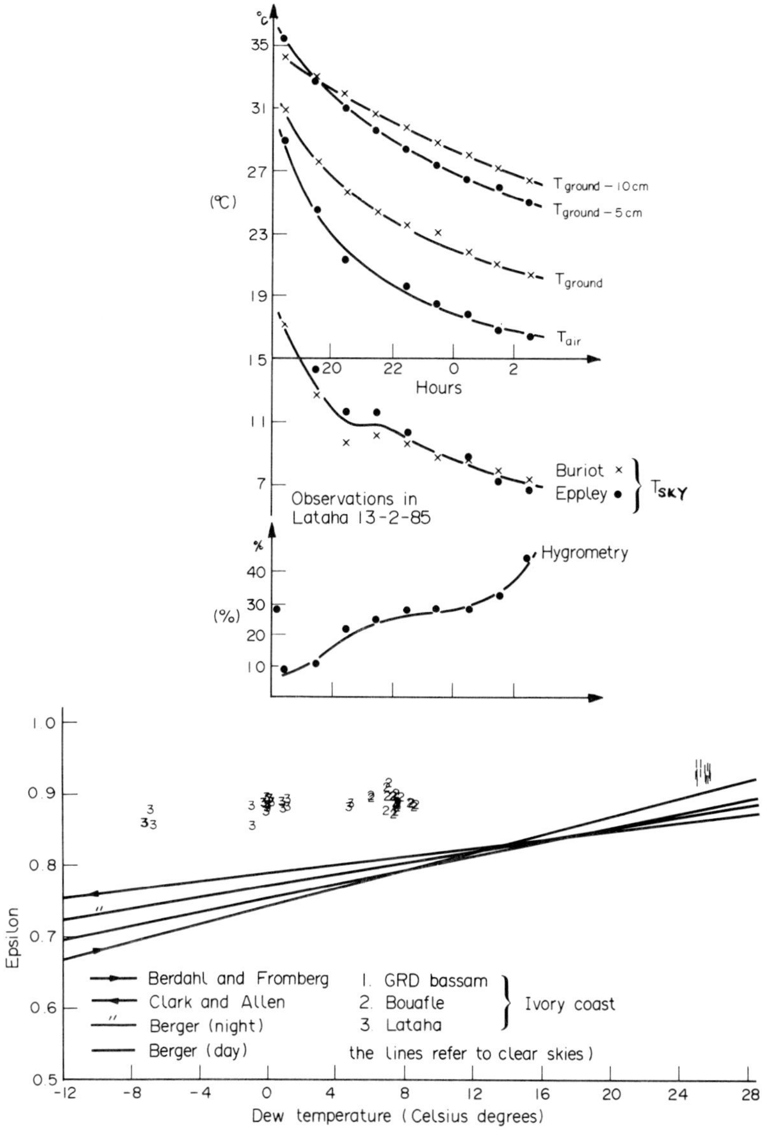

Fig. 6. Observed T_{sky} in hot polluted climate.

losses e.g. respiratory, convective C_{res} and evaporative E_{res} exchanges, skin (through cloth), convective S_{conv}, radiative S_{rad} and evaporative (insensible diffusion S_{dif} and sweat S_{sw}) exchanges.

$$M_L + C_{res} + E_{res} + S_{conv} + S_{rad} + S_{dif} + S_{sw} = 0$$

The various expressions of elementary exchanges are given in Deva (8). Combination of the 6 main parameters (activity Act, clothin insulation clo, main radiant temperature T_{rad}, air temperature T_{air}, air velocity V_a, air hygrometry Hr) for which a null balanc is obtained delimit a comfort zone (when sweat is null) in the six-dimensioned space. For the fixed four parameters clo, Act, v_a and Hr, the comfort lines are entangled in a delimited area in plane T_{rad}, T_{air}.

Two "nodal zones" are defined by lines:

$T_{clo} = T_{skin}$: indifference to cloth insulation (T_{clo}, mean clothing surface temperature)

$T_{clo} = T_{air}$: indifference to air movement.

Beyond these nodal zones the comfort lines start to diverge and define two complementary unstable comfort zones: the radiative cooling comfort zone and a part of the convective cooling comfort zone. This convective cooling zone may be considered as a heating radiative comfort zone characterized by an excess of convective losses. An optimal comfort zone is so pointed out in opposition with two unstable comfort zones (Fig. 7). Sweat possibilities add a supplementary comfortable evaporative adaptation zone.

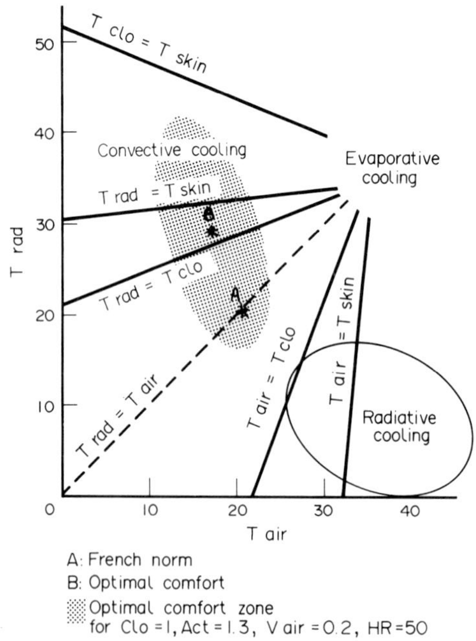

Fig. 7. The radiative cooling zone in the (T_{air};T_{rad}) plane.

In the radiative cooling comfort zone, a cold radiation compensates for hot air. The greater part of heat losses are radiative: convective exchanges are small, and even null in the node zone. Real comfort is obtained without a sweat requirement, but it is unstable: any deviation from a given situation cannot be compensated by a small variation of another parameter. For example any activity shall lead to higher metabolism and excess convective heat exchange. As the air temperature is high, the null balance will never be obtained. On the other hand, lower activity will lead to lower metabolism which will not compensate the lower convective exchange. The radiative cooling zone is only fitted for low activity situations such as sleeping or office work.

The T_{clo} temperature being a function of the radiative and convective exchanges, of the skin temperature and cloth insulation (9), its value may fall below T_{dew}, producing this well known feeling of piercing cold.

Complementary conditions of radiative isotropy must be fulfilled to appreciate the cold radiative comfort zone. A recent study by Fanger et al. (10) fixes the limits for assymetry of the thermal radiative field. In a climatic chamber, with a mean radiant temperature 17.5°C and a vertical radiant temperature asymmetry of 15°C, only 6.3% of the subjects, at rest, felt a thermal sensation of discomfort (head and neck). The air temperature was about 30°C, air velocity 0.2 m/s and the cloth insulation 0.6 clo.

The limits for horizontal gradients for radiant temperature are higher, about 6°C, depending on mans' posture (11). It then appears that the radiant cooling by sky radiation should not really misfit the comfort of people when lying down.

b) Experiments in the radiative cooling zone

Experiments were carried out by J.C. Deval in a climatic chamber of the Centre d'Etudes Bioclimatiques in Strasbourg. Four subjects stayed for half-days in four thermal ambiences (Table 3). After a neutral initial phase the subjects were submitted to heterogeneous climatic conditions they were not aware of. They were able to act on the radiant temperature after a while but did not know the parameter they were acting on. They were questioned every 5 minutes about their feelings and their evaluation of the environment. A set of parameters was recorded every minute (Fig. 8).

It may be observed (Fig. 9), that the subjects adjusted the mean radiant temperature at low values compatible with the predicted comfort lines in plane $T_{air} - T_{rad}$.

An unexpected result, pointed out in this experiment is that low relative humidity (15% to 25%) associated with high air temperature was never felt to be unpleasant.

Table 3. Classification of the typical experimental conditions (In the calculations and in the resulted curves, experimental measured data have been taken into account).

	T_{air} (°C)	T_{rad} (°C)	T_{dew} (°C)	V_{air} (m/s)
A	30	10	5	0.2
B	30	10	5	0.5
C	35	8	5	0.2
D	35	8	5	0.5

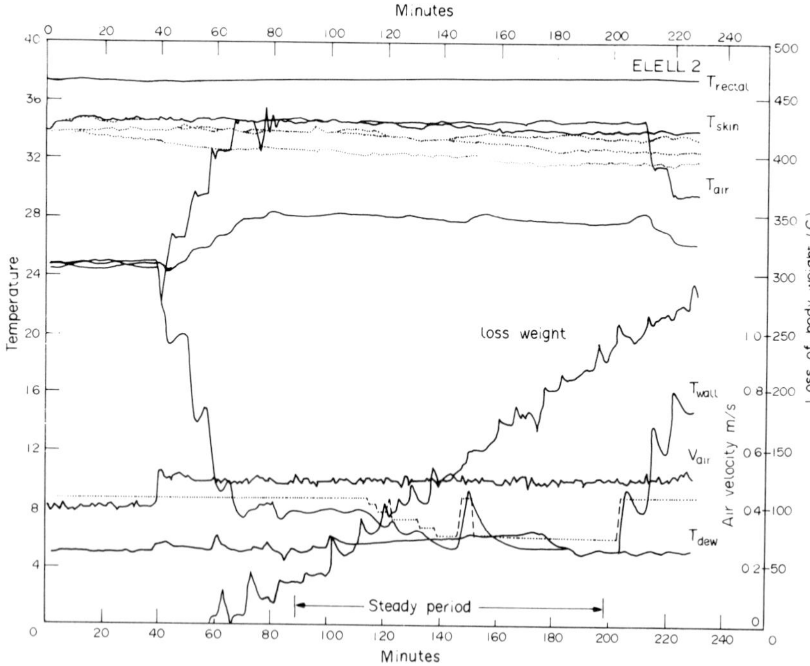

Fig. 8. An example of recorded variables during an experiment.

In fact the evaporative exchanges are functions of the partial vapour pressure of water which is strictly the same as 30°C, 30% humidity, and at 20°C, 55% humidity. On the other hand, the skin temperature, bound to the ambient (or operative) temperature

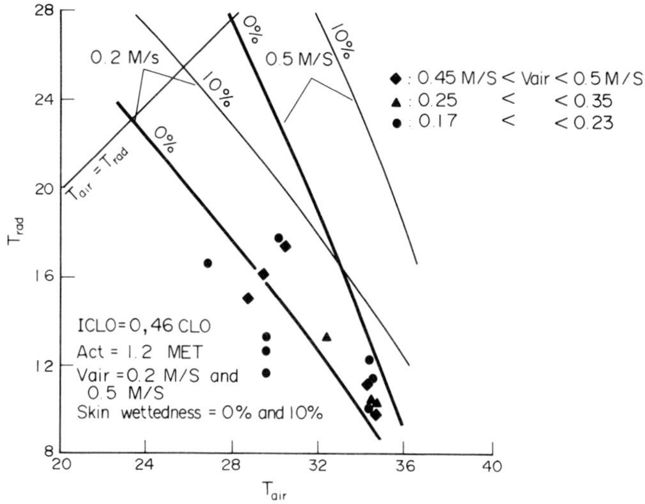

Fig. 9. Predicted lines and measured data.

has been found to be the same as in the optimal comfort zone. Thus, the evaporative exchanges do not appear to be affected by the discrepancy between the air and the mean radiant temperatures. The skin reacts strictly as if it were submitted to a uniform air and radiant comfort temperature. This explains the good agreement between experiments and computed simulations both for skin temperature and skin wettedness (Table 4). The global comfort equilibrium balance has certainly been obtained as none of the subjects drunk more than one glass of water during a 3.5 hours session. Moreover, only one subject experienced a slight respiratory discomfort due to low humidity. This was quite unexpected as usually a dry atmosphere leads to actual breathing discomfort.

c) Thermal and psychological interest of the radiative comfort zone

Sweat requirements for comfort is a strain which involves time for rest. The radiative cooling comfort zone is a real comfort zone without sweat requirements, therefore it is actually and psychologically efficient for low activity applications.

Under hot, dry climates, it may be reached by a simple projection along the T_{rad} axis (Fig. 10) without any change of the three air parameters (temperature, humidity, air velocity).

Under equatorial climates, the daily amplitude of temperature is too small to allow direct access to the comfort zone. When the daily thermal cycle is large enough, the radiative cooling is a transient period between complementary systems, and the climatic

Table 4. Experimental results.

Subject	MH	MH	AME	AME	AME	AME	BP	BP	BP	BP	MH	MH	ELA	ELA	ELA	ELA
Thermal ambiance	A	B	A	B	C	D	A	B	C	D	C	D	A	D	A	C
Session Number	1	2	3	4	5	6	7	8	9	10	11	12	13	14	15	16
$T_{wall\ set}$ (°C)	15.7	12.9	16.3	16.2	6.2	6.2	10.2	12.3	5.5	5.7	10.0	7.5	9.1	7.5	10.9	8.5
T_{air}	26.6	28.8	29.5	29.6	34.4	34.6	29.4	29.1	34.1	34.5	34.4	34.4	29.4	34.4	29.4	34.4
T_{dew}	6.1	7.1	6.4	5.2	4.6	4.7	6.0	7.1	6.1	4.5	6.1	5.8	5.7	5.6	6.7	5.9
T_{wall}	14.8	12.3	15.5	15.2	5.5	4.9	9.3	13.8	5.2	5.5	9.0	6.7	8.2	6.7	10.1	7.7
T_{globe}	25.4	25.9	25.6	26.6	27.1	28.4	24.1	25.9	27.2	27.5	27.5	27.7	23.9	27.9	24.8	27.6
T_{rectal}	36.8	37.3	36.8	36.6	36.3	36.8	36.6	36.9	36.8	37.2	36.7	36.7	37.5	38.1	36.9	37.8
T_{chest}	34.7	35.0	34.1	34.2	33.6	32.9	34.9	35.5	35.1	36.1	35.1	35.9	35.0	35.9	35.1	35.2
$T_{lower\ arm}$	32.8	33.3	33.1	33.8	32.2	34.2	32.2	33.7	32.4	35.3	32.8	32.5	31.7	33.8	32.1	33.4
T_{thigh}	33.3	33.5	32.0	31.8	31.4	32.4	31.5	32.3	30.9	32.8	32.1	33.1	31.2	33.3	31.4	31.9
T_{leg}	33.8	33.4	32.6	33.0	32.4	33.5	31.9	32.6	30.7	32.5	32.7	34.1	32.0	32.2	31.1	31.1
T_{skin}	34.0	34.1	33.2	33.3	32.7	33.3	33.1	33.8	32.7	34.4	33.5	34.3	32.9	34.3	32.9	33.3
V_{air}(m/s)	0.19	0.49	0.19	0.48	0.21	0.45	0.20	0.45	0.17	0.28	0.25	0.30	0.22	0.40	0.19	0.23
T_{rad}	16.6	15.0	17.7	17.5	10.4	10.0	12.6	16.2	10.1	10.4	13.2	11.4	11.7	11.4	13.3	12.2
Hygrometry (%)	27	26	24	22	16	16	24	26	18	15.5	17.5	17	23	17	25	17
Sweat rate g/h	58.2	49.8	54	61.4	-	99.3	46.2	-	51.8	64.2	42.5	49.2	61.9	104.0	64.6	75.0
Measured total E_{ev} (W)	39.6	33.9	36.7	41.8	-	67.5	31.4	-	35.3	49.0	29.6	33.5	42.1	70.7	44	51
Predicted total E_{ev} (W)	30.8	39.3	40.9	38.9	43.2	56	24.3	-	48.2	60.8	48.5	46.6	47	64.2	46.2	53

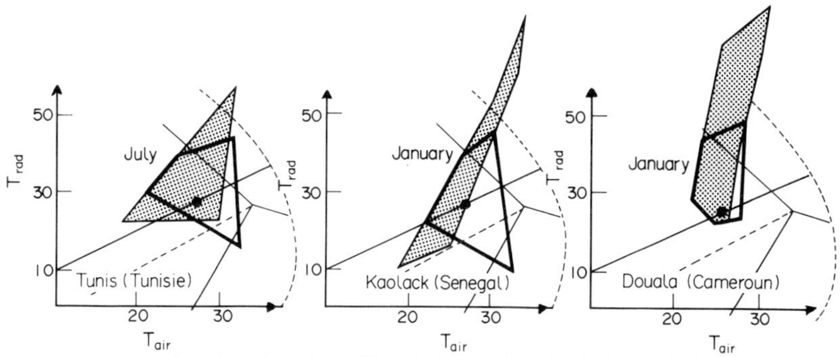

Outside daily cycles, with protection against direct solar radiation
Proposed daily cycles (——) and desirable situation (•)

Fig. 10.

cycle may be used directly, modified by projection or smoothed by thermal inertia. At sunrise, the air temperature is generally low and well adapted for human comfort. The thermal inertia of the house will help to keep comfort during the morning.

At midday, radiative cooling, though its efficiency could be optimal for zenithal views, becomes actually impossible as the sun, nearly vertical, has an effect still more efficient. High air temperatures and solar radiation are unavoidable and the only answer to climatic stresses is rest in dark environment (siesta).

During the second part of the afternoon, the air temperature is at its maximum but the zenithal angle of the sun increases rapidly. Radiative protection is useful and sky radiation becomes more efficient than diffuse solar radiation. During that time and still after sunset, the whole environment is overheated and the houses are uncomfortable. The only possibilities of comfort are found outdoors under direct radiative exchange with the sky vault, the only cooling source yet available. Later on, the air cooling associated with radiative cooling enables indoor comfort and cooling thermal storage for the following day.

Radiative cooling should be associated with complementary cooling processes like evaporative (mainly at midday) or convective (mainly after sunset) effects.

III - CONCEPTION OF A RADIATIVE SYSTEM ADAPTED TO LOCAL CLIMATIC CONDITIONS

Which cooling strategy shall we choose with such characteristics of sky radiation and of human body thermal comfort? Sky radiation should be considered as a "high quality refrigerating energy" which comes from a low energy diluted source in an especially narrow spectral band. Therefore, the natural "within reach" cooling sources (as convection, evaporation, earth cooling source) and the heating protections should be used when available, and the system should be optimized, taking into account the local climatic conditions and the comfort needs.

Daily cycles for air temperature and hygrometry determine the sky clearness, the radiative cooling power of the atmosphere and the situation of the dew point. The former set the maximum energy which can be lost. The latter indicates if an open or closed (relatively with the atmosphere) evapo-condensative system can be set up. Latitude has also to be known, as well for conception of the best adapted solar protection, as for taking into account the day-night duration: in the Mediterranean zone, 8 hours of night in summer have to be opposed to 16 h of sunshine. In the tropical zones day and night last about 12 h each throughout the year.

It must be remembered that radiative cooling needs both large radiating surface and large optical aperture (Abbe principle). The radiating power is low (less than 6 w/m² °C), still lower at low temperature, and often the temperature decrease is cut off by atmospherical convection and dew condensation.

a) Anticonvective protection, heat transfer with a covered radiator

To lower the radiator temperature, a cold selective coating is more efficient than black painting as it limits the energy gains outside the atmospherical window. However, in this case, the optimized cooling level is obtained with a limited emission efficiency and convection must then be prevented with infrared transparent covers which should also be selective. The cooling power rate becomes very much lower as soon as a lower level of cooling temperature is desired.

The net radiative heat loss W of a covered radiator at equilibrium (12, 13) is the result of 4 terms:

- the balance of radiative emission between the sky w_s and the radiator w_r modulated by the transmittance t_c of the cover and the emissivity ε_r of the radiator, with a corrective term depending on the reflectance of the cover r_c and of the radiator r_r.

- a similar term to express the exchanges between the radiator and the cover, in which the cover emissivity ε_c replaces the transmittance.

- two terms corresponding to the conduction-convection gains between radiator and cover and with the surrounding air:-

$$W = \int \frac{t_c \cdot \varepsilon_r}{1 - r_r \cdot r_c}(w_r - w_s) + \int \frac{\varepsilon_c \cdot \varepsilon_r}{1 - r_r \cdot r_c}(w_r - w_c)$$

$$= h_{r,air}(T_r - T_{air}) + h_{r,c}(T_r - T_c) = 0$$

these, r, c indices refer respectively to the sky, the radiator and the cover.

The first two terms are to be integrated in wavelength and in direction. The temperature T_c of the cover is determined from the cover heat balance, assuming its heat capacity is null.

For ideal cover and ideal back insulation we get:

$$t_c = 1$$

$$h_{r,air} = 0$$

$$T_{c0} = \frac{h_{r,c}.T_r + h_{c,a}.T_{air}}{h_{r,c} + h_{c,a}}$$

and $T_r = T_{c0} - \int \frac{\varepsilon_r}{h_{r,c}}.(w_r - w_s)$ (1)

In equation (1), it appears that an ideal cover has only a convective action: $h_{r,air}$ is replaced by $h_{r,c}$ which is smaller and lowers T_r.

For $t_c < 1$, ε_c differs from zero and $T_c > T_{c0}$. In equation (1), w_s is modified in $t_c.w_s + (1 - t_c).w_c$ which is greater than w_s. T_r tends to T_c and therefore increases.

This gives the limit beyond whcih a cover is not useful. Let us take:

$$h_{r,c} = \frac{1}{3} h_{c,a}; \quad \varepsilon_r = 1;$$

$$w_r - w_s = k(T_r - T_s) \text{ with } k \approx 6 w/m^2.°C$$

then we get

$$T_c \quad T_{air} - 1.5\ °C$$

$$T_r \quad T_{air} - 7.5\ °C$$

Owing to the partial transparency of the cover T_r tends to approach T_{air}, this explains that, according to Berdahl et al. (12) "about 5°C below air temperature... a white painted surface produces more cooling than a pvf surface".

As a conclusion, anticonvective selective covers are needed only when a temperature more than 5°C below air temperature is looked for. In such cases, daily solar protection is necessary. The residual cooling power is very low and gives to such systems very few practical applications (the situation is similar to that of solar collectors, but in this case, the related energy is more important and allows valuable applications for single and double glazed systems).

b) Concentration of the radiant effect

When a large cooling power is requested, it is necessary to concentrate the cooling energy collected on larger areas onto a smaller surface. Many natural systems exist acting either convectively (patios and peristyles collecting the cooled air of adapted pitches of the roofs, a situation often found in towns and valleys), radiatively (cool environment of whitewashed or selective surfaces) or by water thermigration (natural heatpipes, capillary transfer, diode roof).

The radiation coming from a selective body B at temperature T of infrared emissivity ε (selective body) is mainly of the form (radiosity expression)

$$A.\varepsilon.\sigma.T^4 + A.F.(1 - \varepsilon).\varepsilon_{sky}\sigma.T_{sky}{}^4$$

- a term of direct transport of radiant energy from B to the observer P, where A is the solid angle of view of the body from P, and a second term for radiant energy of the sky partially reflected by B in the direction of P where F is the angle factor between the sky and the body. This second term corresponds to a partial concentration of sky cool radiation. Moreover the temperature T itself is affected by radiative exchange of body B with the sky vault.

c) Radiative cooling associated with other cooling systems

Wind should be used to enhance cooling every time the radiator is warmer than the ambient air. Protection against forced convection should be insured in the reverse case (14). Association with convection has two main interests: first of all it lets the radiator work most of the time about or below the air temperature secondly it throws out efficiently extra heat from internal walls or thermal storage. If large convective coefficients (with draught effects) may be obtained (greater than 20 W/m² °C) and large exchange surfaces may be used, one has to avoid the problems, such as dust or noise, associated with convection.

Associating evapo-condensation in an open system (Lebanese porous walls, micro-porous media like lime paintings...) allows a natural condensative mass storage of low temperature heat. Water evaporation is used to limit the overheating caused by exposure to the sun. This process is controlled and delayed by capillary forces and remains efficient as long as the storage dew bulb temperature stays below the external air temperature. Such systems are used either with natural dew condensation under Mediterranean climates, or by artificial wetting in places where water is available (greens, irrigated roofs, roofponds...). In places where water is scarce and when the dew point is not reached, closed systems may be utilized (diode roofs).

Protection against solar radiation or convective gains may be obtained by screens (walls, trees...), upside down conical roofs (which protect against the vertical sun and allow a large solid angle of sky view around the radiator), vertical (or tilting) blades upon the radiator, polyethylene white films (transparent in the infrared), selective white paintings (white lime, titanium oxyde). The coatings of selective systems should be renewed frequently as they are very sensitive to dust and ageing. Table 5 gives approximate values of solar absorptivity and infrared emissivity for usual selective coatings.

Table 5. Approximate coefficients for solar-infrared selective coatings

coating	α_O	ε_{IR}
titanium oxyde, white paint	0.19	0.94
white paint	0.2-0.4	0.8-0.95
acrylic white	0.26	0.90
lime stone	0.25	0.95
white marble	0.47	0.95
ice	-	0.97
water	-	0.96
snow	0.20	0.82
sandstone	-	0.83-0.90
white Mgo	0.14	0.73
white Tabor	0.05-0.1	0.90

Radiators, protected from solar radiation, and therefore, working only at night time, may be black in the whole spectrum as well as those without thermal inertia only used at night. A large emissivity, stretched over the whole infrared spectrum , is required for exchanges with a heat storage of which the temperature level is over or about the air temperature. For temperatures below the air temperature, the absorptivity must be low all over the spectrum except for the atmospherical "window band" (Fig. 11). In this case a minimum value of storage temperature associated with weak exchange powers is required.

Let us quote a few examples of natural radiative cooling systems which are the result of local cheap availability and often of "secular know-how":

- the white lime painted dwellings in which, not only the roof, but also the vertical walls, the paving stone streets and the courtyards (Mykonos Cyclade Island, for instance) are

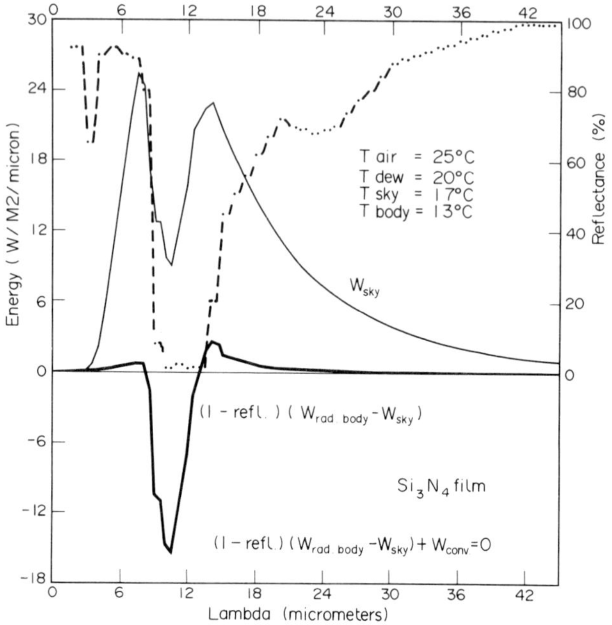

Fig. 11. Example of the net radiation between a selective coating and the sky.

radiative cooling systems. The outer surface temperature may be found quite low even under sunlight.

- the porous Lebanese walls where a partial radiative cooling on which the humidity of the sea-breeze condenses at the end of the night.

- the situation in a deep valley (Ourika, near Marrakesh) where the whole of the surroundings act like a radiator which feeds a cool breeze at nightime and prevents direct solar radiation at sunrise and sunset.

- patios and closed courtyards which act in a similar way during nighttime. In daytime the effusive marbles or sandstones and the selective mosaics (metallic oxydes) contribute to a low hemispherical radiative emission. They also protect against any excessive effects of wind.

- pebble bed roofs, which have been studied by Givoni (15). A series of experiments have been carried out in Israel to investigate various techniques of storage of night radiative cooling. Pebble bed storage, concrete roofs or waterponds were studied experimentally together with operable insulation. Average values of the roof temperature (17°C) were observed to be only one degree over the wet bulb point. The indoor air temperature could then be maintained at about 18 °C with outdoor average temperatures of about 21 °C. These systems are attached to the availability of evaporable water and the need for an operated insulation system.

- lawns, gardens of some Indian palaces (or Alhambra in Spain...) in which alternate radiative cooling and radiative heating are associated with condensation (dew) and evaporation to maintain a natural greenness and comfortable conditions for life.

- statues in the parks, pebble heaps in the desert or even domes in towns, may initiate chimneys-effects by their differential temperature with the ground. They may be used to create natural convective motions of large amplitude.

- ice has even been produced traditionally by radiative methods in countries such as Iran or Chile (in Atacama desert for desalination of brackish water). Under temperate climates such production of ice (silver thaw) is observed on streets with air temperatures over zero (5-8°C).

IV - THE DIODE ROOF

The system is a closed one; evapocondensation is used to enhance external heat exchange by a heatpipe effect. Convective stability blocks the reverse heat exchange when outdoor temperature is higher than indoor temperature. The system provides an efficient storage of cooling.

It has the cooling efficiency of the waterpond roof with operable insulation but adds the advantage of being a purely passive system. It consists of a special pebble bed roof adapted to specific climatic conditions:

- in hot dry tropical climates, for which it is intended, the dew point may not be reached easily. Then, the evapo-condensative possibilities which may improve a radiative system, in a Mediterranean site, should be used in a closed system to avoid the water loss.

- a closed system involves the existence of a supplementary heat exchange. The following consequence is a decrease of efficiency which is compensated under such climates by a colder sky temperature, a longer night and a good diurnal protection against outside overheating.

- because many hot dry countries are countries with reduced financial possibilities, the system has been chosen to be simple, cheap, and almost natural, even though many variants are possible.

a) Description

The system is composed of a flat corrugated iron roof (ceiling of the room) covered with inflated plastic sacks (Fig. 12). The sacks are partially filled with pebbles and a small quantity of water (heat storage). Their upper external surface (radiator) is coated with cold selective painting (white titanium).

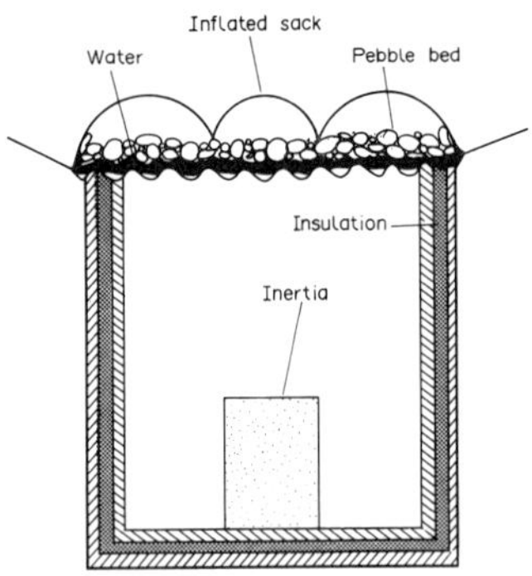

Fig. 12. The experimental cell.

This simple principle may be improved in simple ways:

- external protections against solar direct irradiation while still keeping large overview of the sky.

- low emissivity coating on the pebbles to reduce the radiative transfer during daytime (aluminium painting) and external treatment to diminish their porosity and improve the trickling of condensed water.

At night time, and if its solar absorptivity α is near 10 %, the temperature of the radiator falls below the ceiling temperature. Water vapour inside the sacks condensates. Drops slide along the curved radiator or fall by gravity. This heatpipe effect carries heat outwards and cold water inwards. Heat exchange is improved by radiation between the two humid internal surfaces ($\alpha \cong 0.9$). This heat supply is dissipated by radiative and sometimes external convective exchanges. When the radiator temperature at dawn becomes higher than that of the ceiling, no condensation may occur and the convective stability of air enclosed in the bags constitutes an efficient insulation which reduces heat transfer to a minimum value.

The theoretical resolution by an explicate finite difference method allowed the efficiency of this system for a set of conditions to be studied (16).

These computations show the importance of such a system for tropical dry countries. The mean temperature of the inside test-cell may be kept a few degrees over the minimum nocturnal outdoor air temperature throughout the day. The first parameter of influence is the solar absorptivity α of the radiator. This value should be lower than 0.2. The maximum indoor temperature being obtained at sunset, associated ventilation with an important air renewal rate is suggested during the first part of the night. This may lower the average temperature of 3 or 4 degrees.

b) Experimental studies

A cubic test-cell (2.5 m each side) with a small window (0.65 m x 0.40 m) has been built just beseides the laboratory in Nice. The corrugated iron ceiling, larger than the living area (4.3 m x 4.3 m), was covered with inflated plastic sacks, partially filled with pebbles (550 kg) and a small quantity of water (60 kg). Their external surface was cold selective painted (white titanium).

Figure 13 shows the recorded temperatures. Though the efficiency should be greater in a hot dry country compared with Nice (Mediterranean site), the computed simulations were still confirmed (17).

Laboratory experiments have been carried out to determine the optimum conditions for such systems. The efficiency is higher as working temperatures (evaporator and condensor) are also higher. Observations showed that a 8 cm air layer is the maximum thickness (18). The apparatus and the experimental results are described in Figures 14 and 15.

c) Comments

Thermal comfort not only depends on the instantaneous climatic conditions, but also on the average ones on account of the physiological adaptation of human body. Consequently a "preferred" ambient (operative) temperature is observed (9), generally below the lower sweat limit at rest (<28°C). In hot countries,

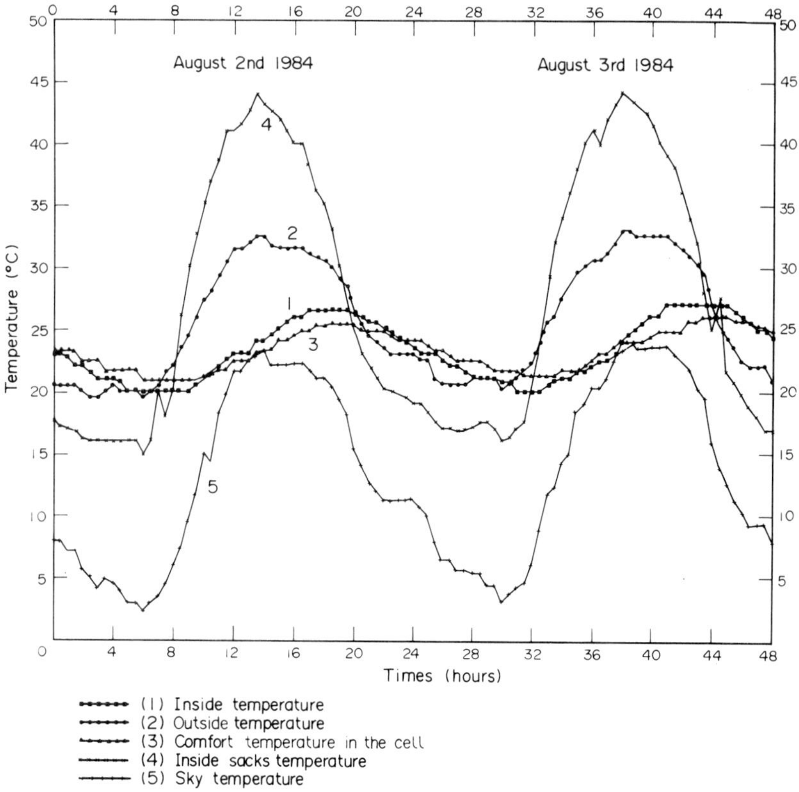

Fig. 13. The thermal behaviour of the diode roof cell, as observed in Nice (France).

sensitivity to coldness seems larger than sensitivity to heat (19), and often the minimum daily temperature is felt to be too cold, even if at 23 °C. Therefore, to maintain the indoor temperature a few degrees over the minimum outdoor temperature seems a good goal. It has been achieved in this experiment.

CONCLUSION

Cooling depends, at the same time, on knowledge about cold resources, the physiology of human body, efficiencies of thermal exchanges, and even sociological aspects. Tradiational systems and ways of life seem to have used most aspects of cooling possibilities. Present knowledge, modern materials and automatic control systems allow more efficient systems. But cooling needs should not lead to expensive or complicated solutions which do not integrate into a local context.

Sky radiation is an advantageous solution as it avoids the inconvenience attached to mass transfer e.g. dust, noise or damp.

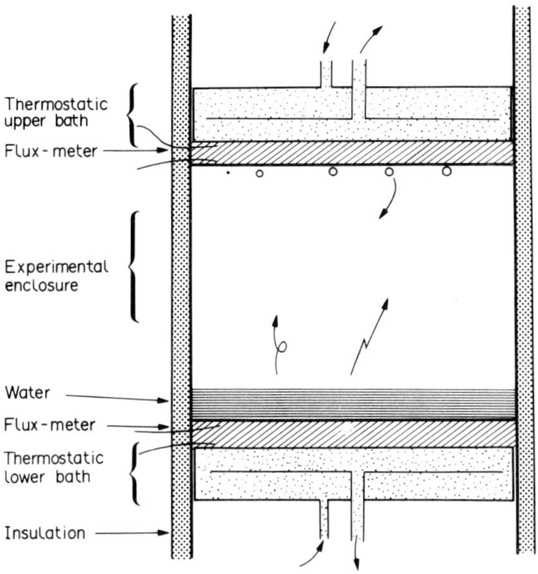

Fig. 14. Skeleton diagram of the laboratory appartus.

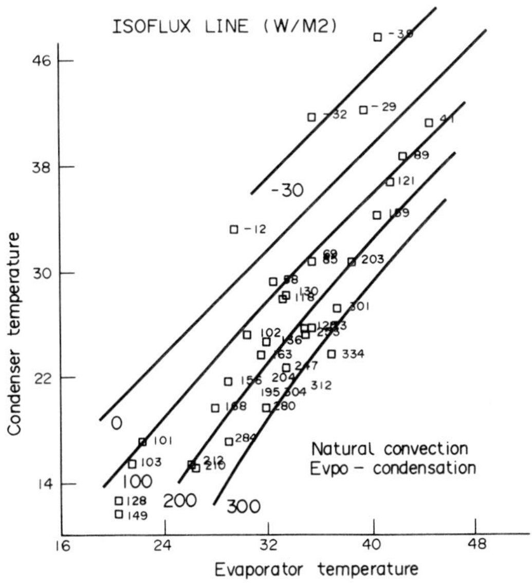

Fig. 15. Experimental results on the heatpipe effect.

But it has to be associated with solar protection, a large optical aperture towards the sky, and a clear celestial vault. It finally appears as an efficient system during some periods of the day, but it must also be used in association with the other available cooling sources.

REFERENCES

1. X. BERGER, D. BURIOT, F. GARNIER; 1984, Solar Energy 32, 725-733. About the equivalent temperature for clear skies.
2. I. DONET, X. BERGER; 1984, Colloque A.F.M.E. Météorologie et Energies renouvelables, Sophia Antipolis, mars. The atmospheric radio-soundings as a way to determine the sky radiative temperature.
3. P. BERDAHL; 1982, Solar Energy 29, 299-314. The thermal radiance of clear skies.
4. K. Ya. KONDRAT'YEV; 1965, Pergamon Press. Radiative heat exchange in the atmosphere.
5. B. CUBIZOLLES; 1985, Contrat A.F.M.E.-I.R.Bat.-C.N.R.S., Développement d'un modèle de calcul de la température de ciel.
6. P. BERDAHL, M. MARTIN; 1981, A.S.-I.S.E.S. Int. Passive and Hybrid Cooling Conference, Miami. Thermal radiance of skies with low clouds.
7. G. CLARK; 1981, A.S.-I.S.E.S. Int. Passive and Hybrid Cooling Conference, Miami. Passive-Hybrid comfort cooling by thermal radiation.
8. J.C. DEVAL; 1984, Rev. Phys. Appl., 19, 513-531. Le confort thermique en climat tempéré.
9. K. CENA, J.A. CLARK; 1981, Elsevier. Bioengineering thermal physiology and comfort.
10. P.O. FANGER, B.M. IPSEN, G. LANGKILDE, N.K. CHRISTENSEN, S. TANABE; 1985, Energy and Buildings, 8, 225-236, Comfort limits for asymmetric thermal radiation.
11. D.A. McINTYRE; 1980, Applied Science publishers. Indoor Climate.
12. P. BERDAHL, M. MARTIN, F. SAKKAL; 1983, Int. J. Heat and Mass Transfer, 26-6, 871-880. Thermal performance of radiative cooling panels.
13. X. BERGER, M. SCHNEIDER; 1981, A.S.-I.S.E.S. Int. Passive and Hybrid Cooling Conference, Miami. New techniques in natural cooling.
14. A.A. SAYIGH; 1976, Solar Heating and Cooling Forum, Miami. Summer night cooling in Saudi Arabia.
15. B. GIVONI; 1981, A.S.-I.S.E.S. Int. Passive and Hybrid Cooling Conference, Miami. Experimental studies on radiant and evaporative cooling of roofs.
16. C.N. AWANOU; 1984, thèse de doctorat de spécialité, Université de Nice, Etude de réfrigération passive par toiture diode.
17. X. BERGER, M. SCHNEIDER, J.C. DEVAL, F. KIENO, C.N. AWANOU, I. DONET; 1984, P.L.E.A. Mexico, Thermal comfort in hot dry countries and radiative cooling by diode roof.
18. F. KIENO; 1985, thèse de doctorat de spécialité, Université de Nice, Etude théorique et expérimentale d'un dispositif caloduc pour climatisation en pays chaud et secs.
19. X. BERGER, J.C. DEVAL, F. GRIVEL; 1985, Convention A.F.M.E. n 4 330 2326, Le confort thermique vecu en climat chaud.

CHAPTER 8

NOCTURNAL COOLING

P. Boon-Long

Department of Mechanical Engineering
Chiang Mai University,
Chiang Mai, Thailand

1. Introduction

In temperate climates, a phenomenon usually observed during clear nights in the Spring or Autumn is that, even though the ambient temperature is above freezing, ice forms on shallow water ponds, frost occurs on grass, and fruits suffer freezing injury. Farmers have known about this phenomenon for years, and they have correctly identified that it was due to the clear "cold" sky. This phenomenon has been turned into advantage in ancient times, for example to make ice naturally in Iran and India.

The fact that an object near the ground, be it a blade of grass or a water surface, can be colder that the surrounding air at night, implies that some heat transfer mechanisms other than convection with the air must be at work. This heat transfer mechanism is radiation; the object cools by radiative exchange with the sky, which is much colder than the surrounding air. The steady-state temperature of the object is that at which this radiant heat loss is equal to the heat gain from the air by conduction or convection. This steady-state temperature may therefore be lower than the freezing point of water, while the ambient air temperature is above it.

Consider the transient cooling of a ground-level surface at night, initially at a temperature T_i higher than ambient temperature, T_a (Fig. 1). In the beginning, the surface will cool by radiation to the sky, by convection to the surrounding air, and by evaporation if the surface is wet. Once T_r falls below T_a, convective cooling will turn into convective heating. If T_r drops further below the ambient wet bulb temperature (T_w), evaporation will cease. At this point, the only means of cooling is radiation to the colder sky, which has an apparent temperature T_s. This radiation cooling theoretically can persist until the object's temperature reaches the sky temperature, which could be several tens of degree C below the ambient air temperature. In practice, the sky temperature can never be reached, as convection and conduction heat gain become more prominent as T_r

decreases below T_a. However, a surface temperature 15-20 C below ambient has been achieved in practice, through radiative cooling only.

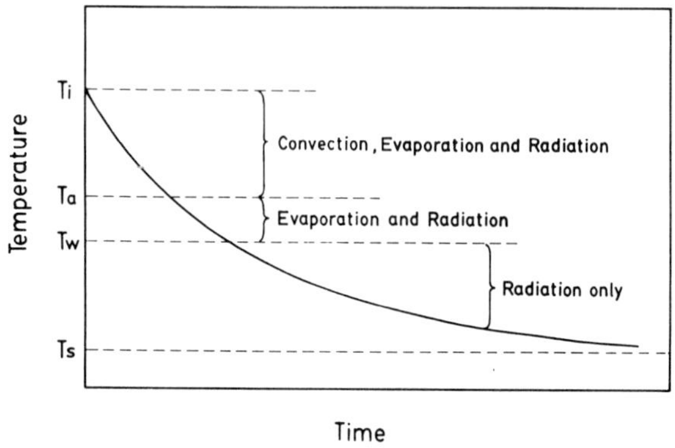

Fig. 1. Cooling of a surface with an initial temperature T_i, in air at T_a and wet-bulb temperature T_w, and with sky temperature T_s.

The sky (through the atmosphere) is thus an infinite heat sink several tens of degree C colder than ground air temperature. Actually, the sky or outer space itself has a temperature close to absolute zero; however, the sky as we see it is much warmer (250-300 K), due to the effects of the earth's atmosphere, particularly the water vapour, carbondioxide and ozone gases in it. These atmospheric constituents absorb and emit radiation in the thermal wavelengths, which increases the apparent sky temperature. In order to take advantage of the sky as a radiant heat sink, we must therefore understand the characteristics of the radiation emitted from the atmosphere.

Consider a ground object which is exchanging heat with its surroundings at night. Assuming that its internal temperature is uniform throughout ($\nabla T = 0$), and that there is no internal heat generation, at any instant its temperature will be changing according to

$$m\, c \frac{dT}{dt} = q_r + q_c + q_e \qquad (1)$$

Our task is to determine each term on the right hand side of eq. (1). Once this is done, a nocturnal cooling system can be analyzed and its performance predicted. The terms q_c and q_e may be estimated from available empirical correlations. It is the term q_r that we shall focus our attention on first.

2. Radiative Cooling

Radiative cooling of an object near the ground occurs because there is a difference between the object's temperature and the sky's apparent radiant temperature. This temperature difference will lead to radiative heat exchange according to

$$q_r = \phi\varepsilon_r(\lambda,\theta)[1-\varepsilon_s(\lambda,\theta)]i_b(\lambda,T_a)-\phi\varepsilon_r(\lambda,\theta)[i_b(\lambda,T_a)-i_b(\lambda,T_r)] \quad (2)$$

If the radiative properties can be assumed to be diffuse and gray, then

$$q_r = \varepsilon_r(\sigma T_r^4 - \varepsilon_s\sigma T_a^4) \quad (3)$$

From the above equations, in order to determine q_r, two quantities must be known: the characteristics of the sky radiation, and the characteristics of the radiating surface. The former may be either directly measured or estimated from meteorological data. The latter may be determined experimentally, using appropriate equipment.

2.1 Sky (Atmospheric) Radiation

What is sometimes called sky radiation in the thermal wavelengths actually originates from the atmospheric constituents. The main constituents which contribute to this radiation are water vapour, carbon dioxide, and ozone. We shall therefore refer to this radiation as atmospheric radiation.

The intensity of atmospheric radiation varies with wavelength, zenith angle, and amount of the relevant atmospheric constituents.

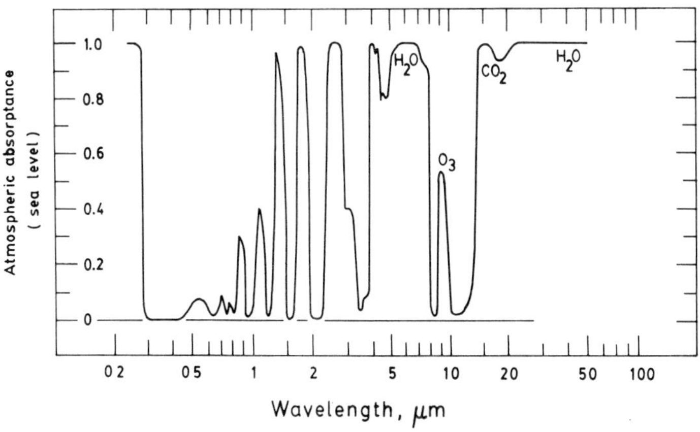

Fig. 2. Absorption spectrum of the atmosphere [1].

Wavelength Dependence. A typical absorption spectrum of the atmosphere is shown in Fig. 2. The spectrum is dominated by absorptance of water vapour, with carbondioxide and ozone contributing in the region around 15 μm and 9.6 μm, respectively. The water vapour absorptance is due to vibrations (centred around 6.3 μm) and to rotations (beyond 20 μm) of the water molecules.

Note, however, that there is a marked drop in absorptance (or conversely, a marked increase in atmospheric transmittance) in the range 8-14 µm. This wavelength is commonly called the atmospheric window, through which radiation from ground surface could pass to the colder sky beyond. It also happens that this window coincides with the wavelength region of peak intensity of radiation emitted from objects at typical ground temperatures. Numerous studies have thus been conducted, to take advantage of this window and maximize the cooling rates, to naturally cool ground objects.

Angular Dependence. The intensity of atmospheric radiation also depends on angle of measurement away from the zenith, as shown in Fig. 3. The radiation intensity has a maximum at the horizon, and a minimum at zenith. (In fact, along the horizontal direction, the atmosphere appears as an almost completely black body.) Therefore, all other things being equal, a surface in a horizontal position will have the maximum cooling rates. Practically, radiative cooling rates do not vary significantly within a cone defined by an angle of about 45° - 60° from the zenith.

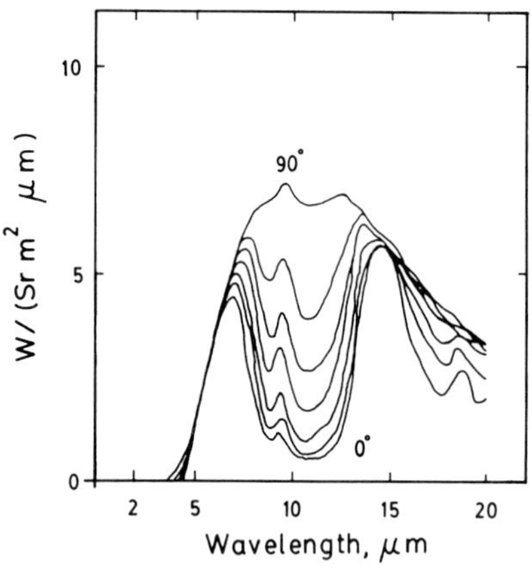

Fig. 3. Atmospheric radiation intensity at different angles away from zenith, measured at Elk Park Station, Colorado. The angles in sequence are 0°, 60°, 75.5°, 82.8°, 86.4°, 88.2° and 90° [2].

Atmospheric Constituents. Water vapour is the main constituent which is responsible for most of the atmospheric radiation. The amount of water vapour in the air is therefore the determining factor on the intensity of atmospheric radiation. Fgiure 4 shows this effect. The amount of water vapour in the air is represented by the vapour pressure, e, or the dew point temperature, T_{dp}, or the amount of precipitable water, L*. The lower the water vapour

*The vapour pressure may be related to the dew point temperature T_{dp} and the depth of condensing water L, as [4,5]

$$L = 0.17\ e$$

$$e = 0.61078 \exp \frac{17.2694(T_{dp} - 273.16)}{T_{dp} - 35.86} \qquad (T_{dp} \text{ in K})$$

content, the lower the atmospheric radiation intensity, particularly in the 8-14 μm window. If the vapour pressure is sufficiently low, a second transparency window opens in the region of 17-22 μm. This window is much less significant on radiative cooling than the first one, since its absorptance is higher, and its wavelength range is not in the peak radiation intensity range of surfaces at normal ground temperatures.

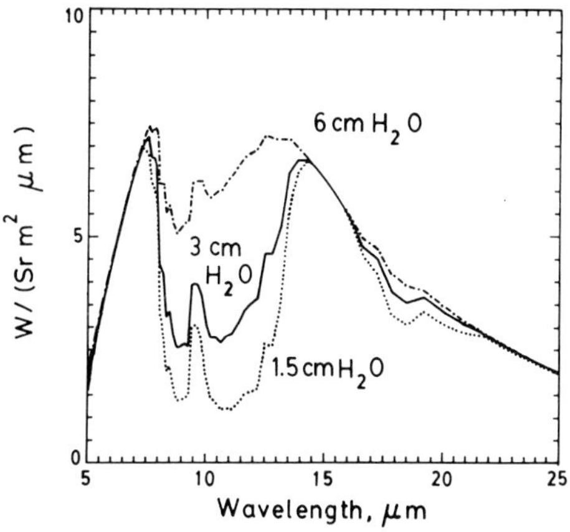

Fig. 4. Atmospheric radiation intensity at different water vapour contents [3].

The other two atmospheric constituents, carbondioxide and ozone, play a less significant role, particularly in more humid climates. Only in dry climates are their effects pronounced; in tropical or humid climates their effects tend to be masked by the strong water vapour presence. Also, ozone is present mainly at high altitudes, and therefore its effects are important only in high-altitude radiation exchanges.

The calculation of atmospheric radiation is usually made by assuming that the atmosphere is at ambient dry bulb temperature, and with an averaged emissivity ε_s. The magnitude of atmospheric radiation is therefore given by

$$R = \varepsilon_s \sigma T_a^4. \tag{4}$$

Alternatively, the atmosphere could be considered a black body (ε = 1.0) at some temperature T_s (sky temperature) below ambient temperature. In this case the atmospheric radiation can be calculated from

$$R = \sigma T_s^4. \tag{5}$$

Comparison between the two equations gives the relationship between ε_s and T_s as

$$T_s = \varepsilon_s^{1/4} T_a. \tag{6}$$

The above of course is made with the simplifying assumption that the emissivities are independent of wavelength, and therefore applies only to gray surfaces.

The calculation of atmospheric radiation is now reduced to determining the quantity ε_s. In the following, we shall discuss the methods of obtaining this quantity. Two main methods are possible: 1) calculation, using measured temperature and data on atmospheric constituents and their known spectral emissions, and 2) empirical correlations based on direct measurements of atmospheric radiation and relevant meteorological variables such as temperature, humidity and cloud cover.

2.1.1 Calculation of Atmospheric Emissivity

Atmospheric radiation may be calculated from measured data, by dividing the atmosphere into layers 1,2,...n. If τ_{ij} is the transmittance between the levels i and j, the emission from the layer between levels n and n+1 is $i_b(\lambda,T_n)(1 - \tau_{n,n+1})$, where $i_b(\lambda,T_n)$ is the Planck function for the spectral intensity per unit solid angle of a blackbody at temperature T_n. (Note that $\tau = \tau(\lambda,z,\theta)$, and that it is implicitly dependent on atmospheric constituents at height z). The spectral intensity of atmospheric radiation can then be shown to be [3]

$$\bar{R}(\lambda,\theta) = -\int_0^\infty dz\ i_b(\lambda,T(z))\frac{d\bar{\tau}(\lambda,z,\theta)}{dz} \tag{7}$$

The problem is how to determine $\tau(\lambda,z,\theta)$. This quantity may be calculated using publicly available computer programs, if the characteristics of the atmosphere (pressure, temperature, humidity, cloud fraction, etc.) at each height at the selected location are known. These data, called upper-air data, are measured with weather balloons carrying sensors and radio transmitters (radiosondes), which report back the atmospheric characteristics as the balloon ascends through the atmosphere. Such measurements are a normal procedure at most airports.

A widely-used program for calculating atmospheric transmittance, τ, is the LOWTRAN computer code [6]. This program calculates atmospheric transmittance as well as atmospheric radiation from upper-air data, and includes the effects of absorption, scattering, and aerosols. It also contains typical model atmospheres, which may be used without inputting upper-air data, if preferred. These model atmospheres include 5 seasonal atmospheres: tropical (15°N), midlatitude summer (45°N, July), midlatitude winter (45°N, January), subarctic summer (60°N, July), subarctic winter (60°N, January), as well as the 1962 U.S. Standard Atmosphere. The atmospheres are specified in terms of vertical profiles for temperature, and densities of water vapour, CO_2, O_3 and other gases. Further information can be found from ref. [6-9]. Errors in LOWTRAN results have been noted [8,9].

2.1.2 Empirical correlations.

<u>Clear-sky correlations</u>. Several empirical correlations exist,

the majority of which relate total atmospheric emissivity, ε_s, with some form of water vapour content in the atmosphere and/or ambient temperature. One of the most comprehensive and recent ones is by Berdahl and Martin [10].

$$\varepsilon_0 = 0.711 + 0.56\left(\frac{T_{dp}}{100}\right) + 0.73\left(\frac{T_{dp}}{100}\right)^2 \quad (8)$$

for $-13\ C < T_{dp} < 24\ C$,

in which ε_0 is the clear-sky emissivity and T_{dp} is the dew point temperature, in °C. Practically the same values of emissivity could also be obtained from

$$\varepsilon_0 = a + b\sqrt{e_a} \quad (9)$$

where $a = 0.564$, $b = 0.059$, and e_a is in mb.

Alternatively, Idso [11] gave the following relation in terms of vapour pressure and dry-bulb temperature:

$$\varepsilon_0 = 0.70 + 5.95 \times 10^{-5}\ e_a \exp(1500/T_a) \quad (10)$$

(T_a in K).

Spectral and angular emissivity. Martin and Berdahl [12] proposed an equation to relate spectral directional sky emissivity to total hemispherical sky emissivity:

$$\varepsilon_s(\lambda,\theta) = 1 - (1-\varepsilon_s[\tau(\lambda)/\tau_{av}]e^{b(1.7-1/\cos\theta)} \quad (11)$$

where the parameters b and $\tau(\lambda)/\tau_{ave}$ are obtained from Table 1, for different wavelength ranges. This equation is based on extensive measurements at six locations in the U.S.

Cloud effects. The presence of clouds, especially low, opaque ones, will increase the amount of atmospheric radiation. The magnitude of the increase may be up to 40% over the clear-sky value [5]. These effects may be taken into account in the form of an additional emissivity [13].

$$\varepsilon_s = \varepsilon_0 + (1 - \varepsilon_0)n\ \varepsilon_c\ \Gamma, \quad (12)$$

where n is obtained from observed weather data. The value ε_c is approximately 1.0 for low, opaque clouds; for other types of clouds, ε_c may be found from Fig. 5. Γ is found from Fig. 6, in which it is seen that the temperature of the cloud base must be known. This temperature may either be deduced from upper-air measurements, or estimted from fixed "typical" atmospheric lapse rate (temperature decrease as a function of height). If a lapse rate of 5.6 C/km is used, the resulting Γ vs. cloud base height is shown in Fig. 7.

Table 1. Values at least squares fit parameters for use with eqn (11) to predict spectral sky emissivity [12].

	$b = A\varepsilon_S + B$		$\tau(\lambda)/\tau_{av} = C\varepsilon_S + D$	
	A	B	C	D
no filter	1.493	-0.867	1.124	0.600
8-14 μm	1.792	-1.113	1.807	1.034
8.8 μm	1.281	-0.771	5.119	-1.192
9.6 μm	1.305	-0.715	5.321	-1.609
11 μm	1.778	-1.159	3.174	0.452
15 μm	-5.778	5.258	0.041	-0.007
17-22 μm	-0.691	1.653	-1.549	1.298

$$\varepsilon_s(\lambda,\theta) = 1 - (1 - \varepsilon_s)[\tau(\lambda)/\tau_{av}]e^{b(1.7 - 1/\cos\theta)}$$

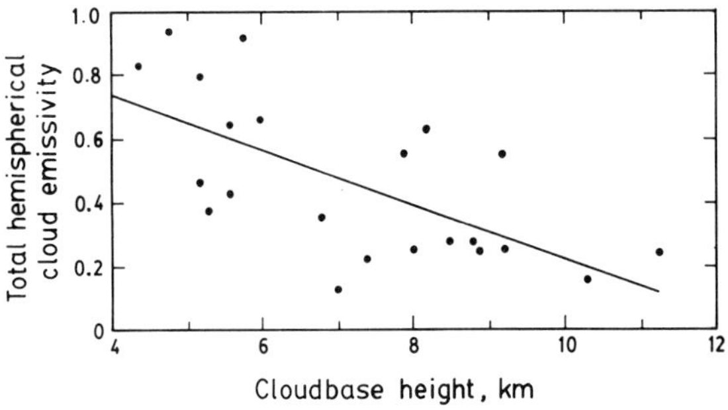

Fig. 5. Averaged hemispherical emissivity for cirrus (high) cloud systems [14].

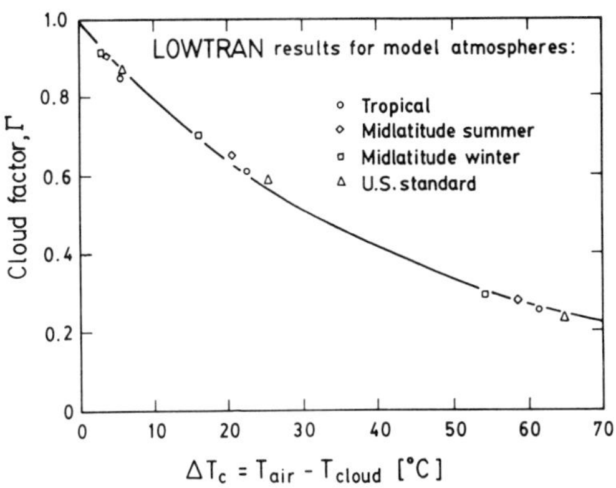

Fig. 6. Cloud factor Γ as a function of cloud base temperature depression [13].

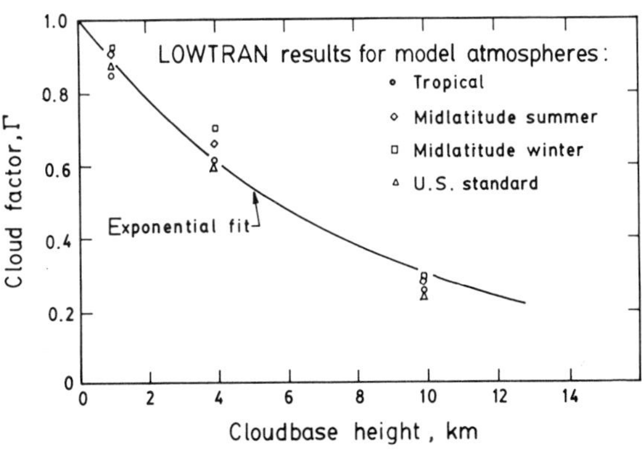

Fig. 7. Cloud factor Γ as a function of cloud base height [13].

If several cloud layers are present, the total contribution to atmospheric emissivity may be estimated from [13].

$$\varepsilon_s = \varepsilon_0 + (1 - \varepsilon_0)\sum_i n_i \, \varepsilon_{c,i} \Gamma_i \qquad (13)$$

It should be noted, however, that high clouds will contribute much less to the total atmospheric radiation than low clouds, since high clouds are at lower temperatures, and usually not as opaque. Therefore, the main contribution will come from the first cloud layer, and any error in estimating the heights or temperatures of the higher clouds would be relatively small.

2.2.3 Measurement of Atmospheric Radiation

A device commonly used for measurement of atmospheric radiation is the pyrgeometer (infrared radiometer). This device is similar to the pyranometer used for measuring solar radiation intensity, except that the pyrgeometer dome cover is made from special infrared-transparent materials. (Ordinary glass cannot be used, since it is relatively opaque to long-wave radiation). The pyrgeometer contains a thermopile to sense incoming radiation, and usually has a built-in circuit to compensate for emission from the sensor surface. This emission is taken to be of the form $R_p = \varepsilon_p \sigma T_p^4$. In some cases, the user has an option of measuring T_p directly, and compute the quantity σT_p^4 separately.

This method avoids the error due to instability of the battery used in the compensating circuit, estimated to be about 0.02 emissivity units in the current Eppley pyrgeometer [3].

A major source of error in pyrgeometer measurement is the solar heating of the dome cover. Ideally, the cover should reflect all radiation in the solar wavelengths. The actual dome material transmits and absorbs some solar radiation, and reradiates the absorbed energy to the sensor surface. Conduction heat transfer within the airspace between dome and sensor also contributes to heat the sensor. Together, the overall effect is that the pyrgeometer's output could be up to 30-35 W/m² too high, on a clear day with the sun overhead [3]. This error decreases with decreasing solar intensity and higher wind speed, which help reduce the temperature of the dome cover. At night, of course, this error ceases to exist.

What a pyrgeometer will measure is the total downward atmospheric radiation; in some cases where net radiation from a surface on the ground is needed, two pyrgeometers may be used, one facing up and one facing down over the surface of interest, the difference between the two readings will be the net radiation. Alternatively, a net radiometer (which has two radiating surfaces, upward and downward) may be employed.

Instead of using a pyrgeometer, a simple heat flux meter may be constructed, for measurement of atmospheric radiation and net radiation from ground surfaces at night. A type of this flux meter is shown in Fig. 8, in which heat is supplied to the

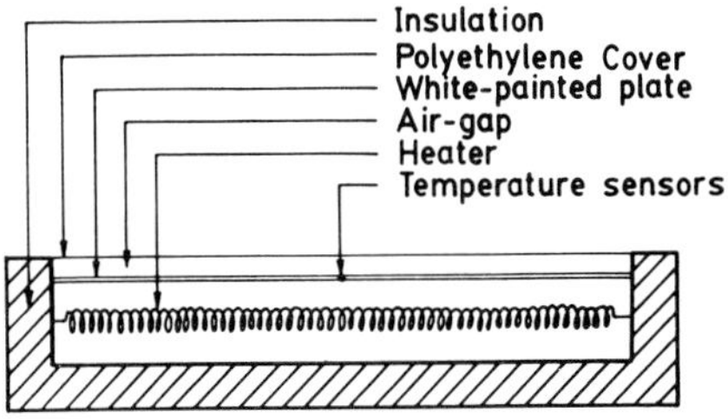

Fig. 8. Heat flux meter, electrical resistance type.

device by electrical resistance wire. The surface temperatures are measured at several points, and the net radiant heat flux from the device may be calculated from

$$q_{net} = q - q_c \tag{14}$$

where q is taken to be the electrical power input, and q_c can be calculated from standard heat transfer correlations.

This net radiant heat flux must then satisfy the equation (assuming for simplicity that the transmittance of the cover is unity)

$$q_{net} = \varepsilon_r(\sigma T_r^4 - \varepsilon_s \sigma T_a^4) \tag{15}$$

from which the atmospheric emissivity, ε_s, may be determined, given surface characteristics. The net radiant heat loss, q_{net}, can then be calculated at other radiator temperatures, with proper corrections for conduction and convection heat transfer.

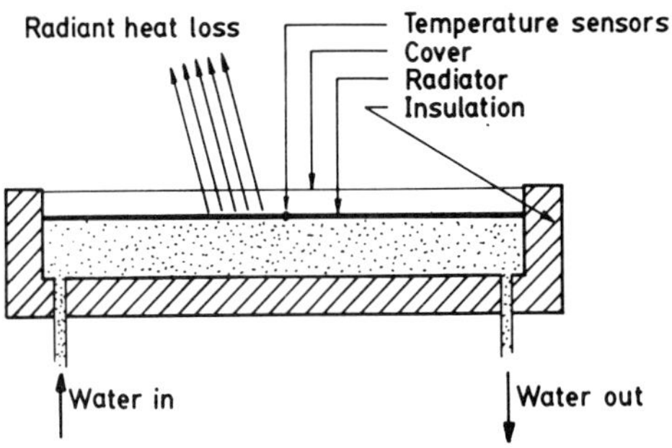

Fig. 9. Heat flux meter, water flow type.

Alternatively, a heat flux meter utilizing water flow may be employed, as shown in Fig. 9. In this case, q_{net} can be found from

$$q_{net} = \dot{m}\, c\, (T_{w,i} - T_{w,o}), \qquad (16)$$

where q_{net} includes the effects of non-radiative heat transfer, which must be properly taken into account. The magnitude of atmospheric radiation can then be calculated from eq. (15).

In both cases, if good precision temperature sensors are used (error in the order of ± 0.5 °C), and proper care is taken in th design and operation of the device, the resulting values of atmospheric radiation obtained could be within 5-10% of pyrgeometer readings, in the absence of solar radiation. If one is interested in net radiant cooling rates from ground surfaces, these methods offer direct measurement of radiation from the actual radiating surfaces under consideration, and with probably more accuracy than using pyrgeometer outputs to calculate net radiation from the surfaces.

2.2 Radiating Surfaces

For cooling purposes, the 'ideal' radiating surface should have ε_r = 1.0 in the 8-14 µm atmospheric window, and zero elsewhere, to take maximum advantage of the primary atmospheric window. (In dry climates where the effect of the 17-22 µm secondary window is significant, an ideal radiating surface should theoretically have high emissivity in this wavelength range as well, but this will not be considered here.)

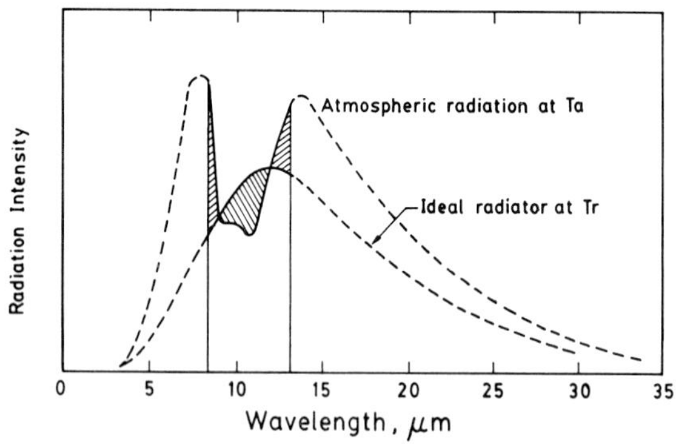

Fig. 10. Radiation absorbed and emitted by an ideal radiator [15].

Consider radiative exchange between an ideal radiator and the atmosphere, as shown in Fig. 10. If it is assumed that only the primary atmospheric window is present, the ideal radiator will derive maximum advantage from this window, since it will radiate only in this wavelength range, and will not absorb any radiation from the atmosphere outside this range. Its equilibrium tempera ture will therefore be lower than that of the atmosphere; this temperature will in fact be that at which the radiator's emitted radiation in the 8-14 µm band is equal to radiation it absorbs from the atmosphere in the same band. (No radiative exchange occurs outside this band.)

A black surface will be less satisfactory for radiative cooling than an ideal surface, since a black surface will absorb atmospheric radiation outside the 8-14 μm band as well. Therefore its equilibrium temperature will be higher than that of the ideal radiator.

However, if the 17-22 μm secondary window is present, a black surface can radiate in this wavelength range, whereas the so-called ideal radiator as we define it cannot. Therefore it may happen that a black surface performs better than a supposedly ideal surface, in dry climates.

Radiator Materials. Many spectrally-selective surfaces have been developed, to maximize radiative cooling in the atmospheric window. One of the first, developed by Catalanotti et al. [15], was aluminized polyvinyl fluoride plastic (PVF). Its spectral characteristic is shown in Fig. 11. Other materials subsequently developed include aluminum coated with 1.3 μm of silicon oxynitride ($SiO_{0.6}N_{0.2}$) [17], MgO and LiF on metal foils [18], and many others. Commercial white paint with high titanium dioxide (TiO_2) contents also is considered a good radiator surface, although not very spectrally selective.

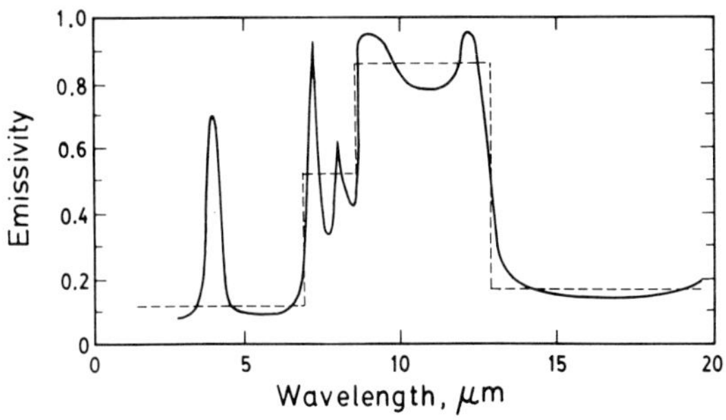

Fig. 11. Spectral emissivity of aluminized PVF [15].

Gases also have been identified as potential radiators for cooling purposes [19]. The advantages of gaseous radiators have been cited as being relatively cheap, easy to transport, may be used as coolant medium directly, and also having the ability to be mixed together to obtain desired characteristics. Examples of gaseous radiators are ethylene (C_2H_4), ethylene oxide (C_2H_4O), and ammonia (NH_3).

Radiator Cover. In order to reduce non-radiative heat gain into the radiator, in addition to side and back insulation, the radiator surface is usually covered with an infrared-transparent screen, leaving a small air gap. The most common material used for radiator cover is polyethylene film, whose transmittance is shown in Fig. 12. The stagnant air layer trapped between the radiator surface and the screen will serve as a thermal resistance. Theoretically, no convection current will be present, since the air layer is in a stable configuration (colder lower boundary and warmer upper boundary). The air layer will therefore be a very effective insulation. In practice, however, convection currents usually develop in the larger radiators due to wind-induced fluttering of the flexible cover [16], and significantly reduce the conductive insulation effect. Therefore the cover needs to be reinforced for increased rigidity. Nilsson et al. [20] reported results from V-corrugated high-density polyethylene foils (Fig. 13), designed to reduce convective effects even in the presence of winds. A transmittance of 0.73 in the infrared and a thermal resistance of 1.1 m^2 K/W were reported.

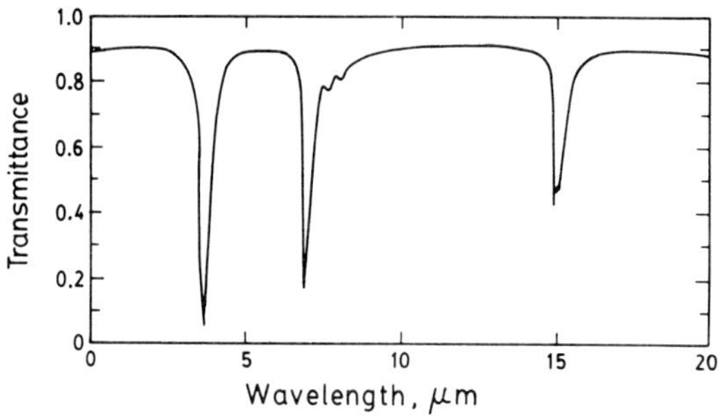

Fig. 12. Transmittance of 10 μm polyethylene film [15].

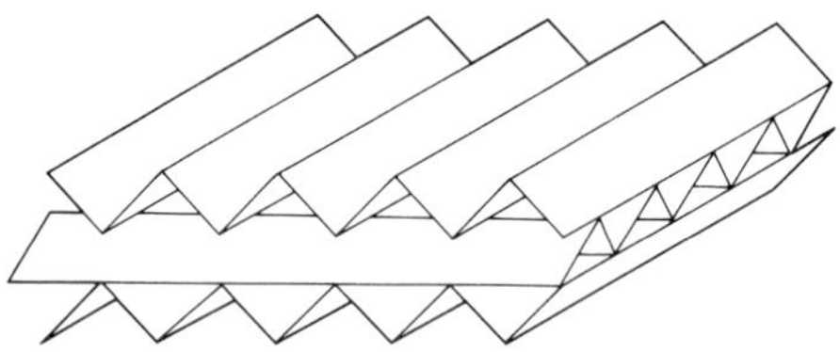

Fig. 13. A V-corrugated polyethylene cover [20].

In the presence of the cover, the net cooling rate from the diffuse radiator could be written as [25],

$$q_{net} = \phi \frac{\tau_c(\lambda,\theta)\varepsilon_r(\lambda,\theta)}{1 - \rho_r(\lambda,\theta)\rho_c(\lambda,\theta)} \; i_b(\lambda,T_r) - i_b(\lambda,\theta,T_s)$$

$$+ \phi \frac{\varepsilon_c(\lambda,\theta)\varepsilon_r(\lambda,\theta)}{1 - \rho_r(\lambda,\theta)\rho_c(\lambda,\theta)} \; i_b(\lambda,T_r) - i_b(\lambda,T_c)$$

$$+ h_{ra}(T_r - T_a) + h_{rc}(T_r - T_c) \quad (17)$$

The cover temperature, T_c, has to be found from the following heat balance equation on the cover, assuming that the cover has no heat capacity:

$$\phi\ \varepsilon_c(\lambda,\theta) + \frac{\tau_c(\lambda,\theta)\varepsilon_c(\lambda,\theta)\rho_r(\lambda\ \theta)}{1 - \rho_r(\lambda,\theta)\rho_c(\lambda,\theta)}\ i_b(\lambda,T_c) - i_b(\lambda,\theta,T_s)$$

$$+ \phi\ \frac{\varepsilon_c(\lambda,\theta)\varepsilon_r(\lambda,\theta)}{1 - \rho_c(\lambda,\theta)\rho_c(\lambda,\theta)}\ i_b(\lambda,\ T_c) - i_b(\lambda,T_r)$$

$$= h_{rc}(T_c - T_r) + h_{ca}(T_c - T_a) = 0 \qquad (18)$$

Materials for Day-Time Radiative Cooling. One interesting application of radiative cooling is for cooling in the day time, under direct sunlight. Consider a structure shown in Fig. 14. A selective film is used as the roof of the structure. The objective is to keep the temperature inside the structure lower than ambient, through radiative cooling. In order to achieve this objective, the characteristics of the selective film must be: high transparency to infrared radiation, high solar reflectance on its upper surface, and high solar absorptance on its lower surface [21]. A film which fulfills these requirements will allow the surface underneath it to cool by infrared radiation to the sky, yet prevent solar radiation from entering and being trapped in the cooled volume. Initial results from a double polyethylene film [21], and a selective film using 100-μm polyethylene charged with 2.4% of TiO_2 on its upper side and 0.5% of carbon black on its lower side, have been reported [22]. The results of both investigations indicated that the temperature inside the cooled volume varied from ambient to 2-3°C below ambient temperature. Further development of this material could have a far-reaching impact in applications in hot regions of the world.

Dew Formation. In more humid climates, where the dew point temperature of the ambient air is relatively high, it is possible that through radiative cooling the temperatures of the radiator and/or the infrared-transparent cover would fall below the ambient dew point. In such cases, dew will form on the surfaces. Since water has a high absorptivity in the infrared, this dew film will reduce the net radiant cooling rate of the radiator. This dew formation is a self-accelerating process: the water film on the cover becomes the radiating surface itself, and, due to its small thermal mass, cools rapidly, which causes more condensation onto the surface [23,24]. The cooling performance of the radiator is then drastically reduced. This phenomenon may render ineffective the use of selectively-radiating surfaces in humid climates.

Radiator Efficiency. It is useful to define radiator efficiency as the ratio of the actual radiative cooling rate of a surface over some idealized cooling rate:

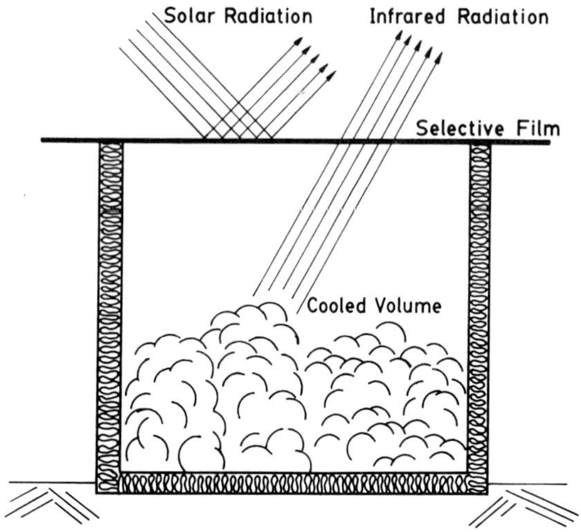

Fig. 14. Light selective structure for radiative cooling under direct solar radiation [21].

$$\eta = \frac{q_{net}}{q_i} \tag{19}$$

Berdahl et al. [25] defined q_i to be the cooling rate which occurs from an uncovered black radiator (or a covered black radiator with cover transparency = 1.0) which is at ambient temperature,

$$q_i = (1 - \varepsilon_s)\sigma T_a^{\,4} \tag{20}$$

A dimensionless temperature difference is also defined as

$$\delta = \frac{4(T_a - T_r)}{(1 - \varepsilon_s)T_a} \tag{21}$$

Plots of η vs. δ showed that they are relatively insensitive to changes in atmospheric conditions such as humidity, temperature and wind speed, and suggested that they could be used in the same way as the now-standard efficiency curves for solar collectors.

2.3 Radiative Cooling Rates

Net Radiative Cooling Rate Calculation. Using equations (8), (9), or (10), one can calculate the atmospheric emissivity in a locality of interest, based on humidity and temperature records. Effects of clouds can then be added, using equations (12) and (13). Then, with known or assumed emissivity of the radiator, one can calculate the magnitude of radiative cooling flux at various radiator temperatures with eq. (3). The net cooling flux at any particular radiator temperature is finally estimated by subtracting non-radiative heat gain into the radiator from the radiative cooling flux. Estimates of this non-radiative heat transfer are given in Table 2 [25], or they can be calculated from standard heat transfer equations.

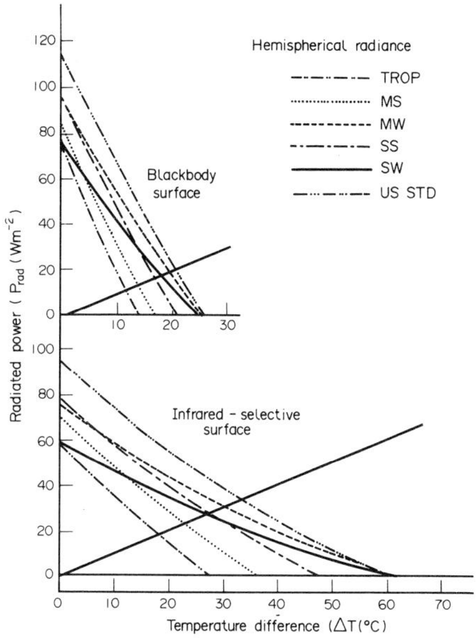

Fig. 15. Radiative cooling rates calculated with LOWTRAN 5. The model atmospheres are: tropical (TROP), midlatitude summer (MS), midlatitude winter (MW), subarctic summer (SS), subarctic winter (SW), and the 1962 U.S. standard (US STD). The slant line represents a conductive heat gain coefficient of 1 W/m^2K [7].

Some representative radiative cooling rates are shown in Fig. 15. The curves were calculated with the LOWTRAN 5 computer code, for all six model atmospheres, and for the black and the ideal selective radiators [7]. Radiative cooling fluxes up to 110 W/m^2 are predicted at ambient temperatures. Note that the black surface performs better than the ideal surface, near ambient temperatures. The reason is due to the secondary atmospheric window, as previously described. This is further illustrated in Fig. 16 [26], for two different climates: dry and humid (as measured by Bell et al. [2] in Colorado and Florida, U.S.A., respectively). Consider the humid case first. At $\Delta T = 0$ ($T_a = T_r$), the ideal radiator and the black radiator produce the same radiative cooling rate, as they both exchange radiation with the atmosphere only in the 8-14 μm range. (No radiative exchange occurs outside this range, since the secondary window is insignificant in

Table 2. Non-radiative heat transfer coefficients [25]

Heat transfer coefficient	Symbol	Heat flow direction	Value (W/m^2K)	Remarks
Radiator-to-air (back losses)	h_{ra}		0.3	10 cm plastic foam
Radiator-to-cover	h_{rc}		0.9	2.7 cm air gap
	h_{rc}		$1.3(T_r-T_c)^{1/4}$	laminar convection T_r-T_c in C
Cover-to-air	h_{ca}		5.4	forced convection with windspeed of 2 m/s length scale = 1 m
	h_{ca}		17	forced convection with windspeed of 10 m/s

this case.) However, in a dry climate, where the 17-22 μm secondary window is significant, the black surface can also radiate in this secondary window; therefore its radiative cooling rate is higher than that of the so-called ideal radiator. As the radiator temperature drop (T_a - T_r increases), the benefit of the black surface over the ideal one decreases. The radiant flux from both surfaces will be equal at some radiator temperature at which the black surface's cooling benefit gained in the secondary window equals the atmospheric radiation it absorbs outside the primary window. Thereafter, the ideal surface will perform better.

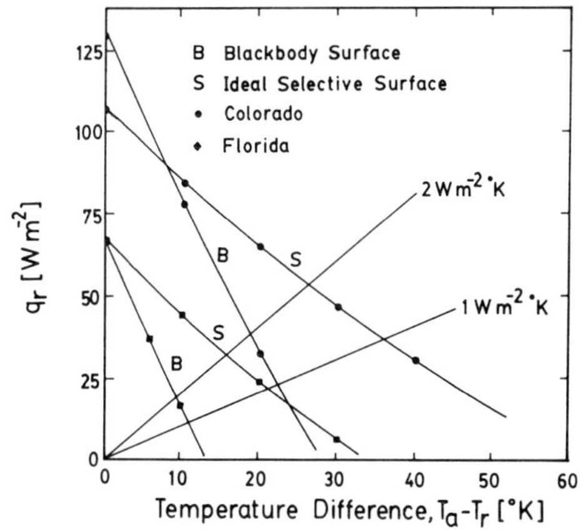

Fig. 16. Radiative cooling rates calculated from data at two locations [26].

Several studies [7,25,27] have indicated that in applications where the desired radiator is near ambient temperature (T_a - T_r < 5°C), a black surface is preferred to an ideal surface, for reasons of performance as well as cost. Simple white paint may in many cases be superior to a supposedly selective surface such as aluminized PVF [25].

In practice, there will be non-radiative heat gained into the radiator, which will reduce the net cooling rates, and increase the radiator's equilibrium radiative temperature. If this non-radiative heat gain is 1 W/m²K, the equilibrium temperature depression for an ideal radiator will be reduced from 32°C to 21°C, and for a black surface from 14°C to 11°C, in the Florida case. However, it should be recognized that near this equilibrium temperature, the useful net cooling rate is very small.

Error in Net Cooling Rate Calculation. It is important to recognize that net radiative cooling is the difference between two relatively large numbers, the radiating surface's emission and the atmospheric radiation (eq. 3). Using some typical values of ε_r = 0.9, ε_s = 0.85, and T_r = T_a = 293 K, the surface's emission is $\varepsilon_r \sigma T_r^4$ = 355 W/m², and the atmospheric radiation is $\varepsilon_s \sigma T_a^4$ =

334 W/m². The net radiative cooling flux is therefore 56 W/m². If the calculated or measured value of atmospheric radiation contains a 5% error (which is not easy to achieve), the absolute error in atmospheric radiation value is 17.8 W/m². This represents an error of around 32% already in the calculated net radiant cooling flux. Other uncertainties in emissivity values of the radiator and in temperature measurements will compound this error further.

3. Evaporative Cooling.

Evaporative cooling is another potentially important mechanism in nocturnal cooling, particularly in situations where radiative cooling is obstructed by clouds, or in the early stage of cooling when the water or (wetted radiator) surface is well above the ambient wet bulb temperature. The magnitude of evaporative cooling may be estimted from [28]

$$E_a = 0.35\ (1 + 0.54\ U_2)(e_a{}^* - e_a), \qquad (22)$$

where E_a is in mm/day, and e_a^* and e_a are in mm Hg.

4. Actual Systems

In an actual system utilizing nocturnal cooling, three major components are required:

i) the radiator
ii) a cold - storage mass (for example a body of water)
iii) the cooled volume.

The cold storage mass, which is cooled through the radiator during the night, will act as a heat sink for the cooled volume during the day. There must therefore be a means of heat transfer between the radiator and the cold-storage mass, and between the cold - storage mass and the cooled volume. The former will operate only at night, whereas the latter has to operate throughout the entire day.

Roof Pond Systems. In the majority of actual systems built to date, the radiator and the cold - storage mass are combined in the form of a shallow body of water (roof pond) placed on top of the cooled volume. A set of movable insulation panels is installed over the roof pond; these panels will be opened to expose the roof pond to the sky at night, and closed during the day to prevent heat gain into the pond by solar radiation (Fig. 17). Heat transfer between the water mass and the cooled volume is by natural convection, assisted by ceiling fans when necessary. This arrangement has been developed by Hay and Yellott [29,30]. Hay obtained patents on the system [31-33], and gave variations of it trademarked names such as Skytherm™, Thermopond™, and Thermospace™.

The majority of roof pond systems currently operating are situated in the United States. They were all designed to be able to

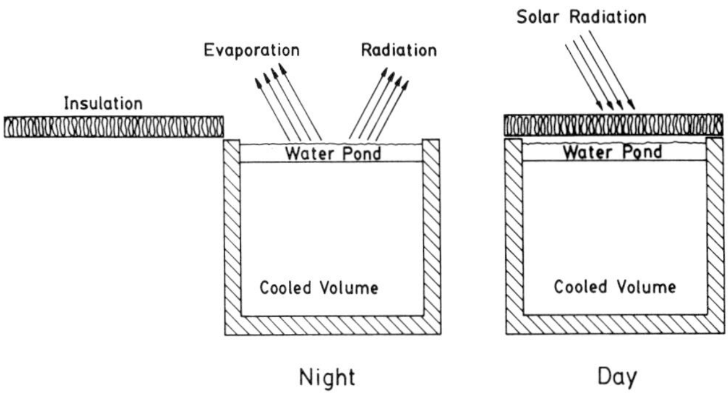

Fig. 17. Roof-pond system.

operate in the heating mode as well as the cooling mode. (The heating mode is effected by simply reversing the operation of the movable insulation panels: opening them during the day to admit solar radiation into the pond, and closing them at night to prevent heat loss.) There are approximately 15 such structures in the U.S. [34]. All of them are designed for human comfort (residential house or school). Details are summarized in Table 3. In general, the water mass is enclosed in plastic bags, and no evaporative cooling is possible, except in some cases where the bags are sprayed with water during certain periods, to increase cooling effect. The structures are limited to a single storey.

The thermal performance of these roof pond structures has been reported to be satisfactory. Temperature inside the buildings is maintained within the human comfort range, with little or no auxiliary cooling needed. The problems most often reported are water leakage from the pond, and faulty operation of the movable insulation.

Nocturnal Cooling for Storage of Agricultural Products. Intensive work has been conducted in the northern area of Thailand, to utilize the nocturnal cooling method for storage of fruits and vegetables. The method is seen to be appropriate for rural areas, due to its low cost, low maintenance, and ability to operate without electricity. Also, the high humidity inherent in the roof pond room is suitable for storage of fruits and vegetables. Two structures, each 25 m^3 in volume, have been constructed, using the open roof pond concept with movable insulation (Fig. 18). Cooling at night occurs by both radiation and evaporation. Heat exchange between the water mass and the room is by natural convection only.

The local climate in this case is warm and humid. In this climate, the water can usually be cooled to the minimum ambient, and the room to within 2°C of the minimum ambient temperature. Diurnal temperature fluctuations in the room were less than 2°C,

Table 3. Summary of Roof Pond Systems in the U.S. (Adapted from ref. 34)

Name	Location/ Year of construction	Floor Area,m^2	Roof-pond Area,m^2	Roof-pond depth,m	Remarks
1. Phoenix Prototype	Phoenix,Az, 1967	11.2	17.2	0.15-0.175	First prototype constructed by H. Hay
2. Atascadero House	Atascadero, Ca., 1973	110.7	102.2	0.22	Enclosed water bags, with inflatable air cell to reduce heat loss in heating mode
3. Sherwood House	Winters,Ca. 1975	111.5	26.8	0.30	
4. Mobile/ Modular Home II	Los Alamos, NM, 1977	101.0	83.2	0.15	Primary for heating
5. Farallones Institute Cabin	Occidental,	26.5	5.4	0.23	Contains rockbed thermal storage
6. New Maxico State U. House	Las Cruces, NM, 1977	171.9	120.8	0.15	
7. Bruder House	Phoenix, Az., 1976-79	167.2	148.6	0.28	Provision for water spraying on top of water bags, for evaporative cooling
8. Skytherm North	St. Paul, Minn.,1979	185.8	92.9	0.45	Heating mode only. Uses double acrylic glazing.
9. San Joaquin House	San Joaquin Valley,Ca, 1978-80	276.9	276.9	0.25	
10. Trinity U. Test Facility	San Antonio, Tx.,1980	74.3	74.3	0.08-0.15	Two identical test facilities.
11. Garrett House	Las Cruces, NM,1980-81	353.0	223.0	0.30	

Table 3. (cont'd)

Name	Location/ Year of construction	Floor Area,m^2	Roof-pond Area,m^2	Roof-pond depth,m	Remarks
12. Pala Passive Solar Project	Pala,Co., 1981	23.8	23.8	0.13	
13. Camelback School	Paradise Valley,Ca., 1982	35.7	35.7	0.25	Provision for water spraying on top of the water bags.
14. Fry House	Yuma,Az., 1978	139.4	139.4	0.13-0.15	Uses stationary insulation panels submerged in water Water is circulated over the insulation at night.
15. Taliesin West	Scottsdale, Az.,1982-83	44.6	40.4	Two layers, 0.3-and 0.15	Similar to (14).

of the minimum ambient temperature. Diurnal temperature fluctuations in the room were less than 2°C, and relative humidity was maintained between 85-95%, which is suitable for storing most fruits. The optimum water depth, in terms of lowest water temperature attainable while still having sufficient water mass to serve as an effective heat sink, is found to be about 0.10 m.

One room was used to store several types of fruits, with different treatments such as waxing and low-pressure storage in combination with the tempreatures obtained in the roof-pond room. It was found that pear of the Pien Pu variety and persimmon of the P-1 variety gave the best results. Their storage life was extended from one week to several months inside the cool room [35]. An economic analysis of the system indicated that it was very promising.

In this application, the thermal mass of the stored products is roughly on a par with the roof-pond water. Therefore, provided that loading and unloading of products is done on a partial basis, the product mass could substitute for some of the thermal mass of the water. (This is in contrast to roof-pond applications in residential buildings, where the building's contents usually have much smaller thermal mass). Also, if the pond water could be cooled to only around the minimum ambient temperature, as is true in this case, then it would be advantageous to also

Fig. 18. An open roof pond room for fruit storage in northern Thailand.

cool the product directly with the night air by forced ventilation, at such times when the ambient temperature is near minimum.

Another application was to use the roof-pond room as a shed for cultivating Champignon mushrooms. In Thailand, these mushrooms can be grown only in the winter, when the ambient temperature is sufficiently low. The use of the roof-pond method will help extend the production season considerably, and increases the farmer's income. Work is continuing in this aspect.

Nocturnal Cooling for Precooling of Agricultural Products. The cooled pond water can also be used in another application, precooling of fruits and vegetables to remove field heat. This process involves the removal of a large amount of heat from the product in a relatively short time, and is normally achieved through expensive large-capacity refrigeration units. Work has been conducted in Thailand to utilize the cooled roof pond water at night in a packed spray-tower, to forced-air cool products in rural areas. The arrangement is shown in Fig. 19. Air is drawn through the ventilated product boxes, cooled in the packed tower using water from the roof pond, and returned to the room. The fan is driven by a small engine if no electricity is available. The system uses energy for water pumping and for air circulation- but none for cooling. This method was found to be effective for cooling several kinds of products. A precooler having sufficient cooling capacity for 1,000 kg of products has been developed [36].

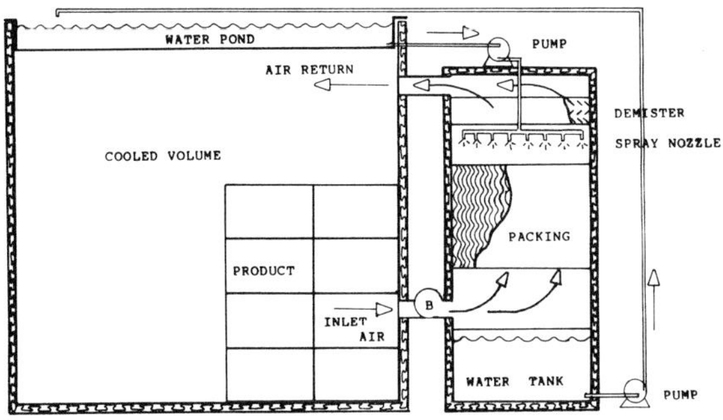

Fig. 19. Forced-air precooling of agricultural products, using nocturnally-cooled water.

5. General Assessment and Future Research

Nocturnal cooling has been shown to be effective in cooling residential buildings, and cold storage for agricultural products. The main limitation of the technique is that it is strongly dependent on the local climate. Therefore, to utilize the technique effectively, one must keep three factors in mind: location of usage, season of usage, and user's requirements in terms of temperature and humidity. For example, for comfort cooling, one can estimate the temperature which may be obtained through nocturnal cooling from the foregoing discussions, and compare it with the human comfort requirements. If they match, either for the entire year or only for certain periods, then nocturnal cooling could be employed (with back-up systems for the rest of the year, if needed). For fruit cooling, on the other hand, one needs to know the product's storage requirements instead.

The major problem which remains in the practical utilization of nocturnal cooling is the lack of better radiator materials. This aspect should be the focus of future research efforts. To date, the use of currently-available materials in nocturnal cooling systems has limited their temperature to the minimum or slightly above the minimum ambient. If this temperature could be lowered 5-10 °C further, a much wider application would result. For example, in Thailand the minimum temperature achievable with the method is around 8-15 °C. This temperature range is suitable only for a few selected fruits (with some also needing complementary treatments such as waxing and hypobaric storage). If the temperature range could be lowered to around 0-2 °C, practically all fresh fruits and vegetables could be stored, using standard processes.

Better materials must therefore be developed for the radiator surface, to maximize cooling in the atmospheric window, and minimize absorption outside it. Better cover materials are needed. Also, more work should be done on the light-selective film for daytime cooling under direct solar radiation. In all cases, the materials developed must be durable, resistant to deterioration under adverse climatic conditions, and relatively inexpensive.

6. Conclusions

In this chapter, we have discussed the phenomenon of nocturnal cooling, with emphasis on radiative cooling through the 8-14 μm atmospheric window. It has been shown that the magnitude of radiative cooling can be large enough for many practical purposes, but that it depends on the local climate, a dry and clear-sky climate being very favourable for it. Several types of wavelength-selective radiator surfaces have been developed, to take advantage of the atmopsheric window. A number of full-scale nocturnal cooling systems already exist, but they are situated in very selected locations, and are designed for very specific purposes. More research work must be conducted, to make the method more widely applicable.

Refereneces

1. Granqvist, C.G. 1984. Spectrally selective surface coatings for energy efficiency and solar applications. The Physics Teacher, September, pp. 372-383.
2. Bell, E.E., Eisner, L., Young, J. and Oetjen, R.A. 1960. Spectral radiance of sky and terrain at wavelengths between 1 and 20 microns. II. Sky measurements. J. Opt. Soc. Am., Vol. 50, No. 12, pp. 1313-1320.
3. Berdahl, P. and Fromberg, R. 1982. The thermal radiance of clear skies. Solar Energy, Vol. 29, No. 4, pp. 299-314.
4. Berger, X., Buriot, D. and Garnier, F. 1984. About the equivalent radiative temperature of clear skies. Solar Energy, Vol. 32, No. 6, pp. 725-733
5. Kimball, B.A. and Idso, S.B. 1982. A model of thermal radiation from partly cloudy and overcast skies. Water Resour. Res., Vol. 18, No. 4, pp. 931-936.
6. Kneizys, F.X., Shettle, E.P., Gallery, W.O., Chetwynd, J.H., Jr, Abreu, L.W., Selby, J.E.A., Fenn, R.W. and McClatchey, R.A. 1980. Atmospheric transmittance/radiance: Computer code LOWTRAN 5, U.S. Air Force Geophysical Laboratory,

Hanscom AFB, Mass., AFGL-TR-80-0067. Available from NTIS, U.S. Dept. of Commerce, Springfield, Va. 22151.

7. Eriksson, T.S. and Granqvist, C.G. 1982. Radiative cooling computed for model atmospheres. Appl. Opt., Vol. 21, No. 23, pp. 4381-4388.

8. Ben-Shalom, A., Barzilai, B., Cabib, D., Devir, A.D., Lipson, S.G. and Oppenheim, U.P. 1980. Sky radiance at wavelengths between 7 and 14 μm; measurement, calculation and comparison with LOWTRAN 4 predictions. Appl. Opt., Vol. 19, pp. 838-843.

9. Stössel, F.P. and Bolle, H.J. Mediterranean temperature distribution derived from METEOSAT data by means of atmospheric corrections based upon LOWTRAN 5. Inst. fur Meteorologie und Geophysik, Universität Innsbruck, Schopfstrasse 41, Austria.

10. Berdahl, P. and Martin, M. 1984. Emissivity of clear skies. Solar Energy, Vol. 32, No. 5, pp. 663-664.

11. Idso, S.B. 1981. A set of equations for full spectrum and 8- to 14-μm and 10.5- to 12.5-μm thermal radiation from cloudless skies. Water Resour. Res., Vol. 17, No. 2, pp. 295-304.

12. Martin, M. and Berdahl, P. 1984. Summary of results from the spectral and angular sky radiation measurement program. Solar Energy, Vol. 33, No. 3/4, pp. 241-252.

13. Martin, M. and Berdahl, P. 1984. Characteristics of infrared sky radiation in the United States. Solar Energy, Vol. 33, No. 3/4, pp. 321-336.

14. Platt, C. and Dilley, A. 1981. Remote sounding of high clouds. IV: Observed temperature variations in cirrus optical properties. J. Atmos. Sci., Vol. 38, p. 1069. Quoted by [13].

15. Catalanotti, S., Cuomo, V., Piro, G., Ruggi, D., Silvestrini, V. and Troise, G. 1975. The radiative cooling of selective surfaces. Solar Energy, Vol. 17, pp. 83-89.

16. Johnson, T.E. 1975. Radiation cooling of structures with infrared transparent wind screens. Solar Energy, Vol. 17, pp. 173-178.

17. Eriksson, T.S., Lushiku, E.M. and Granqvist, C.G. 1983. Optical materials and process technology for energy efficiency and solar applications. Proc. Int. Soc. Opt. Engr., Vol. 428, pp. 106-111.

18. Berdahl, P. 1984. Radiative cooling with MgO and/or LiF layers. Appl. Opt., Vol 23, pp. 370-372.

19. Lushiku, E.M. and Granqvist, C.G. 1984. Radiative cooling with selective infrared-emitting gases. Appl. Opt., Vol. 23, No. 1, pp. 1835-1843.

20. Nilsson, N.A., Eriksson, T.S. and Granqvist, C.G. 1985. Infrared-transparent convection shields for radiative cooling: initial results on corrugated polyethylene foils. Solar Energy Materials.

21. Addeo, A., Nicolais, L. and Romeo, G. 1980. Light selective structures for large scale natural air conditioning. Solar Energy, Vol. 24, pp. 93-98.

22. Andretta, A., Bartoli, B., Coluzzi, B. and Cuomo, V. 1981. Selective surfaces for natural cooling devices. Journal de Physique, Colloque C1, supplement no. 1, Tome 42.

23. Givoni, B. 1982. Cooling by longwave radiation. Passive Solar Journal, Vol. 1, No. 3, pp. 131-150.

24. Mostrel, M. and Givoni, B. 1982. Wind screens in radiant cooling. Passive Solar Journal, Vol. 1, No. 4, pp. 229-238.
25. Berdahl, P., Martin, M. and Sakkal, F. 1983. Thermal Performance of radiative cooling panels. Int. J. Heat Mass Transfer, Vol. 26, No. 6, pp. 871-880.
26. Landro, B. and McCormick, P.G. 1980. Effect of surface characteristics and atmospheric conditions on radiative heat loss to a clear sky. Int. J. Heat Mass Transfer, Vol. 23, pp. 613-620.
27. Michell, D. and Biggs, K.L. 1979. Radiation cooling of buildings at night. Appl. Energy, Vol. 5, No. 4, pp. 263-275.
28. Penman, H.L. 1948. Natural evaporation from open water, bare soil, and grass. Proc. Roy. Soc., Ser. A, 193, 120-146.
29. Hay, H.R. and Yellott, J.I. 1969. Natural air conditioning with roof ponds and movable insulation. ASHRAE Trans., Vol. 75, part 1, pp. 165-177.
30. Yellott, J.I. and Hay, H.R. Thermal analysis of a building with natural air conditioning. ASHRAE Trans., Vol. 75, part 1, pp. 178-190.
31. Hay, H.R. U.S. Patent 3299589 (January 24, 1967).
32. Hay, H.R. U.S. Patent 3450192 (June 17, 1969).
33. Hay, H.R. U.S. Patent 3563305 (February 16, 1971).
34. Marlatt, W.P., Murray, K A. and Squire, S.E. 1984. Roof pond systems, Energy Technology Engineering Center, Energy Systems Group, Rockwell International, prepared for the U.S. Department of Energy, contract no. DE-AM03-76SF00700.
35. Boon-Long, P. and Sornsrivichai, K. 1985. Night cooling for storage of pears and persimmons. Ambeint Energy, Vol. 5, No. 4, pp. 187-192.
36. Boon-Long, P. 1987. Roof-pond system for precooling/cold storage of agricultural products. To be presented at the Solar World Congress, 13-18 September 1987, Hamburg.

CHAPTER 9

ICE-MAKING FOR RURAL APPLICATIONS

R.H.B. Exell

Introduction

Among the applications of solar energy still in the experimental stage refrigeration is attractive because the need for cold is greatest when the sun is hottest. Solar powered refrigeration would be useful in the remote rural areas of developing countries where conventional energy supplies are unobtainable or expensive. It could be employed for making ice or for the cold storage of food and medical vaccines. Ice-making may have an advantage over cold storage in that ice is transportable and can be saved for use on cloudy days when a solar refrigerator would not work. Moreover there is always a ready market for ice. On the other hand it would be easier to cool foods to temperatures in the range 0 to 10 °C. Such cooling could be used for the short term preservation of fresh fruit, vegetables, fish and dairy products, or for the long term preservation of dried foods.

As long ago as 1972 Merriam [1] suggested that efforts should be directed towards the development of a solar ice-maker for rural areas. He suggested using the ammonia-water cycle. The unit would be made of mild steel, and would be simple and rugged in design without precision parts. There should be no problem in constructing it in local workshops. The output would be 60 to 70 kg of ice at -10 °C for an input of 10 to 12 m^2 of solar radiation and the services of a full time unskilled operator.

Broadly speaking there are two ways of using solar energy to operate refrigerators. One way is to convert solar radiation into electricity by means of photovoltaic solar cells and then use the electricity to drive a compression refrigerator. The other way is to convert solar radiation into heat in a thermal solar collector and use the heat to drive an absorption refrigerator.

Photovoltaic converters and compression refrigerators are currently being developed for small-scale domestic use and for the preservation of vaccines in isolated areas. Absorption refrigeration systems with thermal solar collectors enjoy a wider range of possible applications, which includes air conditioning, medium scale cold storage, ice manufacturing, and the same small-scale uses as the photovoltaic systems.

The whole subject of solar powered refrigeration thus embraces a variety of solar and refrigeration technologies, all of which are active areas of research and development at the present time. However, all of the systems developed to date are expensive. Efforts are being made to improve efficiencies and bring down costs, but there is little to choose between the various types, and it is still difficult to see which, if any, will ultimately secure appreciable markets.

In view of the relatively large capital outlay required initially for the installation of a solar refrigerator, and of the better technical efficiency obtainable with larger units, it is likely that voltage-size refrigeration systems would be more practicable and more economic than small domestic units. For example, it has been stated that food coolers of ice-makers serving communities of about 100 people would be valuable in the islands of the Pacific Ocean [2].

Accordingly, the Asian Institute of Technology has investigated the possibility of developing a village-size refrigerator with a solar collector area of 20 to 25 m^2 that can provide 0.5 m^3 of food storage cooled below 10 °C, or make about 100 kg of ice per day at a cost that the people for whom it is intended can afford. For simplicity the intermittent ammonia-water absorption cycle has been used because the thermodynamic properties of this combination are well known, and the technology is reliable.

The type of solar collector employed was determined by the fact that in the tropics the proportion of diffuse solar radiation in the total insolation is usually high. A flat-plate collector was therefore used in preference to a concentrator, which would require accurate orientation and tracking. It seems, however, that the use of flat auxiliary mirrors to enhance the heating by direct solar radiation might be worthwhile.

Early Work on Solar Ammonia-Water Refrigerators

A survey of progress from 1936 to 1972 has been made by Swartman et al. [3]. Early work in this field used designs with concentrating collectors, but later more attention was given to flat-plate collectors, beginning with the pioneering work of Chinnappa [4] in Sri Lanka.

A diagram of Chinnappa's unit, which was constructed of welded steel pipe, is shown in Fig. 1. The generating system consisted of a tube-in-sheet collector 1.52 m by 1.07 m painted dull black with three glass covers and containing 9 kg of ammonia-water solution of concentration 0.46 ammonia by weight. On a good day about 1.2 kg of ammonia could be distilled from the solution. For the refrigeration process a water-cooled absorber was used, and minimum temperatures of -12 °C were obtainable. It was concluded that about 1.4 kg of ice could be produced per square metre of collector area on a clear day, and this performance could be considerably improved by a selective radiation coating on the collector surface.

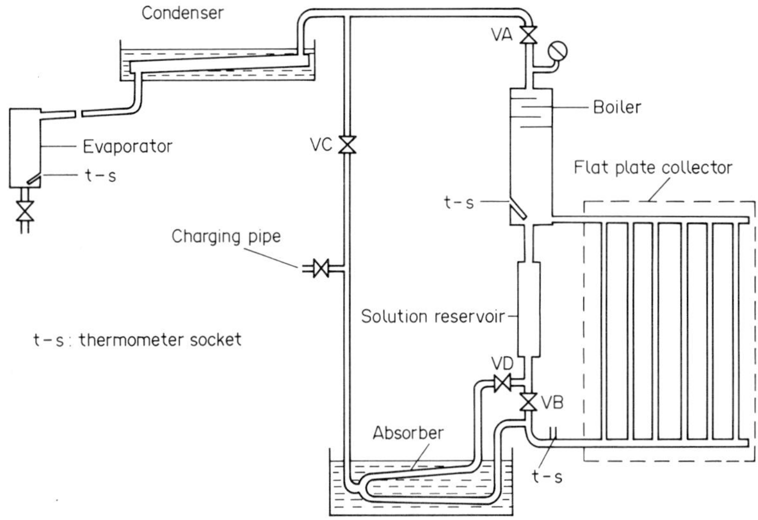

Fig. 1. First solar refrigerator by Chinnappa.

Another solar powered intermittent ammonia-water absorption refrigerator similar to that tested by Chinnappa was studied by Swartman and Swarminathan [5]. Their unit differed from that of Chinnappa in two respects. First, instead of using a boiler and a solution reservoir outside the collector, they used a large header of diameter 152 mm, which provided the required capacity for the solution and a large surface for the evaporation of ammonia. Second, for the refrigeration process they introduced the ammonia vapour from the evaporator into a tube connecting the lower and upper headers of the collector, and the glass cover of the collector was opened to facilitate cooling. Tests showed that, although 1.9 kg of ammonia could be distilled per square meter of collector area, and refrigeration temperatures as low as -12 °C could be reached, the rate of refrigeration was too slow because of slow reabsorption of ammonia vapour into the solution. They proposed using the collector to dissipate the heat of absorption to overcome this problem.

A theoretical analysis of the heat transfer and flow characteristics of tube-in-sheet flat-plate collectors used as generators in solar powered ammonia-water refrigerators was made by Moore and Farber [6]. The work culminated in a collector 1.49 m^2 in area consisting of 25 mm diameter steel pipes running from a 32 mm diameter feeder to a 63.5 mm diameter header. The centres of the 25 mm pipes were spaced 102 mm apart and were soldered to a galvanized iron sheet of thickness 1.0 mm. A single glass cover was used.

Farber's refrigerator, using a continuous ammonia-water cycle with electrically operated circulation pumps, was able to produce 12.5 kg of ice per square metre of collector area per day [7]. This yield is twice what can be obtained using the intermittent cycle, but more components are needed in the construction of the system, and it would not work in locations where electricity is not available.

Basic Concepts

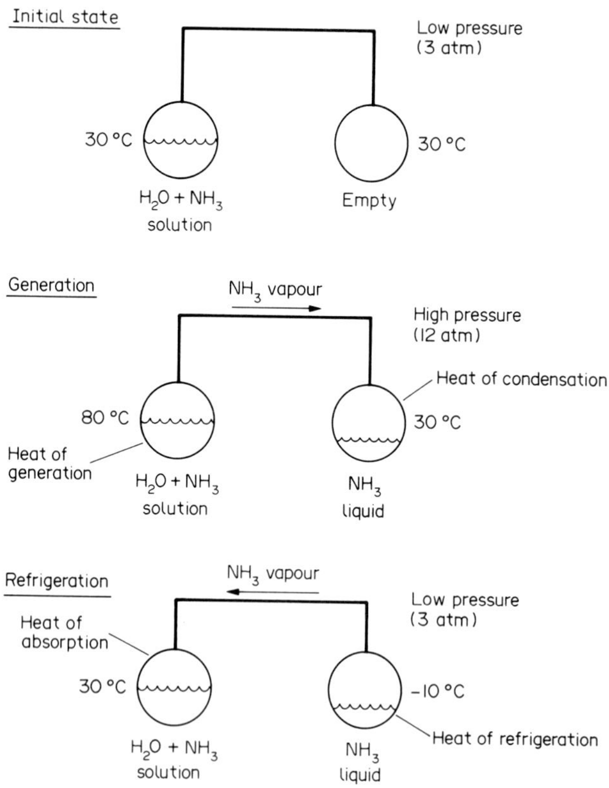

Fig. 2. Principle of ammonia-water absorption refrigeration.

Figure 2 shows the principle of operation of the intermittent ammonia-water absorption refrigeration cycle. Imagine two vessels connected by an overhead pipe. Initially the left hand vessel contains an ammonia water solution, having about 45% of ammonia by weight, at ambient temperature 30 °C. The right hand vessel is empty. First the left hand vessel is heated to about 80 °C while the right hand vessel is kept at 30 °C. This causes the pressure to rise to about 12 bar, and ammonia to boil from the solution in the left hand vessel and condense in the right hand vessel. To produce refrigeration the left hand vessel is now cooled and held at ambient temperature. This reduces the pressure to about 3 bar and causes the liquid ammonia in the right hand vessel to boil at the refrigeration temperature of about -10 °C. The ammonia vapour is reabsorbed into the solution in the left hand vessel. When the liquid ammonia has been used up, the right hand vessel returns to ambient temperature, and the initial state of the system is restored.

During the generation process, heat is supplied at high temperature to the left hand vessel, and the heat of condensation of ammonia is rejected to the environment at ambient temperature from the right hand vessel. During refrigeration, the right hand vessel absorbs heat from its refrigeration load at a low temperature, and the heat of absorption of the ammonia into the solution in the left hand vessel is rejected to the environment at ambient temperature.

The arrangement shown in Fig. 2 would not work in practice because it has two defects. First, during the refrigeration process ammonia vapour reabsorbed into the solution in the left hand vessel does so on the surface of the liquid and increases the ammonia concentration there. Because the strong solution is less dense than the weaker solution below the surface, no mixing takes place and the equilibrium vapour pressure at the surface increases. This raises the temperature of the ammonia in the right hand vessel, and eventually the process stops.

The second defect is that the vapour passing from the left hand vessel to the right hand vessel during generation is not pure ammonia, as desired, but contains 2-3% of water. During refrigeration the presence of this water in the right hand vessel raises the evaporation temperature above the desired value. Moreover, since no water evaporates during this process, more and more water accumulates in the right hand vessel as the cycle is repeated, making the system less and less effective.

An arrangement designed to overcome these defects, and to use solar energy from a flat-plate collector to supply the heat of generation, is sketched schematically in Fig. 3. The vessels, pipes and valves are all made of steel; copper and brass are attacked by the ammonia-water solution and cannot be used.

The ammonia-water solution is contained in the solar collector, and a tall air-cooled rectifier pipe serves to remove water from the generated vapour. The vapour is condensed in a condenser cooled by a tank of water, and the liquid ammonia is collected in the receiver. An evaporator coil provides a heat transfer surface between the liquid ammonia and the refrigeration load, which may be a quantity of water for making ice. There is a drain, closed by a valve C, from the evaporator to the solar collector so that if unwanted water accumulates inside the evaporator it can be returned to the solar collector from time to time. During the refrigeration process the ammonia vapour is directed to the bottom of the solar collector to ensure good mixing and uniform reabsorption throughout the solution.

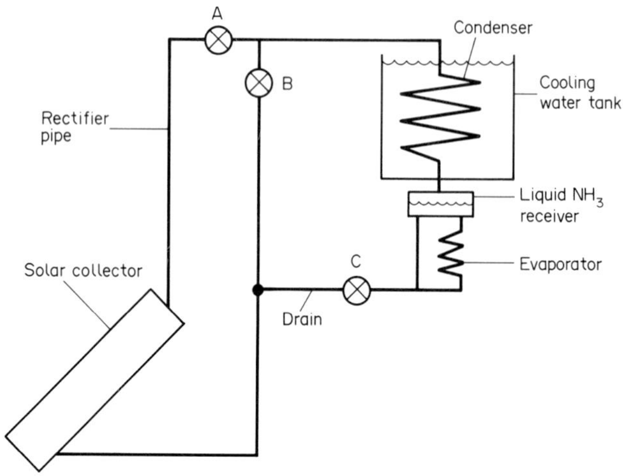

Fig. 3. Scheme for solar refrigerator.

The Ideal Cycle

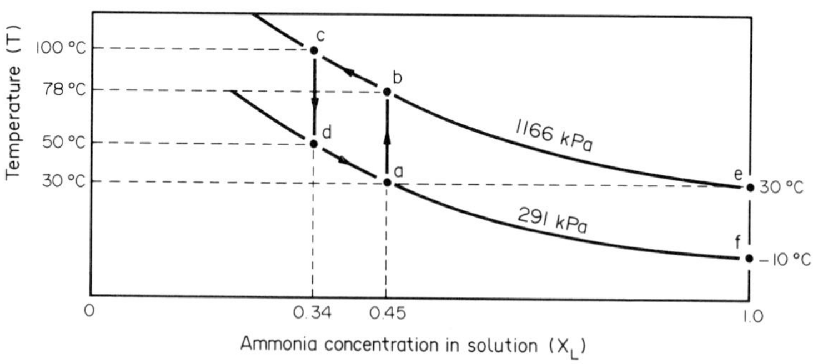

Fig. 4. Ideal thermodynamic cycle.

The ideal cycle for operating the system consists of four processes, which are illustrated on the thermodynamic diagram in Fig. 4. Starting with the solution in the solar collector at ambient temperature and the receiver empty, the four processes are as follows:

(1) a-b: Heating of the solution in the collector with valves A and B closed, during which the concentration X_L remains constant and the pressure P increases.

(2) b-c: Heating of the solution in the collector with valve A open and B closed. During this process, which takes place at

constant pressure, ammonia evaporates from the solution reducing its concentration, condenses at ambient temperature (point e), and accumulates in the receiver.

(3) c-d: Cooling of the solution in the collector with valves A and B closed, during which the concentration X_L remains constant and the pressure P decreases.

(4) d-a: Evaporation of ammonia from the evaporator (point f) with valve A closed and B open. During this process, which takes place at constant pressure, the ammonia vapour is reabsorbed into the solution increasing its concentration.

The heating processes (1) and (2) are carried out during the day under the sun. The cooling processes (3) and (4) are carried out at night with the back insulation removed from the solar collector to allow the heat to escape.

The method of determining the points on the cycle will now be explained. We shall suppose that the ambient temperature is 30 °C and that the required refrigeration temperature is -10 °C. The point f is fixed by the refrigeration temperature -10 °C and the concentration $X_L = 1$. Here the vapour pressure of pure ammonia is 291 kPa. The point a is the intersection of the 291 kPa isobar and the 30 °C isotherm. This gives the ammonia concentration $X_L = 0.45$ of the solution that must be charged into the system at the outset. The point e is fixed by the condensing temperature 30 °C and the concentration $X_L = 1$. Here the vapour pressure of pure ammonia is 1166 kPa. The point b is now the intersection of the 1166 kPa isobar and the line $X_L = 0.45$, which occurs at temperature T = 78 °C. The position of the point c on the 1166 kPa isobar depends on the amount of solar heating that occurs on a particular day, and on the amount of solution the solar collector contains per unit of collecting surface. In a typical design on a sunny day a temperature of 100 °C might be reached. This occurs at a concentration $X_L = 0.34$. Finally, the point d is the intersection of the line $X_L = 0.34$ and the 291 kPa isobar, which occurs at a temperature T = 50 °C. The whole cycle is now established.

It is convenient to have expressions for the mass of liquid m_L in the solar collector, and the mass of ammonia vapour m_V that has been distilled from the liquid, in terms of the liquid concentration X_L at any point and the initial mass $m_L(a)$ and concentration $X_L(a)$ of the liquid at the initial point a. A complete theory must take into account the small amount of water in the vapour. However, to cimplify our calculations we shall assume that the vapour is pure ammonia. Then we have

$$m_L = m_L(a)(1 - X_L(a))/(1 - X_L), \quad (1)$$

and

$$m_V = m_L(a)(X_L(a) - X_L)/(1 - X_L). \quad (2)$$

Note that our thermodynamic system has two chemical components, ammonia and water, and two phases, liquid and vapour. Therefore, by the Gibbs phase rule, there are two degrees of freedom. The two independent variables that we shall use to specify the thermodynamic state of the system are:

X_L, the mass fraction of ammonia in the liquid (dimensionless),

and

T, the temperature (K).

All other properties of the system are functions of X_L and T, for example:

P, the pressure (kPa),
X_V, the mass fraction of ammonia in the vapour (dimensionless),
v, the mass specific volume of the liquid (L/kg),
h_L, h_V the mass specific enthalpies of the liquid and vapour (kJ/kg), and
s_L, s_V the mass specific entropies of the liquid and vapour (kJ/kg K).

Heat Transfers in the Cycle

Let m_{sc} denote the mass of steel in the pipework and the black plate of the solar collector, and let C_s denote the specific heat capacity of steel. Then the differential of the heat entering the steel in any temperature change is $m_{sc}C_s dT$.

We first calculate the heat of generation in the process a-b. Since this occurs at constant volume with no production of ammonia vapour, the differential of the heat supplied to the liquid is $m_L(a)Tds_L$. Therefore, the total heat supplied in the process a-b is

$$Q_{ab} = \int_a^b (m_{sc}C_s dT + m_L(a)Tds_L).$$

For practical calculations the integral may be approximated by

$$Q_{ab} = m_{sc}C_s(T(b)-T(a)) + m_L(a)(0.5)(T(a)+T(b))(s_L(b)-s_L(a)). \tag{3}$$

The values of m_{sc} and $m_L(a)$ depend on the design of the system. In accordance with the designs tested in the Asian Institute of Technology we shall take m_{sc} = 65 kg and $m_L(a)$ = 12kg per square metre of solar collector area, and quote all the results on this basis. We also have C_s = 0.46 kJ/kg K.

Thermodynamic tables give, for T(a) = 303 K and T(b) = 351 K, the values $s_L(a)$ = 0.22 kJ/kg K and $s_L(b)$ = 0.93 kJ/kg K. Substituting in (3) we obtain

$$Q_{ab} = 4221 \text{ kJ}.$$

The process b-c takes place at constant pressure with valve A open, so the differential of the heat supplied to the ammonia-water system in the solar collector is the differential of the enthalpy of the liquid-vapour system with the term for the sensible heating of the vapour omitted. The complete differential of the enthalpy of the liquid-vapour system is

$$d(m_L h_L + m_V h_V) = d(m_L h_L) + h_V dm_V + m_V dh_V.$$

Adding the term $m_{sc}C_s dT$ for the heating of the steel pipes, omitting the term $m_V dh_V$ for the heating of the vapour, and integrating from b to c, we obtain

$$Q_{bc} = \int_b^c (m_{sc}C_s dT + d(m_L h_L) + h_V dm_V),$$

which may be approximated by

$$Q_{bc} = m_{sc}C_s(T(c)-T(b))+m_L(c)h_L(c) - m_L(b)h_L(b) + (0.5)(h_V(b) + h_V(c))(m_V(c)-m_V(b)). \quad (4)$$

Here we have $T(c) = 373$ K, and from thermodynamic tables $h_L(b) = 115$ kJ/kg, $h_V(b) = 1444$ kJ/kg, $h_L(c) = 249$ kJ/kg, and $h_V(c) = 1544$ kJ/kg. Equations (1) and (2) with $m_L(a) = 12$ kg and $X_L = 0.34$ give $m_L(c) = 10$ kg and $m_V(c) = 2$ kg. Substituting in (4) we obtain

$$Q_{bc} = 4756 \text{ kJ}.$$

Summarizing, we find the total heat of generation is $Q_{ab} + Q_{bc} = 8977$ kJ, and this produces 2 kg of distilled ammonia per square metre of solar collector area.

By similar calculations we may find the amount of heat that must be released from the solar collector in the processes c-d and d-a. Thus, replacing a by c and b by d in equation (3) leads to

$$Q_{cd} = -4001 \text{ kJ};$$

and replacing b by d and c by a in equation (4) leads to

$$Q = -4743 \text{ kJ}.$$

Therefore the total amount of heat to be released is $Q_{cd} + Q_{da} = -8744$ kJ, the negative sign indicating that the heat is lost by the solution and the solar collector.

In designing the condenser (see Fig. 3) we must know how much heat is released to the tank of water by the condensing ammonia. Since the process takes place at constant pressure, this is simply the difference between the enthalpy of the ammonia vapour entering the condenser $m_A h_{AV}$ and the enthalpy of the ammonia liquid leaving the condenser $m_A h_{AL}$. Here we have $m_A = 2$ kg, and making the simplifying assumptions that the ammonia is pure and that the whole process takes place at 30 °C, we have from thermodynamic tables of pure ammonia h_{AV} = 1288 kJ/kg and h_{AL} = 142 kJ/kg. The total amount of heat relesaed by 2 kg of ammonia condensing at 30 °C is therefore about 2292 kJ per square metre of collector area. In designing the system one must ensure that there is sufficient water in the tank to prevent its temperature from being raised more than a few degrees by this heat.

We shall suppose that at the point d in the cycle, when valve B opened, the liquid ammonia and the receiver containing it fall rapidly to the temperature -10 °C, which is represented by the transition from point e to point f in Fig. 4. This is a simplification of what actually occurs because the release of the heat of absorption from the solar collector takes time.

Let m_{sr} denote the mass of steel in the receiver and evaporator; then the differential of the heat flow from the receiver to the liquid ammonia is $m_{sr}C_s dT$. The differential of the heat flow into the ammonia is

$$dH_A - V_A dP = d(m_{AL}h_{AL} + m_{AV}h_{AV}) - (m_{AL}v_{AL} - m_{AV}v_{AV})dP,$$

where the subscripts A, L and V refer to ammonia, liquid and vapour respectively. Setting $dh_{AV} = -dm_{AL}$, omitting the terms containing dh_{AV} and v_{AV}, equating the result to the differential of the heat from the receiver, and integrating from state e to state f, we obtain

$$\int_e^f (m_{AL}dh_{AL} + (h_{AL} - h_{AV})dm_{AL} - m_{AL}v_{AL}dP) = \int_e^f m_{sr}C_s dT.$$

Finally, approximating the integrals in the same way as before, we obtain

$$0.5(m_{AL}(e) + m_{AL}(f))(h_{AL}(f) - h_{AL}(e)) + 0.5(h_{AL}(e) - h_{AV}(e)$$
$$+ h_{AL}(f) - h_{AV}(f))(m_{AL}(f) - m_{AL}(e)) - 0.5(m_{AL}(e) + m_{AL}(f))$$
$$(0.5)(v_{AL}(e) + v_{AL}(f))(P(f) - P(e)) = m_{sr}C_s(T(f) - T(e)).$$

All the quantities in this equation are known from the system design and from thermodynamic tables except $m_{AL}(f)$, which may therefore be calculated. Taking m_{sr} = 10 kg and the remaining quantities from thermodynamic tables we obtain $m_{AL}(f)$ = 1.576 kg per square metre of collector.

The useful heat of refrigeration Q_f is the heat required to vaporize the remaining liquid ammonia at constant temperature -10 °C and pressure 291 kPa, which is given by

$$Q_f = m_{AL}(f)(h_{AV}(f) - h_{AL}(f)).$$

Substituting the known values on the right hand side we obtain Q_f = 2046 kJ per square metre of collector. The cooling ratio for this cycle is defined to be $Q_f/(Q_{ab} + Q_{bc})$. Using the results calculated above we obtain a cooling ratio of 0.228.

Heat Transfer Rates in the Solar Collector

The rates at which the processes in the cycle take place are controlled by the energy flow rates to and from the solar collector. In discussing these rates we shall suppose that the solar flat-plate collector has one glass cover at a distance 50 mm from a selective black plate, and there is polyurethane foam insulation 100 mm thick at the back which can be removed at night to allow the system to cool quickly.

The transmissivity $\tau(\theta)$ of the glass cover, and the absorptivity $\alpha(\theta)$ of the black plate, for solar radiation both depend on the angle of incidence θ of the radiation.

The total solar irradiance q_{in} incident on the glass cover is given by

$$q_{in} = I\cos\theta + D,$$

where I is the direct solar irradiance at the angle of incidence θ, and D is the diffuse solar irradiance. The rate q_{abs} at which solar radiation is absorbed by the black plate per unit area is therefore

$$q_{abs} = \tau(\theta)\alpha(\theta)I\cos\theta + \overline{\tau\alpha}D, \qquad (5)$$

where $\overline{\tau\alpha}$ is the mean value of $\tau(\theta)\alpha(\theta)$, which for a collector that is nearly horizontal (as in the tropics) can be found by means of integrals over the hemispherical sky to be about 0.70.

The annual mean values of q_{abs} in the climatic conditions of Thailand for various states of cloudiness calculated from typical values of $\tau(\theta)$ and $\alpha(\theta)$ by the above methods are shown in Table 1. The parameter k is defined to be the ratio of the daily total global solar radiation to the corresponding value under a clear sky. The dependence of q_{abs} on k is not simple because the ratio of direct to diffuse solar radiation, and hence the effect of the angle of incidence θ, changes as k changes.

There are several processes by which a flat-plate solar collector loses heat, as illustrated in Fig. 5. They are as follows (we assume steady state heat transfer):

(1) Convection and radiation q_{up} from the glass cover at temperature T_c to the surroundings at ambient temperature T_a.

(2) Convection and radiation q_{up} between the black plate at temperature T_b and the glass cover.

(3) Conduction q_{down} through the back insulation.

(4) Heat losses through the edges of the collector and through the connecting pipes. These losses may not be negligible, but in good designs they should be small and we shall neglect them here.

(5) Convection and radiation q_{down} from the back of the black plate when the back insulation is removed at night to cool the collector and release the heat of absorption.

Table 1. Mean values of radiation absorbed by a flat plate with one glass cover (W/m^2)

Time of day		k=1	k=0.75	k=0.5	k=0.25
6h	18h	6	6	5	3
7h	17h	101	88	64	34
8h	16h	307	232	156	78
9h	15h	524	374	242	119
10h	14h	699	487	310	151
11h	13h	809	558	354	172
12h		846	582	368	179

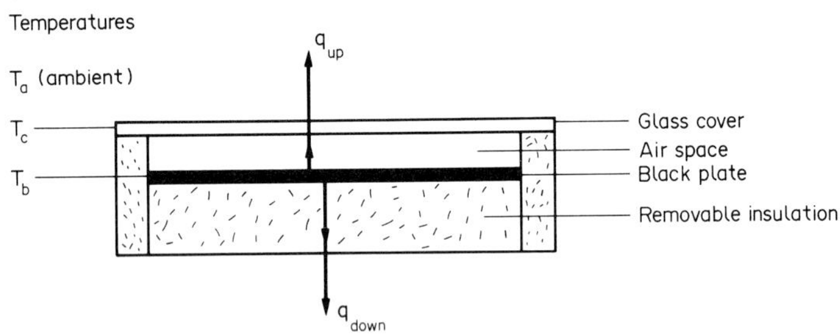

Fig. 5. Temperatures and heat losses in solar collector.

The total heat loss rate q_{loss} from the collector is given by

$$q_{loss} = q_{up} + q_{down}$$

as a function of T with the help of standard equations for the heat transfer in solar collectors [8]. The calculation of q_{loss} must be done numerically.

Some results obtained for ambient temperature 30 °C give for a selective black plate and one glass cover:

$$q_{loss} = 3 + 3.4(T_b - 30) + 0.0064(T_b - 30)^2 \text{ W/m}^2. \qquad (6)$$

where T_b is in degrees Celsius.

When the insulation is removed from the back of the solar collector one obtains

$$q_{loss} = 3 + 10.4(T_b - 30) + 0.0064(T_b - 30)^2 \text{ W/m}^2. \qquad (7)$$

The instantaneous efficiency η of the solar collector is defined by the equation

$$\eta = (q_{abs} - q_{loss})/q_{in}.$$

This depends on the intensities of the direct and diffuse solar irradiance, I and D, on the angle of incidence θ of the direct solar radiation, and on the temperature T_b of the black plate. Consider as an example q_{in} = 800 W/m² with I = 650 W/m², D = 150 W/m², η = 0 at midday, and the black plate at 80 °C such as one might find during generation on a fair day. Then taking $\tau(0)\alpha(0) = 0.83$, and $\overline{\tau\alpha} = 0.70$, we obtain q_{abs} = 644 W/m². Also, for T_b = 80 °C, we have q_{loss} = 189 W/m², and hence η = 0.57.

The overall performance of the system may be defined to be the ratio of the useful heat of refrigeration Q_f to the total solar radiation incident on the collector per day. This equals the cooling ratio multiplied by the average efficiency of the solar collector throughout the day, which may be about 0.4. Taking the cooling ratio 0.288 found earlier we obtain an overall performance of about 0.09. This is close to the overall performance of a large number of solar energy conversion devices that are limited by the second law of thermodynamics.

Modelling the System through Time

Although we have seen how to calculate the total quantities of heat in a particular cycle, and the heat transfer rates in the solar collector that control the processes in the cycle, we still require a method of calculating the response of the system to a given solar input throughout the day. In order to do this we must write down the differential equations for the heat transfers that occur in each of the processes a-b, b-c, c-d and d-a, and integrate them with respect to time. The calculations involved are simple in principle, but detailed and numerous. They must therefore be performed on an automatic computer. Full details cannot be given here, but the basic ideas can be explained briefly.

In the process a-b the differential of the absorbed solar radiation minus the losses from the collector equals the differential of the heating of the collector and solution:

$$(q_{abs} - q_{loss})dt = m_{sc}C_s dT + m_L T ds_L. \qquad (8)$$

Here q_{abs} is given by equation (5) with I, D and specified as functions of time: while q_{loss} is given by equation (6). For simplicity we assume that the temperature T of the solution and the black plate temperature T_b are the same. In real designs these temperatures differ slightly, and the temperature of the solution is not uniform, but we shall ignore these difficulties. The thermodynamic properties of the solution are assumed to be known as functions of X_L and T. The right side of (8) is then written

$$m_{sc}C_s dT + m_L(X_L,T)T(\partial s_L(X_L,T)/\partial T)dT,$$

and the resulting equation when integrated numerically step-by-step gives T as a function of time t. At each step the pressure $P(X_L,T)$ is found and the integration is continued until the condensing pressure 1166 kPa is reached. On a cloudy day the insolation may not be strong enough for this to occur, and then no ammonia is distilled.

In the process b-c the differential equation is

$$(q_{abs} - q_{loss})dt = m_{sc}C_s dT + d(m_L h_L) + h_V dm_V. \quad (9)$$

where m_L, h_L, m_V and h_V are known functions of X_L and T. It follows that the right-hand side can be written in terms of the differentials dX_L and dT. In order to follow the isobar we need $dP(X_L,T) = 0$; so the differentials dX_L and dT must satisfy the equation

$$(\partial P(X_L,T)/\partial X_L)dX_L + (\partial P(X_L,T)/\partial T)dT = 0. \quad (10)$$

Equations (9) and (10) are now solved simultaneously step-by-step to determine how far the process goes. On a good day the final temperature may be high and a large quantity of ammonia may be distilled. It is assumed that when the temperature stops rising, due to the reduction of the solar heating towards the end of the day, the valve A (Fig. 3) is closed and the insulation is removed from the back of the collector to start the process c-d.

At this point we have calculated the amount of distilled ammonia available for refrigeration. For the calculation of the process c-d one uses equation (8) again with q_{abs} small at the end of the afternoon and zero after dark, and q_{loss} given by equation (7). The process d-a is coupled to the refrigeration load. Further development of the model is needed to calculate this process.

The above method of modelling the system has been found to give results that compare fairly well with experimental tests in the Asian Institute of Technology. Table 2 gives some of the result obtained using 4 min time steps in the integrations and assuming uniform states of cloudiness throughout the day. The relationship between the amount of ammonia distilled and the total daily solar radiation with uniformly cloudy skies is approximately linear. In practice the yields of ammonia in experimental tests are variable and usually higher than those calculated in theory. The variability of the experimental yields is caused by the manner in which the cloudiness varies throughout the day. If the weather is fine in the morning and cloudy in the afternoon, the yield of ammonia may be good due to the strong heating of the solar collector during the critical generating period. On the other hand, when the morning is cloudy the collector may not become hot enough by midday to generate ammonia, even if the weather is fine in the afternoon.

Table 2. Ammonia distilled per square metre of collector. H = daily global solar radiation.

Cloudiness, k:	1	0.75	0.5	0.25
H (MJ/m^2):	26.5	19.8	13.2	6.6
Selective plate	3.36 kg	2.04 kg	0.49 kg	0

Early Work in The Asian Institute of Technology [9]

In the first solar powered refrigeration experiments in AIT, started in 1974, a small prototype intermittent ammonia-water unit was developed with a flat-plate collector 1.44 m^2 in area. By 1977 the modified unit was capable of producing 6 kg of ice in one day. It differed from the system shown in Fig. 3 in that a hand operated expansion valve and dry evaporator were used instead of a flooded evaporator. The efficiency of the solar collector was 28 %, the cooling ratio was 0.26, and the overall performance was therefore 0.0726.

During the period 1978 to 1980 a larger unit was designed and tested. It had two flat-plate collectors each of area 2.5 m^2 with movable flat auxiliary mirrors attached to the eastern and western edges to enhance the solar heating. This unit also had a hand operated expansion valve and dry evaporator. On a bright day 14 kg of ammonia could be distilled, and refrigeration the following night gave over 25 kg of ice from water initially at 28 °C. The ice could keep the temperature of a cool storage compartment below 13 °C while the ambient temperature reached 35 °C. A preliminary economic evaluation showed that the cost of ice produced by the system would be 2 or 3 times the city price; this would be acceptable in remote areas.

A Village-Size Solar Refrigerator [10]

The latest system constructed and tested in AIT has a solar collector area 25 m^2 and is designed to provide 0.5 m^3 of food storage space or 100 kg of ice per day under favourable conditions. The principle of operation is shown schematically in Fig. 6. During generation, solar heat causes ammonia to evaporate from the ammonia-water solution in the collector. This ammonia bubble through the liquid P in the seal S, and at night ammonia evaporating from the freezing unit is forced to the bottom Q of the collector by the head of liquid standing in the inner pipe R.

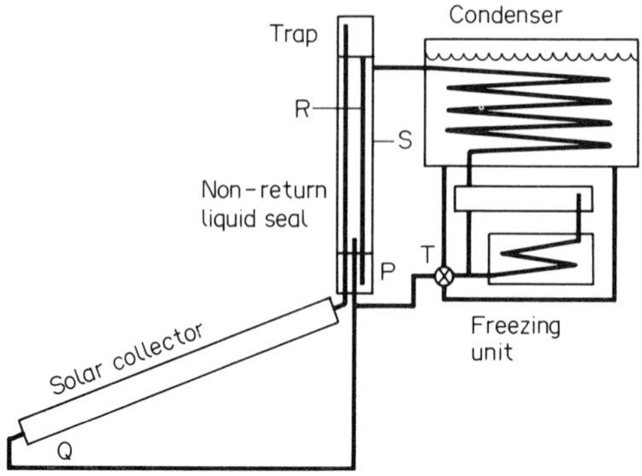

Fig. 6. Design with flooded evaporator and liquid seal.

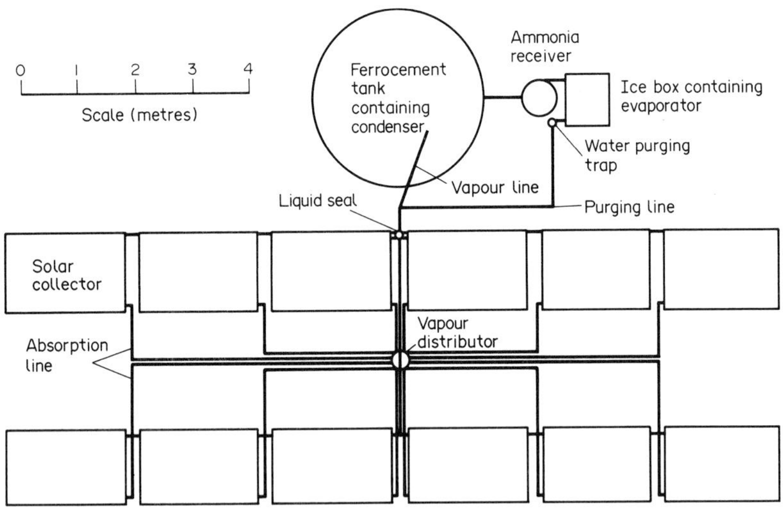

Fig. 7. Plan view of village-size refrigeration system.

Figures 7 and 8 show the plan and side views of the complete system. There are twelve flat-plate solar collectors, each 2.1 m^2 in area making a total area of 25 m^2. The collectors are placed in two rows, each row having six collector panels. The collectors are connected together at the headers. The ammonia-water solution contained in the collectors have a volume of 338 l and a concentration of ammonia 46 % by weight. The insulation covering the underside of the collectors can be lowered to allow the solution to cool rapidly at night.

A liquid seal connected to the header of the second row of collectors acts as a valve to control the flow of ammonia vapour as

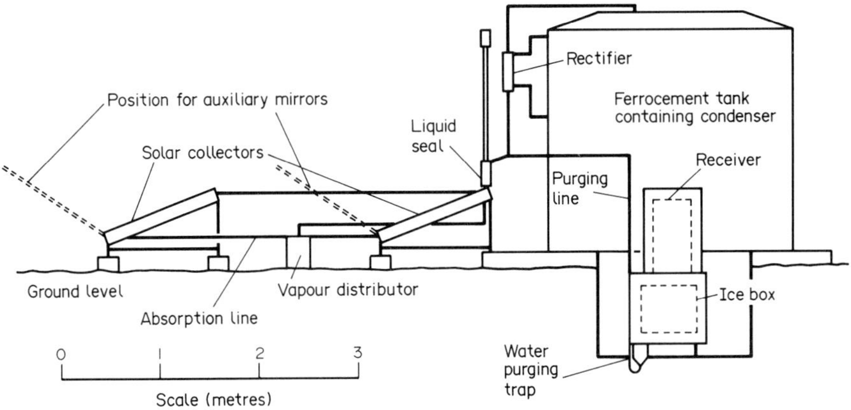

Fig. 8. Side view of village-size refrigeration system.

described above. The rectifier is a vertical tube, connected between the liquid seal and the condenser, to remove unwanted water from the vapour passing from the collectors to the condenser. It is cooled by a water jacket. The condenser is immersed in a large tank of water to remove the heat of condensation of the ammonia vapour.

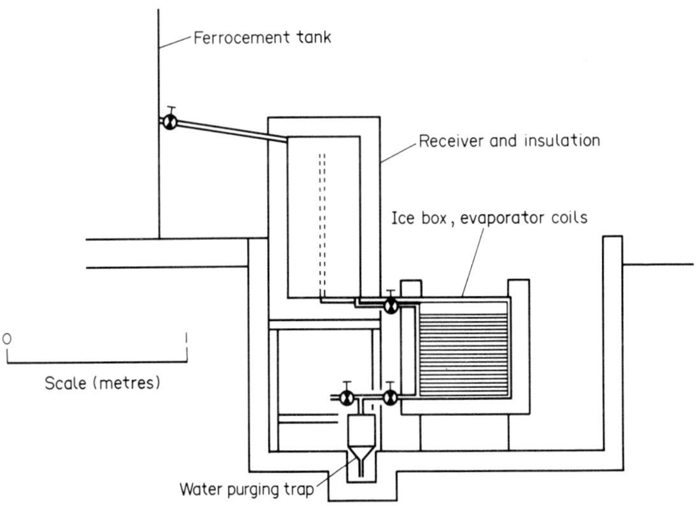

Fig. 9. Details of freezing unit.

Figure 9 shows the refrigeration part of the system. As insulated cylindrical receiver stores liquid ammonia before it runs into the evaporator coils to produce refrigeration. The evaporator coils are immersed in an ice box with brine used as a heat transfer liquid between the evaporator coils and the ice cans. There are five cans, each having a capacity 20 l. A water trap connected to the bottom of the evaporator is used occasionally to return unwanted water from the evaporator to the solar collectors via the liquid seal. This purging operation is the only one performed by hand valves.

Ammonia vapour from the evaporator is directed by the liquid seal to the bottom of the solar collectors for reabsorption into the solution. A distributor, designed especially for this system, ensures that equal amounts of ammonia vapour are sent to the twelve collector panels.

Test runs on the system began in the morning at 8:00 a.m., when the insulated backs of all the collector panels were closed. Ammonia was then distilled from the collectors by solar heat and collected in the receiver. At about 15:30 p.m. the solar radiation becomes insufficient to generate any more ammonia, and the back insulation on the collectors is opened to allow the solution to cool. The amount of ammonia collected in the receiver is noted, and the proper amount of water is filled into the ice cans in order to obtain the maximum quantity of ice.

During the night the cooling of the solution in the collectors is accompanied by the evaporation and cooling of the ammonia, and ice is formed in the ice box. Nothing has to be done by the operator at this time. Early next morning the back insulation of the collectors is closed again, and another test run can be made.

The tests showed that the daily solar radiation falling on the tilted collector must normally be over 7 MJ/m^2 to produce ammonia. In Thailand the annual average global solar radiation on a horizontal surface is about 16 MJ/m^2; and the amount produced by this system on an average day is about 40 kg.

The correlation between the ice production and the ammonia production was poor in our tests, possibly because the correct amount of water to put in the ice cans at the end of the generation was sometimes misjudged. Nevertheless, one can say that 1 kg of liquid ammonia produced 1.7 kg of ice on the average. Therefore, on a day with the average amount of solar radiation the expected amount of ice was 68 kg. This fell short of the target amount of 100 kg. The cause was found to reside in the poor heat transfer rate between the ammonia refrigerant and the freezing water in the ice box so that unfrozen water remained in the ice cans and unused ammonia remained in the receiver at the end of a run. The maximum ammonia that could be evaporated in one night was usually about 31 kg, which produced a refrigeration effect of 38 MJ and was only 61 % of the design amount.

By modelling the heat transfer processes, it has been shown that the performance of the solar collectors was the best possible with this system. In the case of the freezing unit it was found that, taking into account the actual heat transfer rate between the evaporating ammonia and the freezing water, the amount of ice obtained corresponded to the amount of ammonia evaporated. Therefore, there were no appreciable losses in the freezing unit. It is only the slow heat transfer rate that reduces the amount of ice obtained. Once the thermal resistance between the ammonia and the water has been reduced in a new design the full amount of ice (100 kg) will be obtained, and no further improvements will be possible in the system without radical changes in the design concept.

Two simple indicators have been used to evaluate the economic feasibility of the system; the cost revenue ratio, and the pay-back period. The cost of the experimental unit (omitting the expensive instrumentation) was Baht 203,140 (US $ 1.00 = Baht 26 approx.). Of this, 50 % was the cost of the solar collectors. Assuming the actual ice production obtained (a pessimistic figure in view of the possibility of correcting the poor heat transfer in the freezing unit), taking the price of ice slightly greater than that of the ice in Bangkok (as it is in remote places), and calculating the annual production from the known solar radiation statistics in Thailand, one obtained cost revenue ratios varying from 0.4:1, under the best conditions of machine life and interest rates, to 0.8:1 under the worst conditions anticipated. The pay-back period was found to vary from 2 to 6 years, depending on the interest rates and the price of ice assumed. These results suggest that if the cost of the manufacturing system could be reduced, and if the design is corrected, then it will become economically attractive even without radical changes in the design.

A Continuous Solar Ammonia-Water Absorption Refrigerator

If the intermittent ammonia-water absorption cycle could be replaced by a continuous cycle the yield of the system for a given solar input can be increased (perhaps doubled) because smaller amounts of refrigerant and absorbant are circulated and solar energy is not wasted in heating the larger quantities of liquid required in the intermittent systems. A mechanical circulation pump, driven electrically by means of photovoltaic cells, or mechanically by means of a small heat engine incorporated in the system, could be used to maintain the necessary pressure difference between the evaporator and the generator. Such systems have not been developed. (The use of a non-solar electrical supply would not be of interest, since if such a supply were available one would use an ordinary electrical compression refrigerator.)

Another possibility is to use solar heat to operate an absorption refrigerator of the Platen-Munters type, with hydrogen in the evaporator to maintain a constant pressure in the whole system. Early attempts to do this required cumbersome focussing reflectors to raise high temperatures in the solar collectors to drive the system.

However, using the compact non-tracking evacuated-tube solar collectors now on the market, and a commercial Platen-Munz (sic) absorption refrigerator, Uppal et al. [11] have recently devised and tested a system that has no moving parts and is made up entirely of ready made components. The only modification required is a change in the concentration of the ammonia-water solution to match the temperatures of the output of the evacuated-tube collectors; this can be done by the manufacturer during the charging of the refrigeration unit. This development seems extremely promising.

Solid Adsorption Refrigeration Alternatives

Although the ammonia-water system is efficient it has certain practical disadvantages: (1) the problem of rectification (removing water from the ammonia during generation), (2) the corrosiveness of the fluid, and (3) the high pressure and consequent heaviness of the equipment.

The problem of rectification can be avoided by using a solid adsorber for ammonia instead of water. For many years Worsoe-Schmidt has been developing intermittent solar adsorption refrigerators using calcium chloride as the adsorber. In one of his papers [12] he concludes that such systems can now be built which are so simple and reliable that they can be operated for years without maintenance. It is unlikely that they will be competitive with conventionally powered plant, but in many cases a solar driven plant will do a much needed job which in many countries could not be accomplished with a conventional system.

The problem of high pressures can be avoided by using other refrigerants than ammonia. One combination is the zeolite-water system; the concept for using it is shown in Fig. 10 (after Tchernev, [13]). The quantity of water adsorbed by the zeolite is strongly dependent on temperature and only weakly dependent on vapour pressure (Fig. 11). Therefore, during the refrigerati process at night, when the collector containing zeolite is near ambient temperature, the water is readily adsorbed keeping the pressure below 0.55 kPa and making the refrigerant water freeze. During the day the regeneration process is only slightly affecte by the condensation pressure.

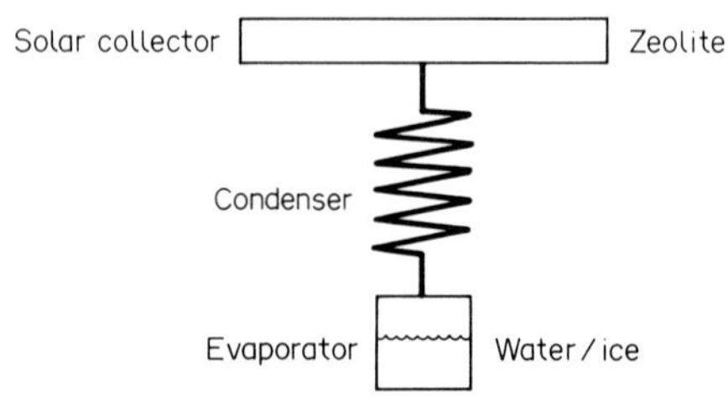

Fig. 10. Solar refrigerator with zeolite.

This system has been used by Tchernev to build small refrigerat of capacity about 100 dm^3, and also by Grenier et al. [14] to build a large cold store of volume 12 m^3. The technology does not show any size effects and is thus well adapted to both larg and small units.

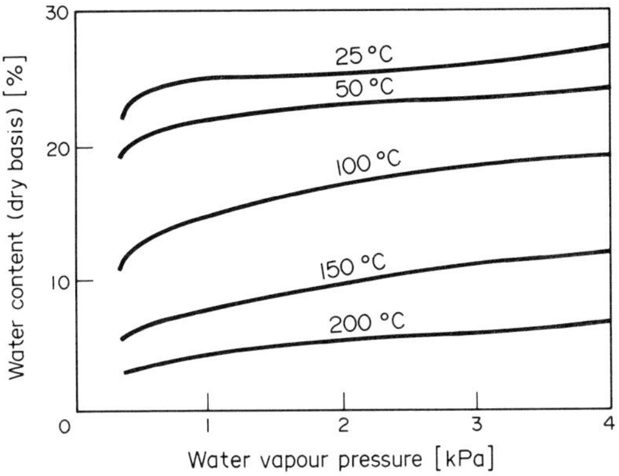

Fig. 11. Adsorption isotherms of water vapour on zeolite 13X.

Another promising absorbent is activated charcoal. Measurements of the adsorbent properties of zeolites 5A, 13X and activated charcoal have been made by Critoph and Vogel [15] for the organic refrigerants R11, R12, R22 and R114 as possible pairs for solar cooling. Charcoal-R22 offered the best cooling ratio (about 0.2), but charcoal-R114 is interesting in that it operates at pressures near atmospheric pressure and units using it would be less prone to leakage.

Another study by Grenier and Pons [16] has compared the thermodynamic performance of zeolite-water and activated charcoal-methanol for application to a solar ice-maker. The results shown in Fig. 12 are for condenser temperature T_C = 40 °C, evaporator temperature T_e = 0 °C, adsorption temperatures T_a = 25 °C and 35 °C, and a range of desorption temperatures T_d. At a desorption temperature of 100 °C, such as one can obtain in a flat-plate collector with a selective surface, the charcoal-methanol combination has a cooling ratio in the range 0.4-0.5 as compared with 0.3 for zeolite-water. It is interesting to note that both activated charcoal and methanol should be manufacturable in developing countries, and that these substances are the most promising for solar cool storage systems.

Conclusions

Summarising, it may be said that solar ice-making is still not commercially available, partly because the technology has not yet been brought to perfection, and partly because even if it had economic viability would be difficult due to the relatively small daily yield of ice for a given first cost of the equipment.

Nevertheless, the following systems appear to be possible contenders for solar ice-making in rural areas (using flat-plate collectors unless otherwise stated):

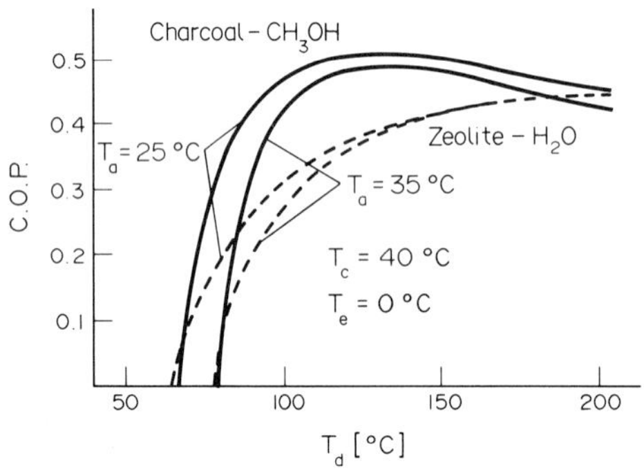

Fig. 12. Comparison between zeolite-water and charcoal-methanol adsorption pairs.

Intermittent ammonia-water absorption.

Intermittent ammonia-calcium chloride adsorption.

Continuous ammonia-water absorption with evacuated-tube collectors.

Zeolite-water adsorption.

Charcoal-methanol adsorption.

The development of the continuous ammonia-water system into a unit of modest first cost, and experiments with locally produce activated charcoal and methanol, would appear to be suitable directions for fruitful research in the future.

References

1. Merriam, M.F. (1972). Decentralized Power Sources for Devel oping Countries. Working Paper Series No. 19, East-West Technology and Development Institute, Hawaii.
2. Johnston, P. (1976). Planning for small-scale use of renew- able energy sources in Fiji. Proceedings of the Meeting of the Expert Working Group on the Use of Solar and Wind Energy U.N. Publication No. E.76.II.F.13, paper NR/ERD/EWGSW/CR.20.
3. Swartman, R.K., Ha, V. and Newton, A.J. (1973). Survey of Solar-Powered Refrigeration. Paper 73-Wa/Sol-6, American Society of Mechanical Engineers, New York.
4. Chinnappa, J.C.V. (1962). Performance of an intermittent refrigerator operated by a flat-plate collector. Solar Energy, Vol. 6, pp. 143-150.
5. Swartman, R.K. and Swaminathan, C. (1971). Solar powered refrigeration. Mechanical Engineering, June 1971, pp. 22-24
6. Moore, G.L. and Farber, E.A. (1967) Combining the collecto and generator of a solar refrigeration system. Paper 67-WA/ Sol-4, American Society of Mechanical Engineers, New York.

7. Farber, E.A. (1970). Design and performance of a compact solar refrigeration system. Paper No. 6/58, International Solar Energy Conference, Melbourne, Australia.
8. Exell, R.H.B. (1983). The theory of a simple solar refrigerator. Renewable Energy Review Journal, Vol. 5, No. 2, pp. 1-17, Asian Institute of Technology, Bangkok.
9. Exell, R.H.B. and Kornsakoo, S. (1981). Design and Testing of a Solar Powered Refrigerator. Research Report No. 126, Asian Institute of Technology, Bangkok.
10. Exell, R.H.B., Kornsakoo, S., Oeapipatanakul, S. and Chanchaona, S. (1984). A Village-Size Solar Refrigerator. Research Report No. 172, Asian Institute of Technology, Bangkok.
11. Uppal, A.H., Norton, B., Probert, S.D. and McVeigh, J.C. (1987). Solar absorption-cycle refrigeration for vaccine storage. Applied Energy, (in press).
12. Worsoe-Schmidt, P. (1983). Solar refrigeration for developing countries using a solar-absorption cycle. International Journal of Ambient Energy, Vol. 4, pp. 115-124.
13. Tchernev, D.I. (1979). Solar refrigeration utilizing zeolites. Proc. 14th Inter. Soc. Energy Conversion Eng. Conf. Amer. Chem. Soc., pp. 2070-2073.
14. Grenier, Ph., Guilleminot, J.J., Mester, M., Meunier, F. and Pons, M. (1984). Experimental results on a 12m solar powered cold store using the intermittent zeolite 13X-H_2O cycle. In: Szokolay, S.V. (Ed.), Solar World Congress, Vol. 1, pp. 353-358, Pergamon Press, Oxford.
15. Critoph, R.E. and Vogel, R. (1986). Possible adsorption pairs for use in solar cooling. International Journal of Ambient Energy, (7, 4, pp. 183-190).
16. Grenier, Ph. and Pons, M. (1984). Experimental and theoretical results on the use of an activated carbon CH_3OH intermittent cycle for the application to a solar powered ice maker. In Szokolav, S.V. (Ed.), Solar World Congress, Vol. 1, pp. 500-506, Pergamon Press, Oxford.

CHAPTER 10

USE OF PASSIVE COOLING IN BUILDINGS

Bo Adamson, DSc, Professor and Head of the
Department of Building Science,
University of Lund, Sweden

1. Passive Cooling

Some decades ago, the concept "passive solar" was introduced as the antithesis to mechanical solar systems which were regarded as "active". A definitiion was laid down which stated that passive solar systems should employ only natural means for the utilisation of solar energy. Very soon, as fans and pumps were used in the systems, the term "hybrid systems" was introduced.

Cooling may concern both rooms and the human body. Cooling of rooms is achieved by cooling the room air and/or a surface in the room. Cooling of the human body takes place by convection and radiation. Convection heat losses from the skin which has a temperature of around 32-34 °C depend on the temperature of the ambient air and also on air velocity. Radiation heat losses depend on a weighted mean temperature of the surrounding surfaces. Cooling can also be achieved by evaporation of perspiration from the skin.

Consequently, passive cooling implies cooling of rooms and of the human body by natural means. Passive cooling comprises solar protection, cooling by wind and stack effects, and also energy management by thermal storage inside the building.

2. Indoor comfort

Thermal comfort is a result of the heat balance of the human body. This is influenced by

activity level
clothing
environment

The parameters describing the environment are temperature, velocity and humidity of the ambient air, and the mean radiant temperature of the environment.

This is not the place for a state of the art review of comfort conditions in hot climates. Good reviews can be found in Markus & Morris 1980, and Meese et al. 1984, which also give extensive lists of references. Only a condensed table for hot climates will be shown here, Table 2.1, which is based on diagrams in Marku Markus & Morris 1980. In rooms with closed windows and doors the air velocity is about v = 0.1 m/s. The air temperature which is acceptable to 80% (70% in brackets) of people is 29 (33) °C if the relative humidity is 20%, and 25 (30) °C if the relative humidity is 80%. If windows are open and cross ventilation sets up an air velocity v = 2 m/s, 80% of people will find 35 (41) °C acceptable if the relative humidity is 20% and 30 (32) °C if the relative humidity is 80%.

3. An approach to a "new" design process

In hot climates, mechanical equipments (active systems) are normally used for indoor climatic control of buildings. The cooling load and the energy requirement are dependent on the building itself and can be changed by varying thermal insulation, window design, window shading devices, the colour of the outside of the building, etc. The degree of load reduction and energy saving are determined by cost-benefit analyses, using marginal load and energy costs and also marginal investment costs.

Another approach to the design process is based on the following:

- mechanical equipments such as cooling machines, fans and pumps are very vulnerable in hot climate - in arid climates they can be damaged by sand and dust, and in wet climates by rust

- mechanical equipments need maintenance and careful operation - when an equipment breaks down, skilled repair personnel is needed

- when a mechanical (active) system breaks down, the indoor climate may become unacceptable, and it may be many days before the system is repaired and the indoor climate is restored to acceptable conditions.

The object of the "new" design process is to eliminate these drawbacks. It has two phases:

1. In the first phase, a good "passive building" (one without mechanical heating or cooling) is designed by utilising knowledge gained from both traditional building and simulation studies.

2. If the passive design does not produce an acceptable indoor climate, heating or cooling is added to an extent just enough to achieve an acceptable indoor climate.

This design process is illustrated in Figure 3.1. Obviously, architects and engineers who are well acquainted with the climate and culture concerned have the necessary knowledge to design a building, but simulation studies of the first design will probably provide a number of new aspects.

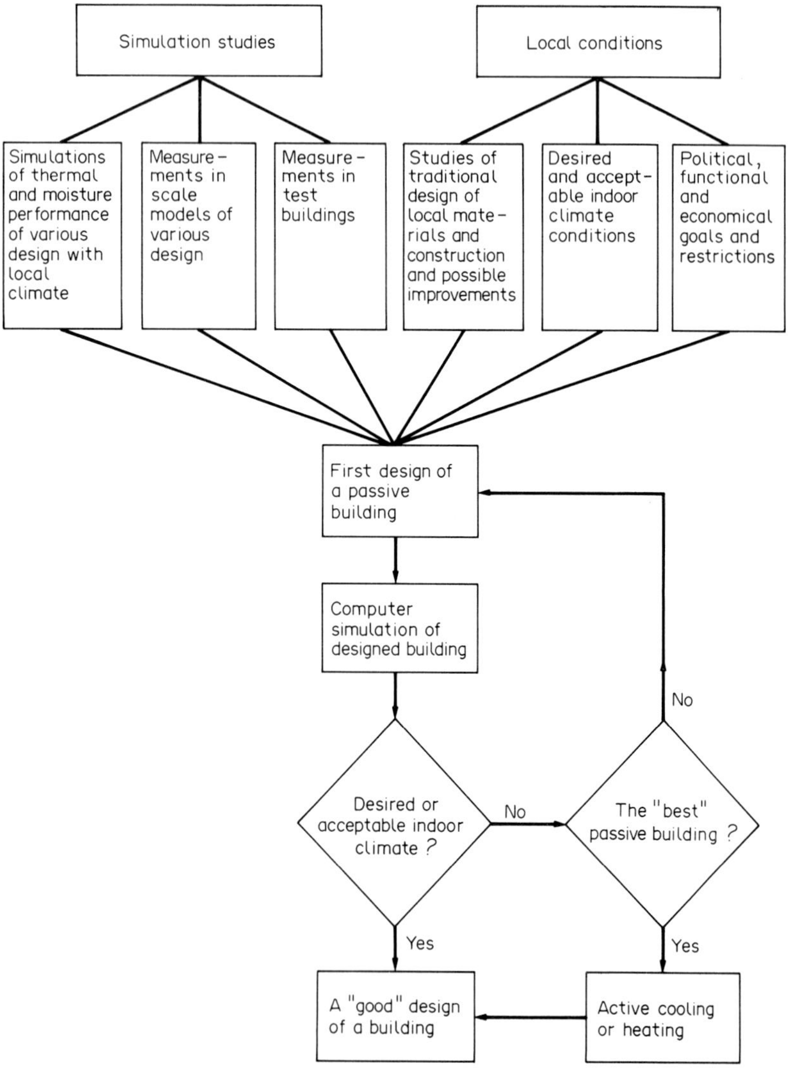

Fig. 3.1. Outline of a "new" design process.

The following sections of this chapter will show how computer simulations of thermal and moisture conditions can provide comprehensive information on the thermal and moisture performance of buildings. This will help in understanding traditional building and will also constitute the basis for good passive design.

4. Computer simulation and parametric studies - a useful design tool

4.1. The accuracy of computer programs

There are hundreds of computer programs available for calculation of the energy consumption for heating, and considerably fewer for cooling. If an accurate calculation is to be made of indoor temperature and moisture conditions, such as the conditions for a given day in a given year, statistics for monthly and yearly cooling loads etc, the demands on the simulation model are extensive:

- solar radiation must be accurately simulated
- physical simulation of the windows must be very accurate
- short wave radiation in the rooms must be accurately distributed
- long wave radiation in the rooms must be accurately calculated
- the convective surface heat flow must be accurately simulated
- the thermal mass must be simulated accurately.

Further, it is an advantage if the simulation program also takes into account

- long wave radiation to the sky
- temperature gradients within the room air
- moisture conditions indoors for a given moisture production
- natural ventilation calculated on the basis of wind and stack effects.

In any case, a computer program is required which calculates temperature and heat flows for time steps of less than one hour. The number of available programs is very low. At the Department of Building Science of the University of Lund, Sweden, we have used the JULOTTA program, Kallblad and Higgs , and the DEROB-LTH program. Both have been subjected to extensive validation studies. Among the limitations of the JULOTTA program must be mentioned the fact that, in spite of accurate calculation of direct and diffuse radiation through windows, the transmitted radiation has a diffuse distribution in the room. Moreover, long wave sky radiation is not normally taken into account. Obviously, the claim for accuracy of a specific part of a simulation model is high if this specific part is subjected to a parametric study, as in the case of windows. If windows are not of particular interest, simulation of windows is not as important as when window effects are studied.

A comparison of computer programs for heat balance calculations has been carried out within the International Energy Agency in 1981.

4.2. Model of the building

The assumed building for which the parametric study is carried out need not be similar to normal buildings in the climatic zone concerned. It may be a schematic building. At the Department of Building Science, University of Lund, the building in Figure 4.1 has been used for parametric studies in different climates. The use of the same building for different climatic conditions provides synergetic effects. It must be borne in mind that the purpose of a parametric study is to produce more knowledge as a basis of building design.

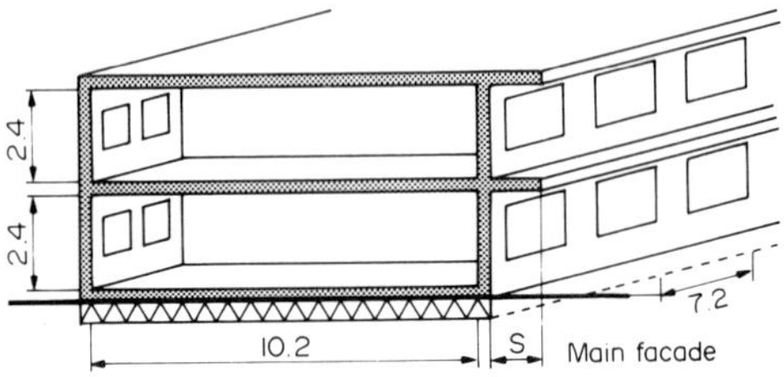

Fig. 4.1. Section of a long building.

The building in Figure 4.1 is a 7.2 m section of a long building of 10.2 m depth. It consists of two storeys and a main facade in which the size and construction of the windows and also the horizontal shading screen above the windows can be altered. In all calculations, the opposite facade is the same.

A number of design alternatives for walls, roofs, window construction and shading, and also different ventilation strategies, have been subjected to analysis. The building and the simulation model is described in Adamson & Rosenlund 1985.

4.3. The parameters to be studied

A parametric study usually starts with a baseline case for which temperatures and possible heating and cooling requirements are calculated for a normal year. This year need not be a "reference year" or a meteorological "normal year". It need only be a year which is considered normal. The parameters which influence room temperatures and possibly also the heating and cooling requirements have been examined in studies for various climates. Such parameters are

- interior walls, fittings and furniture
- "free heat" from occupants and domestic energy
- solar radiation
- ventilation
- window size in the main facade
- window shading in the main facade
- orientation of main facade
- number of panes and insulating glass units
- glass absorption
- overnight and seasonal insulation of windows
- curtains
- outer walls
- roof
- colour of outer surface of walls and roof

- thermal storage capacity of floor between the two storeys
- influence of flooring material
- influence of shading.

4.4. Presentation of results

The presentation of a parametric study is very difficult. The output from the computer shows temperatures (10-20) hour by hour for some interesting days. Very often monthly temperature statistics of room air temperatures are printed. For each month, maximum, average and minimum temperatures are printed. Moreover, a number of heat flows for different surfaces are printed on an hourly or monthly basis. It is possible to present only a very limited part of all data. The output lists from the computer can be examined in various ways and must be kept for further studies. The CPU time on a main computer is normally 15 minutes for a calculation case (on a 386-PC - about 20 minutes). For a parametric study, 50-100 cases are required.

In parametric studies the results are very often condensed and presented in the form of tables. Here, a large number of small diagrams will be shown.

Table 2.1. Maximum room air temperature for sitting people (1.0 met) with light clothing (0.6 clo) as a function of relative humidity RH, air velocity v and percentage of satisfied people. (Markus & Morris 1980).

Relative humidity	Room air temperature, at which					
	70% are satisfied			80% are satisfied		
RH %	v=0.1 m/s	v=0.5 m/s	v=2 m/s	v=0.1 m/s	v=0.5 m/s	v=2 m/s
20	33	39	41	29	32	35
50	30	34	37	27	30	33
80	27	30	32	25	28	30

5. Passive cooling in a hot and arid climate

5.1. Outdoor climatic conditions

The hot and arid climate is typical for deserts. The outdoor climate of a city located in the Sahara desert on latitude 32°N has been chosen to show the influence of various parameters such as ventilation rate, night ventilation, glass area of windows, window orientation, shading of windows, construction of outer walls and roof, and occupation. In the year 1975 the highest temperature (July) was 43.2°C. The lowest temperature (January) was 0.2°C. Figure 5.1 shows the monthly maximum, average and minimum temperatures.

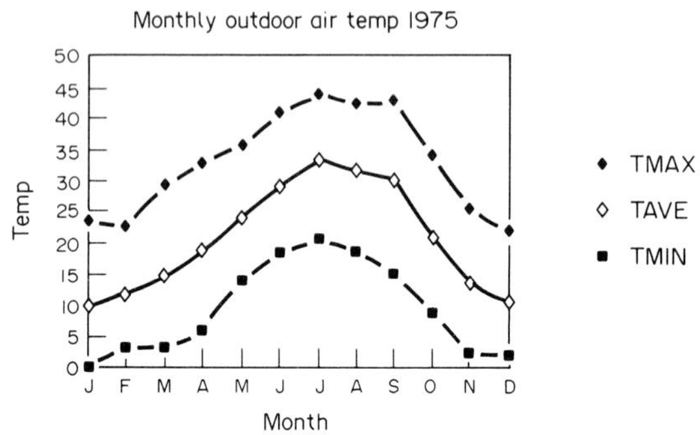

Fig. 5.1a.

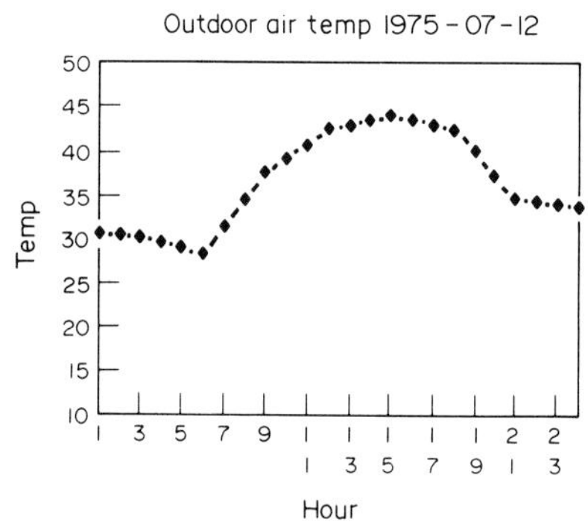

Fig. 5.1b.

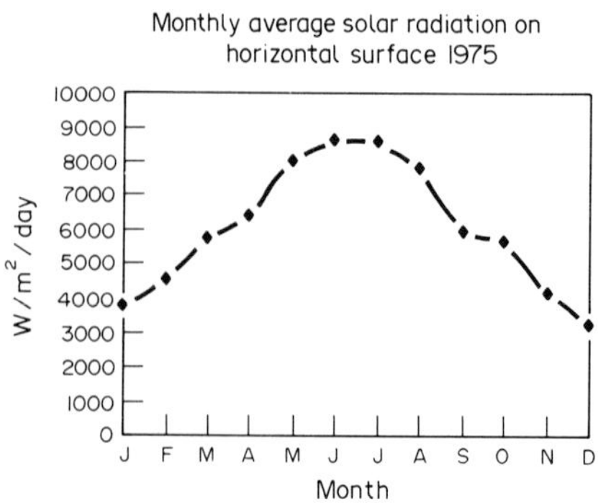

Fig. 5.1c.

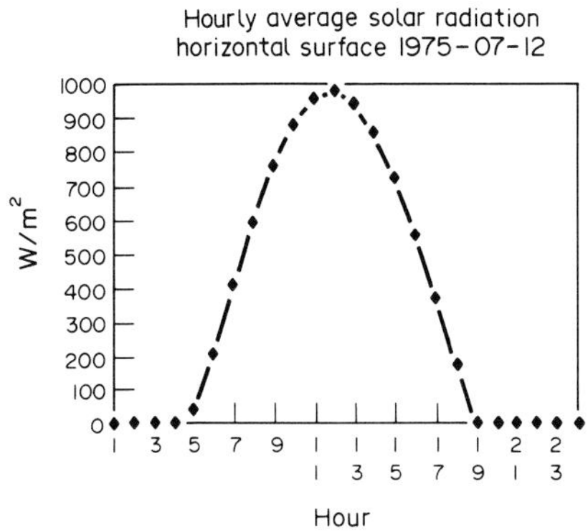

Fig. 5.1d. Climatic conditions for a place in the Sahara desert (32°N).

The temperature on the hottest day, 1975-07-12, is shown in Fig. 5.1b. During that day the maximum temperature is 43.2°C and the minimum temperature is 28.2°C.

The monthly average of solar radiation on a horizontal surface (per day) can be seen in Fig. 5.1c and the daily radiation for the hottest day in Fig. 5.1d. Daily distribution of solar radiation is calculated from the daily sum schematically.

The climate for this particular location is quite typical for the Sahara desert and also for the Arabian desert. The results presented in this section should therefore be representative for these regions.

5.2. The baseline case

A two storey multifamily house, shown in Fig. 4.1, has been chosen as a baseline case for parametric studies. One section of the long building is subjected to analysis. It is assumed that the behaviour of adjacent sections is the same. The section is 7.2 m wide and 10.2 m deep. It may be assumed that each storey constitutes one apartment (73 m² floor area). The following assumptions are made for the baseline case:

- o Glass area A=10.5 m² in main facade
- o Glass area A=4.5 m² in opposite facade
- o Main facade has southerly orientation
- o Single glazing
- o No shading screen above windows, S=0
- o Ventilation rate v=1.1 air change per hour (hygienic ventilation)
- o Light structure in outer walls, k=0.83 W/°C.m²
- o Light structure in roof, k=0.83 W/°C.m²
- o No internal walls or furniture.

Figure 5.2a sets out the monthly maximum temperature (TMAX) and the average temperature (TAVE) for an occupied house. The temperature T50, which is the temperature exceeded 50 hours per month, is also shown. It is seen that TMAX exceeds T50 by 2-3°C in the hot period of the year. Figure 5.2b shows that in the cold period the temperature T-50 below which the temperature drops 50 hours per month is 1-2°C higher than the minimum temperature TMIN.

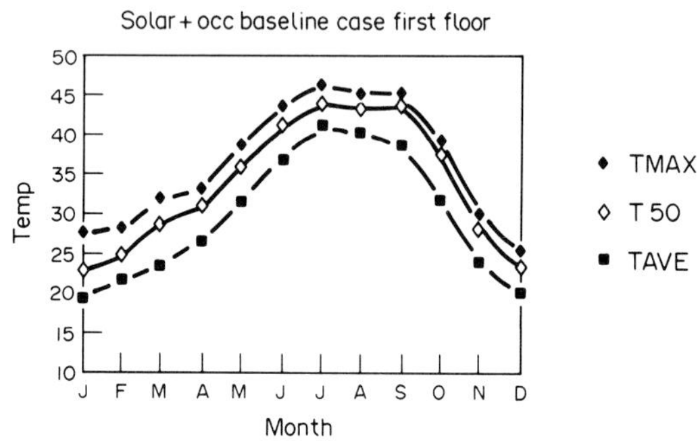

Fig. 5.2a.

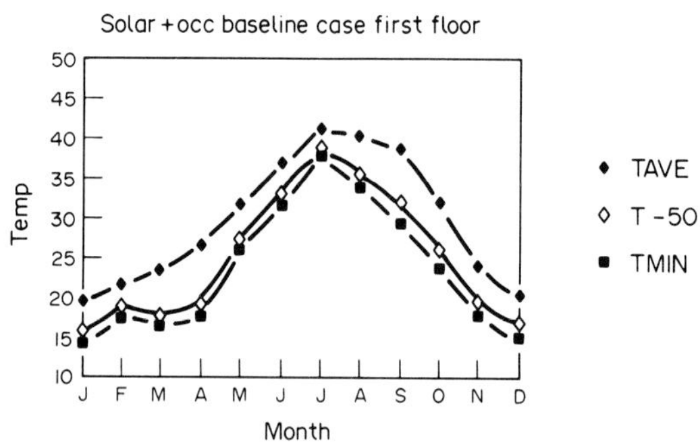

Fig. 5.2b.

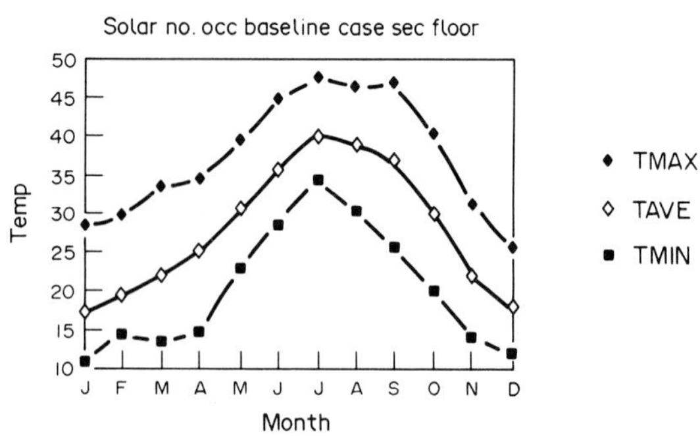

Fig. 5.2c.

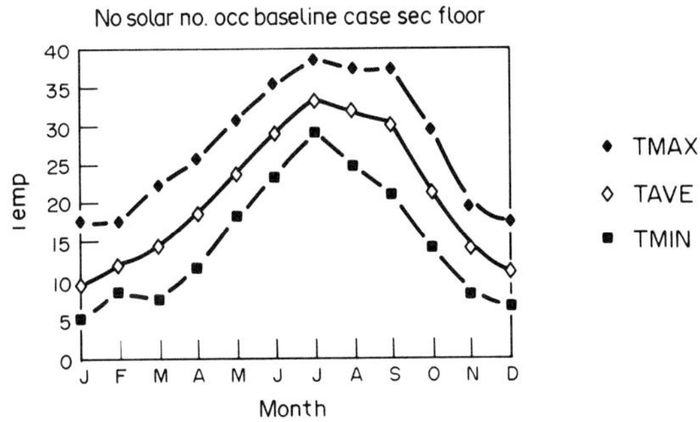

Fig. 5.2d. Monthly temperatures for the baseline case.

Figure 5.2c shows TMAX, TAVE and TMIN for the second storey of an unoccupied building. The maximum temperature is 47.2°C (July). For an occupied building the maximum temperature will be as high as 49.2°C. Without any solar radiation, Fig. 5.2d, the maximum temperature is 38.5°C. In the cold period the minimum temperatures are low, below 15°C. Such indoor climatic conditions are wholly unacceptable.

5.3. Influence of solar radiation and occupation

In Fig. 5.3a the monthly maximum temperature for the second storey is shown for

- o the building subjected to solar radiation and occupancy (S+O)
- o the building subjected to solar radiation but no occupancy (S)
- o the building without any solar radiation or occupancy (No S)

The building is assumed to be ventilated only for hygienic reasons, with v=1.1 ach/h.

It is obvious that solar radiation is responsible for a considerable temperature rise during both hot and cold periods.

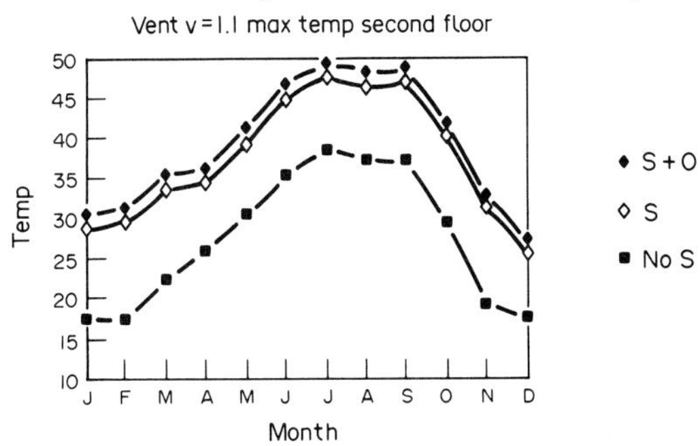

Fig. 5.3a.

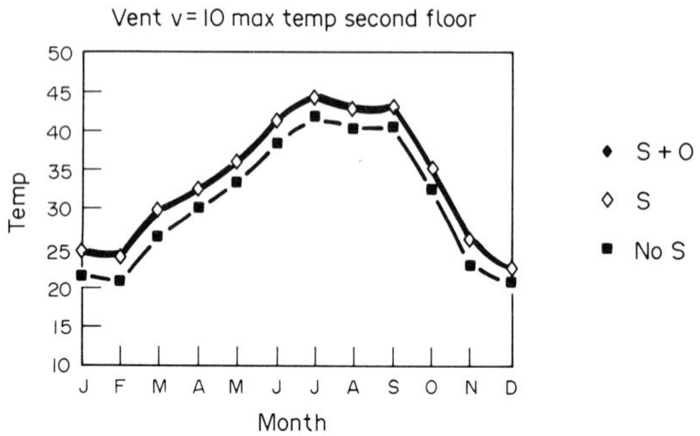

Fig. 5.3b.

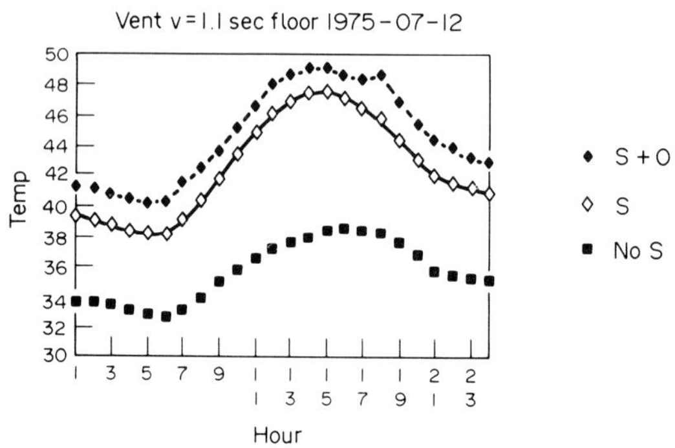

Fig. 5.3c.

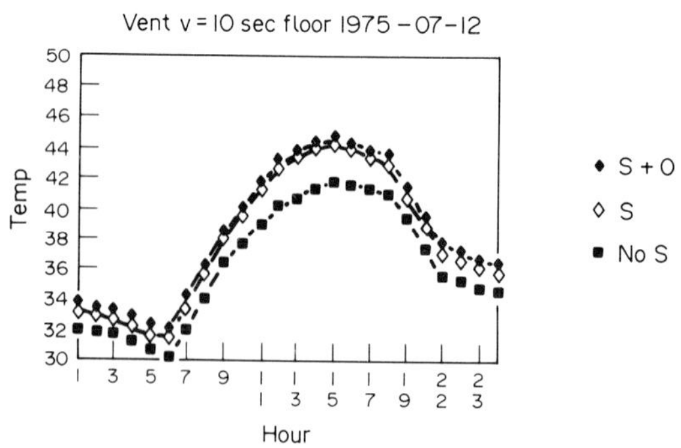

Fig. 5.3d. Influence of solar radiation and occupancy.

The influence of occupants and domestic energy is quite small. It is assumed that occupancy provides "free heat", distributed over the day as follows:

01-06: 350 Wh/h per storey
06-07: 500 Wh/h per storey
07-08: 350 Wh/h per storey
08-17: 200 Wh/h per storey
17-22: 500 Wh/h per storey
22-01: 350 Wh/h per storey

which totals about 8 kWh per day.

Figure 5.3b shows the maximum temperatures for a case with a ventilation rate v=10 ach/h. The influence of solar radiation and occupancy is quite small. Ventilation heat losses dominate the heat balance.

Figure 5.3c and d show the room air temperature for the second storey on the hottest day of the year. These figures confirm the previous conclusions.

5.4. Influence of ventilation rate

Figure 5.4a shows the monthly maximum temperatures on the second storey when the building receives solar radiation and is occupied. The building is assumed to be ventilated at various rates. During the summer an increase in ventilation rate from 1.1 to 10 ach/h will reduce maximum temperatures by 4-5°C. However, when the ventilation rate is changed from v=1.1 to v=10 ach/h, the minimum temperatures will be reduced by up to 10°C (Fig. 5.4b). A high ventilation rate can do a lot in reducing indoor temperatures during the hot period, but produces very low room temperatures in the winter. Consequently, during the winter the ventilation rate should be reduced to what is needed from the hygienic point of view.

If the building is not subjected to solar radiation and occupancy, room temperature is not influenced very much by the ventilation rate, as shown in Fig. 5.4c. With solar radiation and occupancy, room air temperature on the second storey will be very high on the hottest day of the year (Fig. 5.4d). Even when v=10 ach/h the temperature will be as high as 44.6°C which is higher than the ambient temperature that day. Room air temperature will not

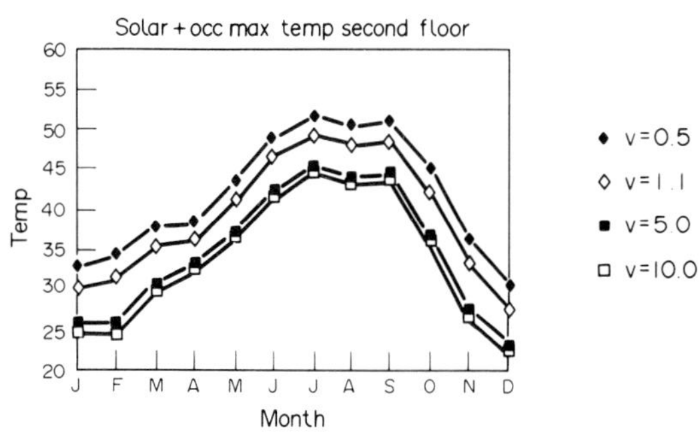

Fig. 5.4a.

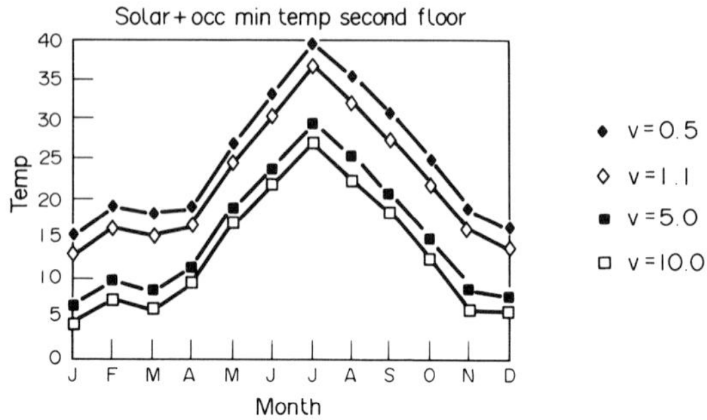

Fig. 5.4b.

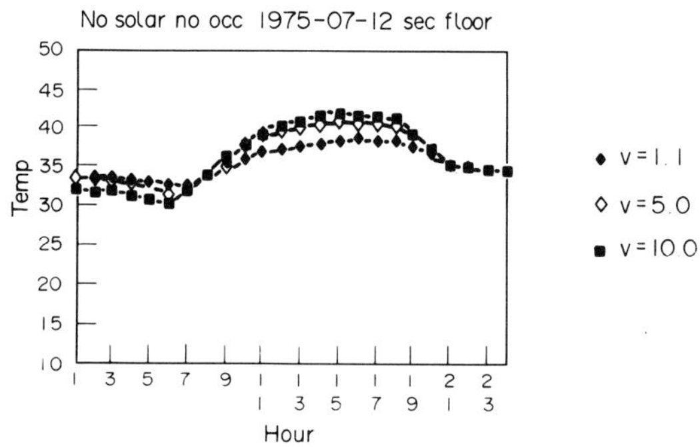

Fig. 5.4c.

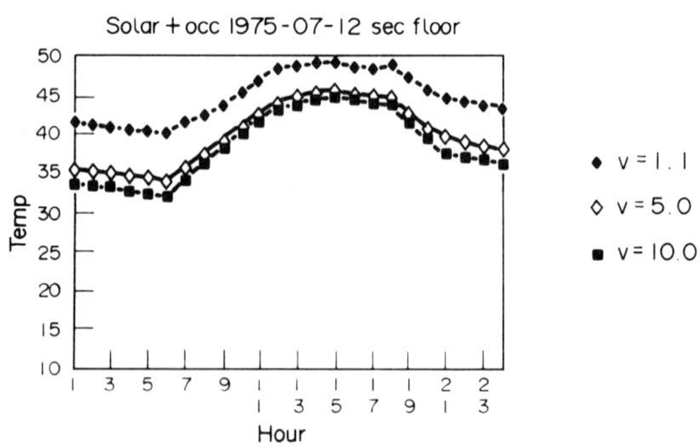

Fig. 5.4d. Influence of ventilation rate.

be decreased very much merely by increasing the ventilation rate; Fig. 5.4d also shows that an increase in ventilation rate from v=5 to v=10 does not reduce room temperature by very much. It must be borne in mind however that all the figures 5.4 relate to continuously ventilated buildings.

5.5. Influence of interior thermal capacity

In all previously discussed calculation cases it had been assumed that there were no internal walls or furniture. It is well known that an interior thermal capacity is necessary in hot and arid climates. This thermal mass is simulated by different numbers of m² of concrete walls. It is assumed that the wall thickness is 200 mm and that both sides of the wall are in convective contact with the room air. Solar radiation through the windows is assumed not to fall on the interior concrete wall.

For the case of no solar radiation and no occupancy, Fig. 5.5a shows that 50 m² concrete wall - 1c(50) - will reduce room air temperature on the first storey by 1.6 °C and 100 m² wall - 1c (100) - by another 1°C. On the second storey, Fig. 5.5b, the reduction due to 100 m² concrete is 3.7°C. Obviously, the reason why the introduction of 100 m² concrete produces a larger temperature drop on the second storey than on the first storey is that the first storey has both a concrete floor and concrete ceiling while the second storey has only a concrete floor. In all the figures it is assumed that the ventilation rate is v≟1.1 ach/h (hygienic ventilation).

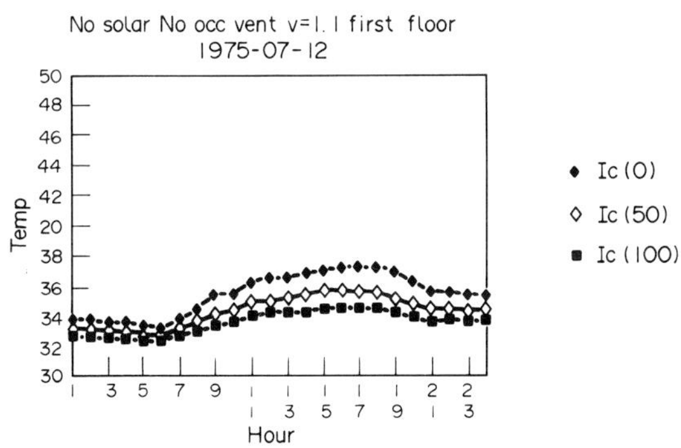

Fig. 5.5a.

If the ventilation rate is increased to v=10 ach/h, the conclusions will be the same but the reduction in the temperature of the room air will be slightly lower.

5.6. Influence of night ventilation

During the night, the cool outdoor air can be used for passive cooling of the building. Night ventilation cools down the interior of the building. During the following day ventilation must be reduced to what is needed from a hygienic point of view

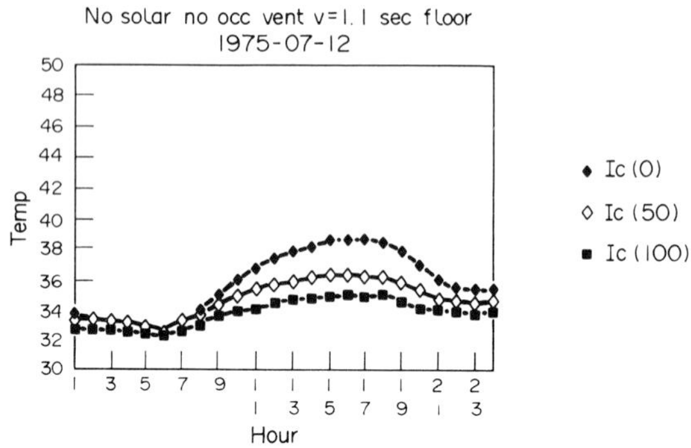

Fig. 5.5b.

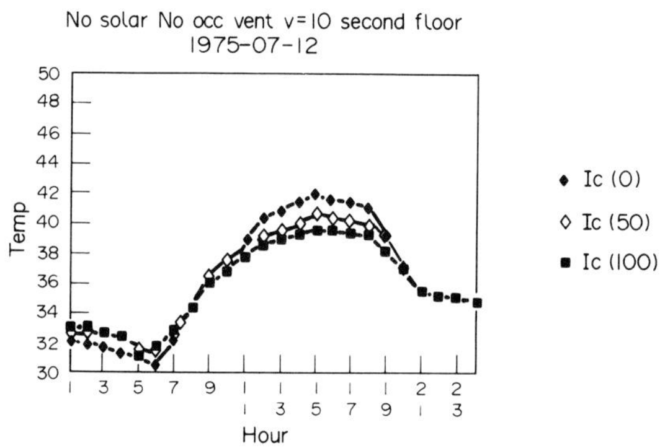

Fig. 5.5c.

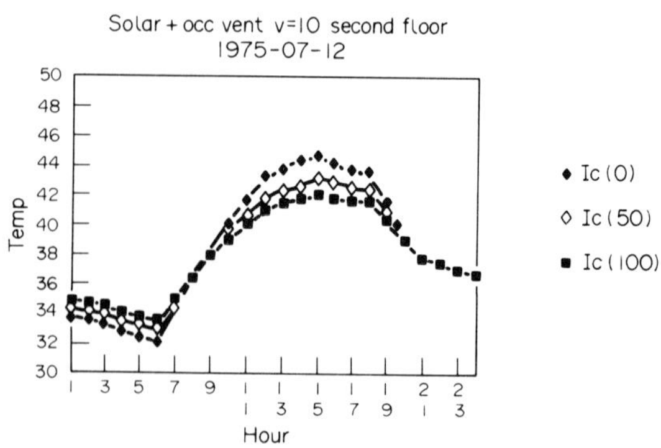

Fig. 5.5d. Influence of interior thermal capacity.

in order to prevent the building being heated up by the hot outdoor air. During the day, the cool interior helps keep down the room temperature.

Figure 5.6a shows the room air temperature on the second storey on the hottest day of the year. It is assumed that the building receives no solar radiation and is unoccupied. The building has an interior thermal mass corresponding to 100 m² of 200 mm concrete-1c(100). There is a considerable temperature drop during the day when the building is ventilated at v=10 ach/h from 2100 to 0600 h, and at v=1.1 ach/h during the day (0600-2100 h). This case is denoted v=10N in contract to continuous ventilation v=10.

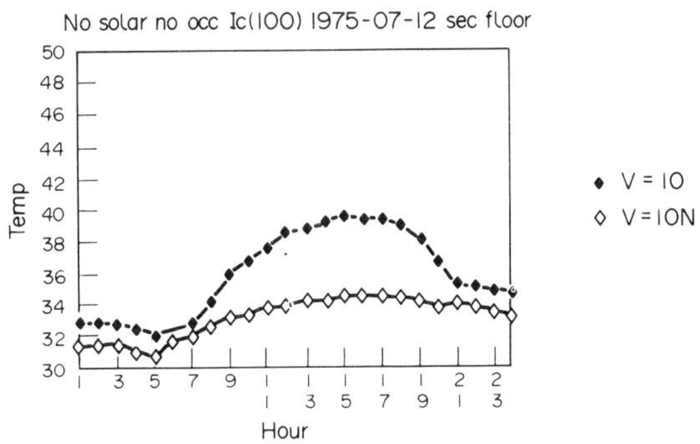

Fig. 5.6a.

If the building receives solar radiation and is occupied, temperature drop is very little as seen in Fig. 5.6b. The reason is that heat gains during the day due to solar radiation and occupancy have a greater influence on room air temperature at a low ventilation rate (v=1.1) than at a high rate (v=10). We should therefore try and reduce solar heat gain through the windows, reduce heat gain through walls and roof by painting them white and perhaps giving them proper thermal insulation.

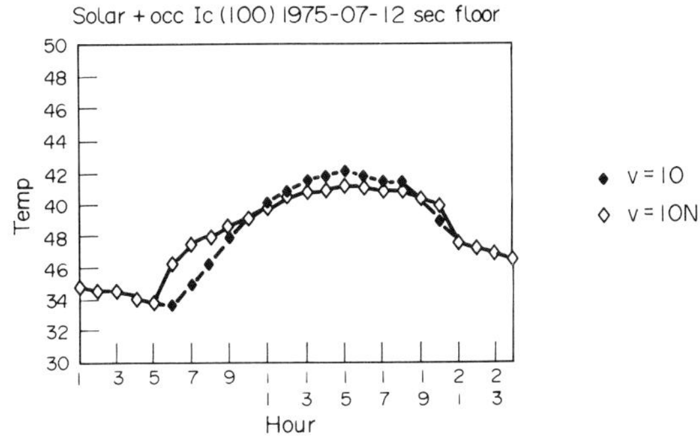

Fig. 5.6b.

In Fig. 5.6c which shows the room air temperature during the hottest day of the year the night ventilation rate has been raised from 10 to 80 ach/h. In this calculation case the windows in the southerly facade are assumed to have a glass area A=6.7 m² and the outer walls and roof are assumed to be white.

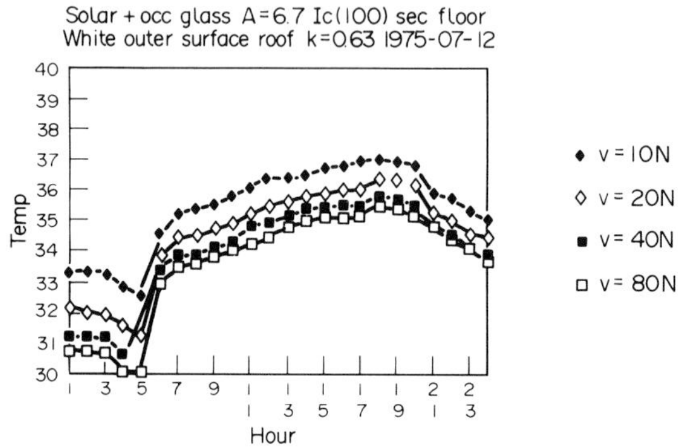

Fig. 5.6c.

In Fig. 5.6d another hot day is chosen. It is seen that for this day room air temperature increases when night ventilation begins at 2100. On this day the outdoor air temperature in the evening is higher (36°C at 2100 h), which causes a rise in temperature when night ventilation begins, particularly at v=40N.

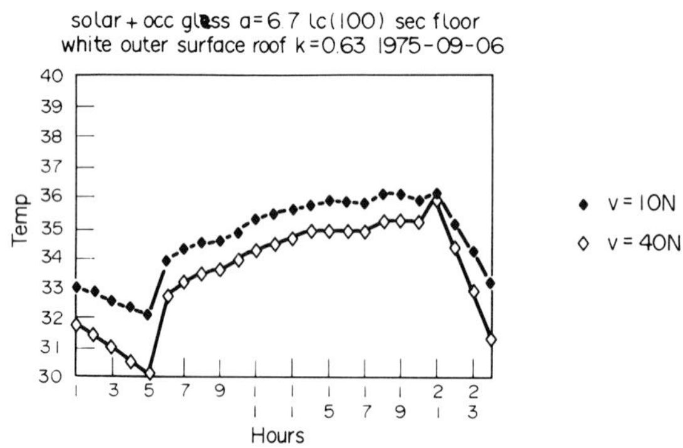

Fig. 5.6d. Influence of night ventilation.

5.7. Influence of the size of south facing windows

As seen from Fig. 5.7a, without solar radiation and occupancy the influence of the size of south facing windows is very small. On the hottest day of the year the temperature difference in the afternoon is no more than 1°C when the south facing glass area is changed from A=6.7 to A=20.9 m². With solar radiation but with the outer surfaces of walls and roof white (Fig. 5.7b), the influence of the glass area is greater but still very moderate for reasonable glass areas (A=6.7 and 10.5 m²). In both cases night ventilation of v=10 ach/h is assumed from 2100 to 0600 h. During the day v=1.1 ach/h is assumed.

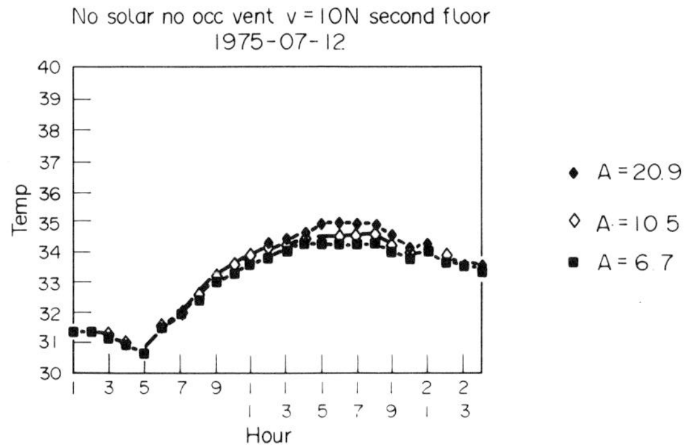

Fig. 5.7a.

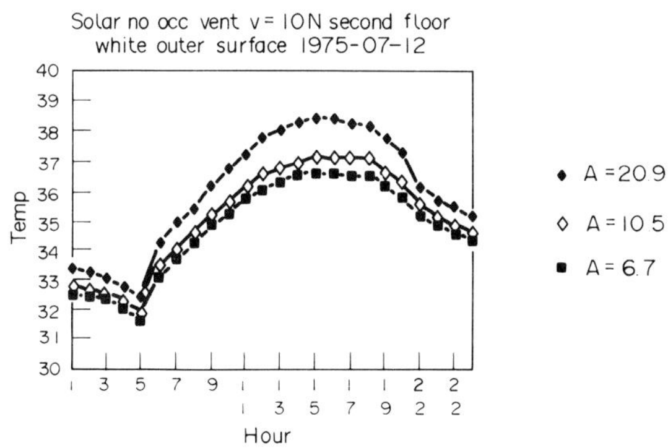

Fig. 5.7b.

Even with a limited glass area (6.7 m² to south and 4.5 m² to north) and white outer surfaces, solar radiation accounts for a 2°C rise in maximum temperature on the hottest day (Fig. 5.7c).

The south facing glass area also influences winter temperatures. During the winter, solar radiation through windows gives the building a considerable heat gain and raises room temperatures. Figure 5.7d shows that the monthly average temperature is raised quite a lot by an increased southerly glass area if the building is ventilated at v=1.1 ach/h (hygienic ventilation). For Dec-Feb, average room air temperature on the second storey is:

A=6.7 m² T_{ave}=17.1°C
A010.5 m² T_{ave}=18.4°C
A=20.9 m² T_{ave}=21.3°C

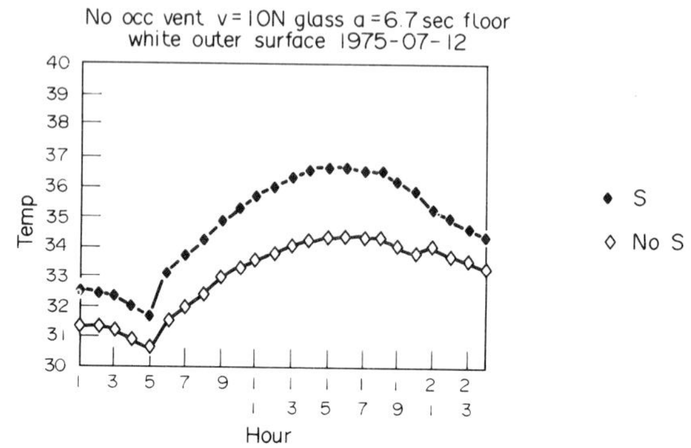

Fig. 5.7c.

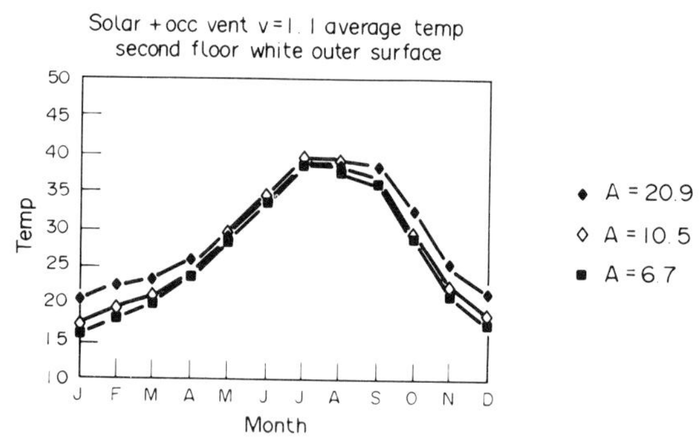

Fig. 5.7d. Influence of the size of south facing windows.

Obviously, the size of southerly glass area has not very much influence on summer temperatures in the building, but it has a favourable effect on indoor conditions in winter. In the summer the sun is at a high altitude and solar transmission through windows is small. In the winter however transmission through the south facing windows is considerable and this helps to heat the building.

5.8. Influence of window orientation and solar shading

It is well known that westerly and easterly windows are disastrous for the indoor climate in hot regions. This is clearly shown in Fig. 5.8a which relates to windows facing west, south and north. For a building with windows in the main facade facing west, the room air temperature on the hottest day of the year is 3°C higher than for one with windows facing south or north. Room temperatures for the last two orientations are very similar. In all three calculation cases the glass area in the main facade is assumed to be A=10.5 m² and that in the opposite facade A=4.5 m². Figure 5.8b shows the monthly average temperature on the second storey of an occupied building. The figure also shows that westerly (and easterly) windows are to be avoided if they are subjected to solar radiation.

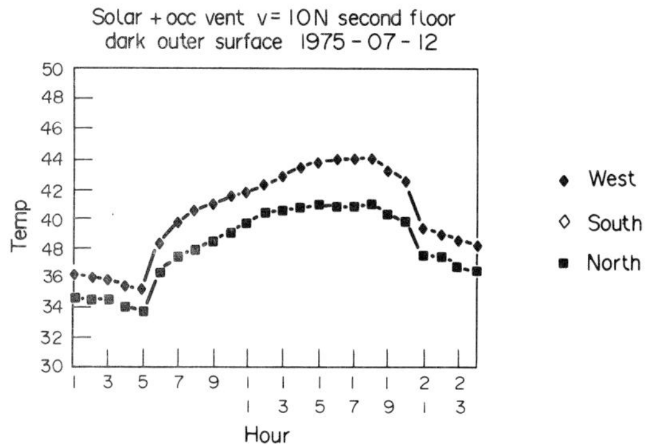

Fig. 5.8a.

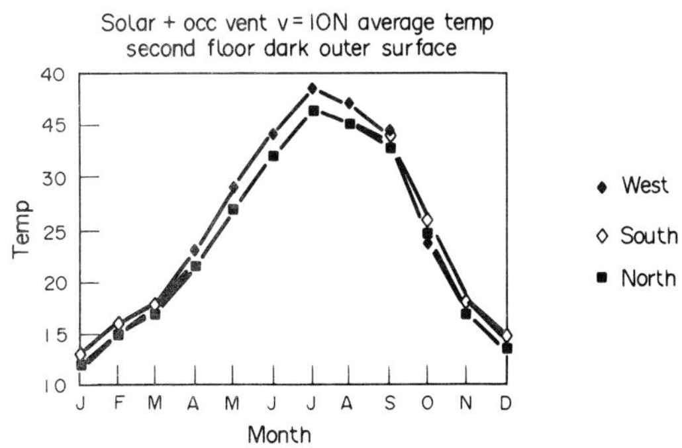

Fig. 5.8b.

As seen in Fig. 5.8c and d, solar shading of south facing windows with a horizontal projecting screen has little value. It will however have been seen from Fig.5.7c that, even for limited glass areas, temperature rise due to solar radiation is worth attention. The reason is that diffuse radiation is not shaded very much by a projecting screen.

During the winter, shading with a horizontal screen projecting from the south facade is a disadvantage since it prevents desirable solar transmission into the building.

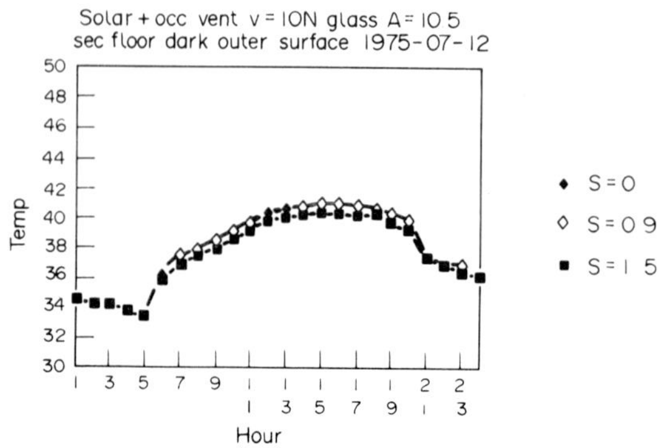

Fig. 5.8c.

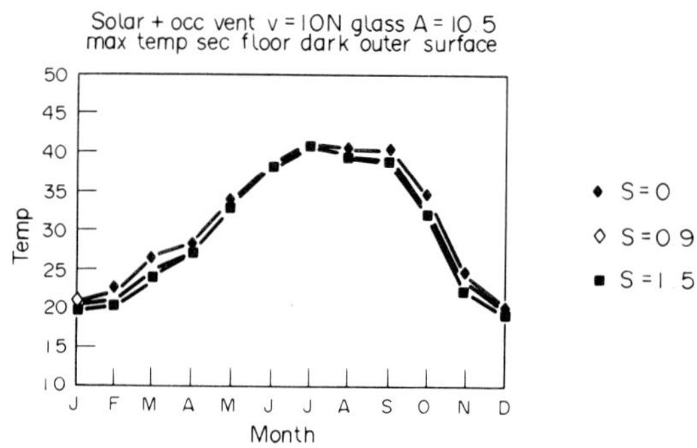

Fig. 5.8d. Influence of window orientation and solar shading.

5.9. Influence of walls and roof (incl. solar absorption on the outside)

In the baseline case and in all previously discussed cases, the walls and roof have been assumed to have a lightweight structure with thermal insulation equivalent to k=0.83 W/°C.m². In this section the wall and roof construction will be changed and the influence of this on room air temperature discussed.

The influence of solar absorption in the outer surfaces of the walls and roof will be studied first of all. Figure 5.9a shows two cases, a building with dark outer surfaces (absorption a=0.8) and the same building with white outer surfaces (a=0.2). The temperature difference on this day is more than 2°C.

For the walls and roof, five alternatives have been studied:

Alt 3: Walls and roof of lightweight structure with k=0.28 W/°Cm²

Alt 4: Light walls of k=0.83 and light roof of k=0.28 W/°cm²
Alt 5: Walls and roof of light structure of k=0.83 W/°Cm²
Alt 7: Walls and roof of heavy structure with 200 mm concrete, k=3.0 W/°Cm²
Alt 8: Walls of 200 mm concrete and roof of 200 mm insulated concrete, k=0.63 W/°Cm².

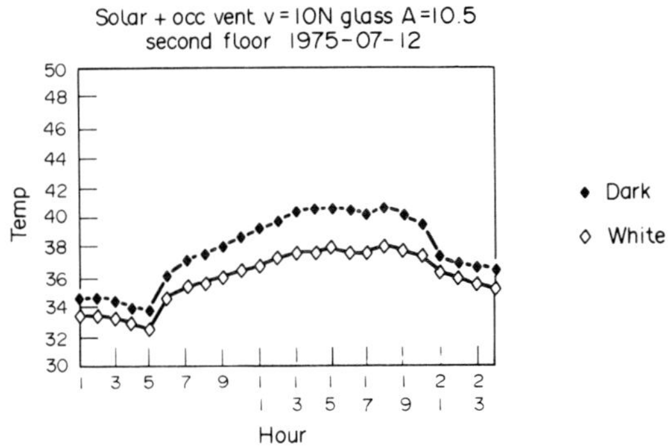

Fig. 5.9a.

For all calculation cases, a building with southerly glass area A=10.5 m² has been assumed. Outer surfaces are assumed to be white, and the building is occupied. Internal mass is simulated with 100 m² concrete, and the building is assumed to be night ventilated with v=10 ach/h in the night and v=1.1 ach/h in the day.

Figure 5.9b and c show that for the second storey the difference between Alt. 3 and 5 is about 1°, and about the same between Alt. 8 and 5. It should be borne in mind that outer surfaces are assumed to be white, which limits the influence of the walls and roof. If the outer surfaces were assumed to be dark, the influence of the structure would be larger.

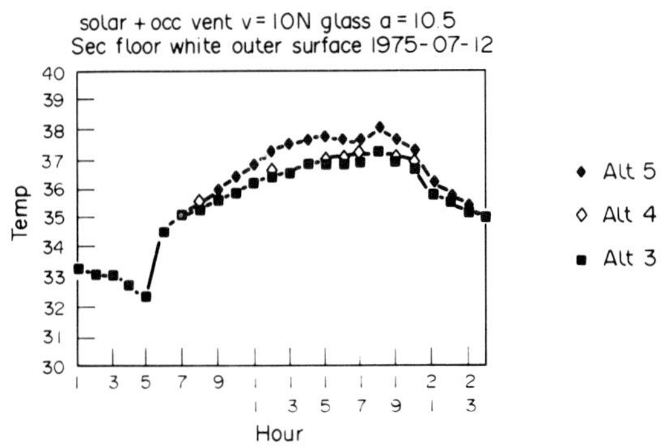

Fig. 5.9b.

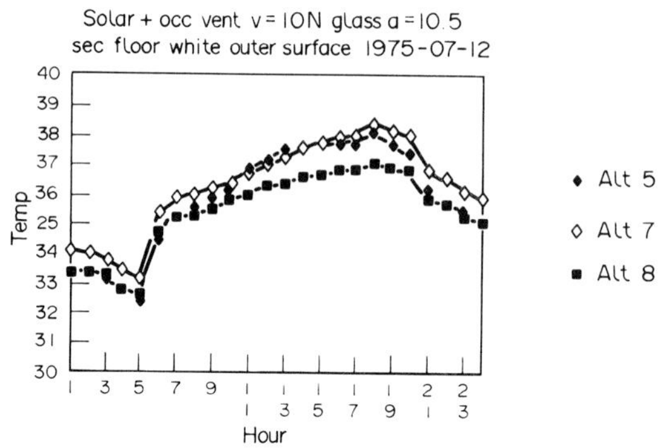

Fig. 5.9c.

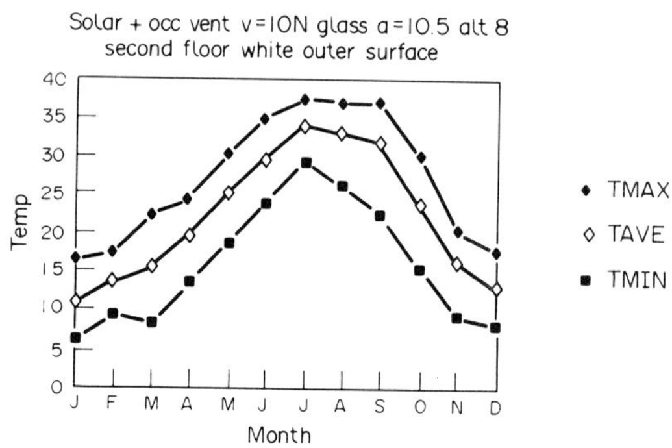

Fig. 5.9d. Influence of solar absorption and construction of walls and roof.

5.10. The "best" passive design

The conclusion of the simulations is that a good passive design for a hot and arid climate should meet the following requirements. The building

- o should have relatively small windows facing south or north, with 7-10 m^2 glass to the south
- o must have a high internal thermal capacity
- o must have night ventilation (v=10-40 ach/h) between 2100 and 0700
- o should have only hygienic ventilation (v=1.1 ach/h) in the "winter" (Nov-Apr) and during "summer" days (0700-2100, May-Oct)
- o must have thermal insulation in walls and roof corresponding to at least 50 mm mineral wool or cellular plastics
- o must have white outer surfaces.

Figure 5.10a sets out the monthly temperatures for a building with outer walls of k=0.83 and roof of k=0.28 W/°Cm². The windows have 6.7 m² glass area facing south and 4.5 m² facing north. The building also satisfies all the above requirements, with night ventilation of v=20 ach/h between 2100 and 0600 h. Figure 5.10b shows the room air temperature on the hottest day of the year.

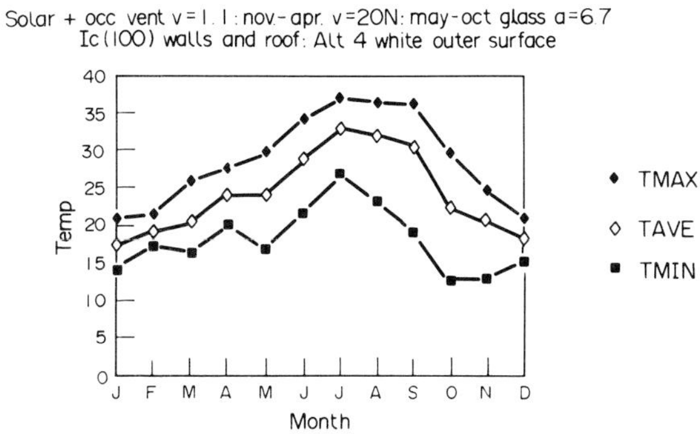

Fig. 5.10a.

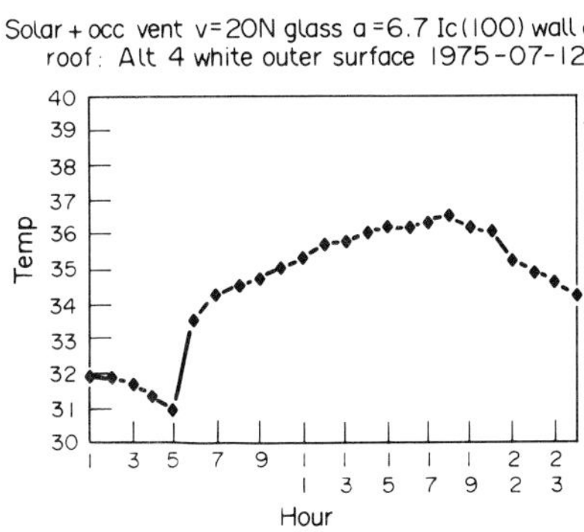

Fig. 5.10b. The "best" passive design in hot and arid climates at the chosen location.

During the year, room air temperature on the second storey will exceed T=35.5°C for 100 hours, and T=34.2°C for 500 hours. Room air temperature is below T=15.2°C for 100 hours and below T= 16.5°C for 500 h.

In a hot and arid climate, humidity is low. The comfort shown above for the summer is therefore questionable since, according to Table 2.1, only 70% are satisfied at v=0.1 m/s and RH=20%. It is still possible to increase night ventilation a little. Neither night radiation to the sky nor evaporative cooling, for instance by sprinkling water on the floor, has been taken into consideration. Increased night ventilation and evaporative cooling will probably reduce room air temperature by 2-3°C.

In practical design, unshaded windows are unacceptable, and some type of lattice is used. In such cases the window area can be increased.

5.11. Traditional design in hot and arid climates

Fig. 5.11. Traditional strategies in hot and arid climate.

Traditional desert cities have narrow streets, often not more than 2 m wide. The streets are often covered and the walls facing onto the streets are of natural sand or stone colour. Solar heat gain by the walls facing the street is very small. The walls of the houses are made of stone or sun dried brick, often 300-400 mm thick. The roofs are made of palm trunks with plaited palm leaves, covered by about 100 mm clay in order to give the room thermal capacity. Thermal capacity has traditionally been used to reduce solar heat gain through the roof - insulating materials were not available in old times. Owing to the heavy structure, rooms cannot be wider than 2.5-2.7 m, and even for such small spans the trunks have a large deflection after some time. If larger rooms are needed, vaults or domes have to be used.

During the summer months, people sleep on the roofs. For reason of privacy and to prevent radiation to the cold night sky, the roofs have one or two rooms of very light construction. The roofs are normally painted white.

In Islamic architecture, more expensive buildings have a courtyard, often with sprinkling water which cools down the air. The cool air is used for gravity ventilation through the building and is exhausted in the upper parts of the building. Ventilation of houses in densely built-up areas is assisted by wind funnels which direct the wind into the building. Outdoor air is sometimes cooled evaporatively inside the supply air shaft.

The results of the parameteric study confirm tradition, but also show that new materials provide better opportunities for climatic control of buildings. Lighter horizontal structures which provide good protection against solar heat gain can be built using insulation materials. Rooms can have larger dimensions than before.

6. Passive cooling in a warm and humid climate with seasons

6.1. Outdoor climatic conditions

Koenigsberger et al, 1974, give a classification of tropical climates. They define the warm and humid climate as that found in Lagos, Dar-es-Salam, Colombo, Singapore, Jakarta and other places near the equator. This type of climate has very little seasonal temperature variation. The daily maximum temperature is 27-32°C and the daily minimum temperature 21-27°C. Relative humidity is high. Some distance from the equator there are more accentuated seasons, and a place located 21°N in the Far East has been chosen for this parametric study. The monthly maximum, average and minimum temperatures are set out in Fig. 6.1a. The monthly maximum temperature TMAX is high even in "winter", but the monthly average temperature TAVE shows a pronounced winter season in Dec-Feb. The monthly average solar radiation on a horizontal surface is shown in Fig. 6.1b. Compared with Fig. 5.1c which refers to a desert climate, average radiation is very moderate.

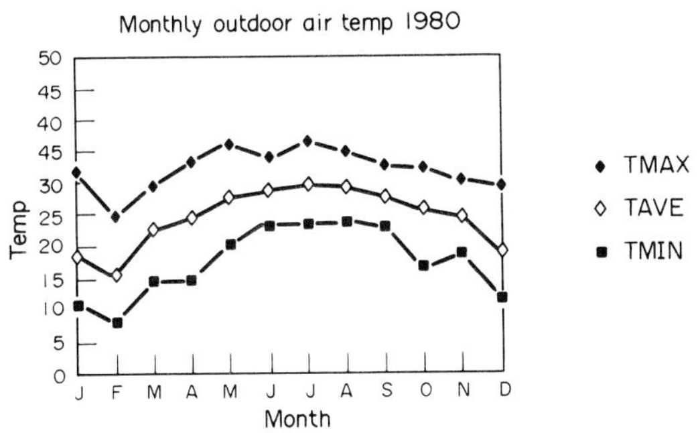

Fig. 6.1a.

Three typical days have been chosen.

a cold and cloudy day, 1980-02-06,
a warm, cloudy and humid day, 1980-03-30
a hot and sunny day, 1980-05-30.

The temperature for these days is set out in Fig. 6.1c. The temperature and relative humidity were as follows:

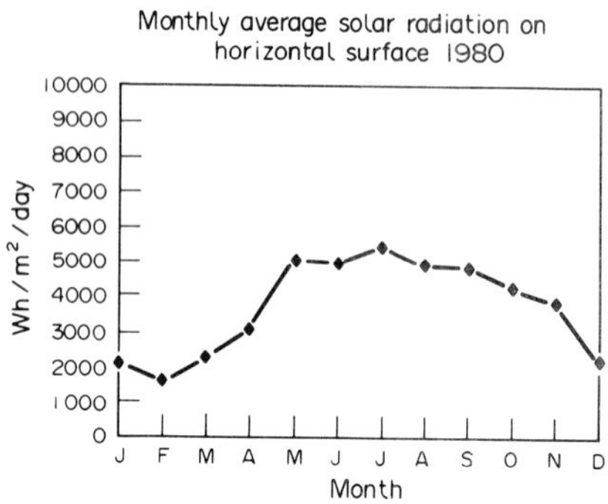

Fig. 6.1b.

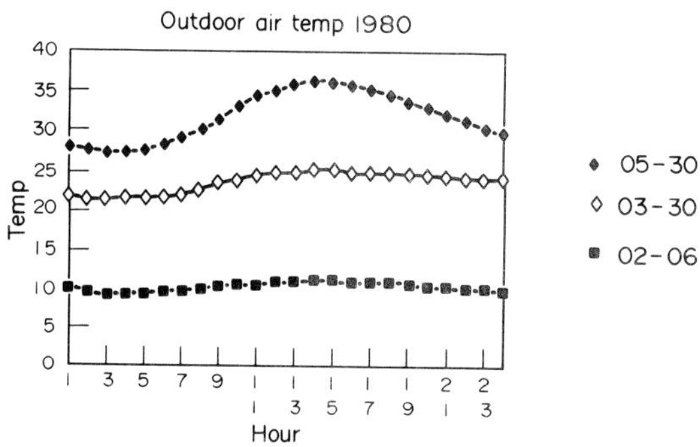

Fig. 6.1c. Climatic conditions for a place in the Far East (21°N).

Day	Temperature °C		Rel. humidity %	
	0600	1500	0600	1500
02-06	9.1	10.8	93	98
03-30	21.4	24.7	97	92
05-30	27.9	35.5	92	54

6.2. Baseline case

The same building as that described in Section 5.2 has been chosen as baseline case for the parametric study relating to indoor climate in a warm and humid region.

Figure 6.2a shows the monthly maximum, average and minimum temperatures on the second storey of an occupied building. For six months, the monthly maximum temperature TMAX is above 40°C. During the summer the monthly average temperature TAVE is high, but in winter it is very moderate. The influence of solar radiation and occupancy on the monthly maximum temperature on the second storey is shown in Fig. 6.2b. Solar radiation accounts for 6-8°C and "free heat" from people, cooking etc for another 2-3°C.

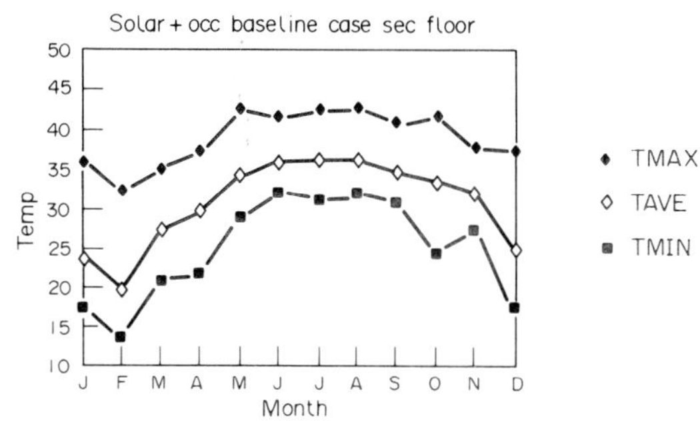

Fig. 6.2a.

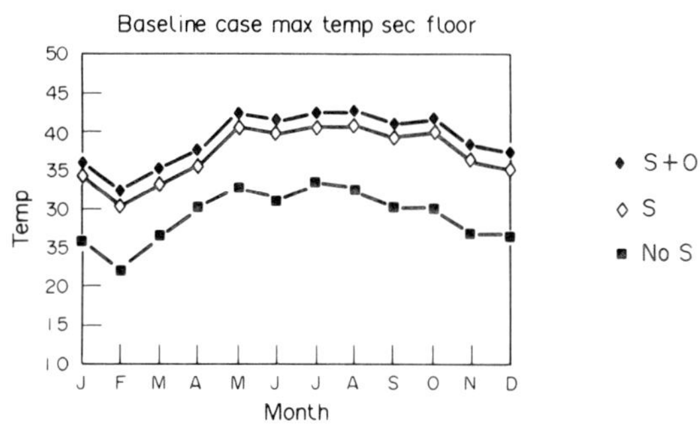

Fig. 6.2b. Monthly temperatures for the baseline case.

In a warm and humid climate, indoor moisture conditions are of great importance. Table 6.1a gives the relative humidity on the first storey of the building on a cold and cloudy day, 1980-02-06. Relative humidity RH is given for three positions and for two times of the day. It is assumed that a kg/h water vapour is added to the room air on each storey. For a family of four persons, 0.2 kg/h is normally emitted by their bodies and, as a daily average, 0.2-0.4 kg/h by cooking, washing, etc. For a=0.4 kg/h relative humidity in the middle of the room is RH=91%, but close to the window it is RH > 100% which indicates condensation.

Relative humidity on a warm cloudy say in March, 1980-03-30, is set out in Table 6.1b. It is obvious that addition of water vapour need be no greater than a=0.2 kg/h for RH = 100% in the middle of the room. Even for this low vapour increment, condensation will occur on the floor.

It is clearly demonstrated that for housing in a warm and humid climate moisture is the most difficult problem.

Table 6.1a. Solar and Occ Baseline case 1980-02-06

Time and location	Temp. in first floor °C	Rel. humidity (%) in first floor with a (kg/h/storey)= a=0.2	a=0.4	a=0.8
0600				
Middle of room	15.4	71	79	87
Close to floor	16.1	68	76	83
Close to windows	11.4	91	(101)	(111)
1500				
Middle of room	15.3	83	91	98
Close to floor	15.9	80	87	95
Close to windows	12.6	98	(107)	(116)

Table 6.1b. Solar and Occ Baseline case 1980-03-30

Time and location	Temp. in first floor °C	Rel. humidity (%) in first floor with a (kg/h/storey)= a=0.2	a=.04	a=0.8
0600				
Middle of room	23.8	91	95	100
Close to floor	23.1	94	99	(104)
1500				
Middle of room	24.8	98	(102)	(107)
Close to floor	23.9	(103)	(108)	(112)

6.3. Influence of ventilation rate

One way of reducing high summer temperatures is to increase the ventilation rate. Figure 6.3a gives the monthly maximum temperatures for no solar radiation and no occupancy. An increase in ventilation rate from v=1.1 to v=10 ach/h reduces maximum temperatures on the second storey by about 2°C. As seen in Fig. 6.3b, maximum temperatures in an occupied building receiving normal solar radiation are reduced by about 5°C when ventilation rate is increased from v=1.1 to v=10 ach/h. The same effect is shown in Fig. 6.3c for the hottest day of the year. The influence of an increase from v=5 to v=10 ach/h is however very small.

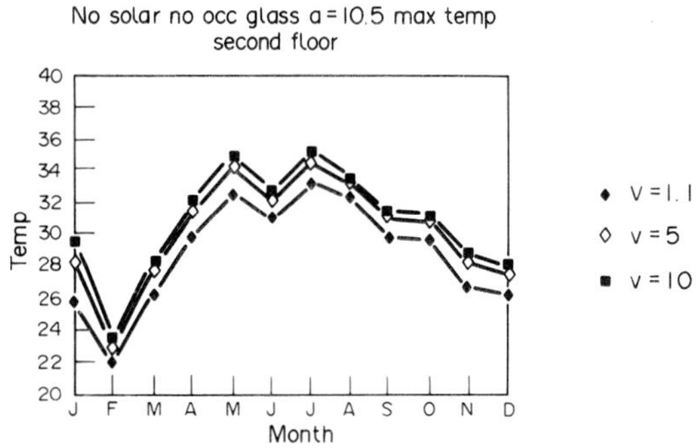

Fig. 6.3a.

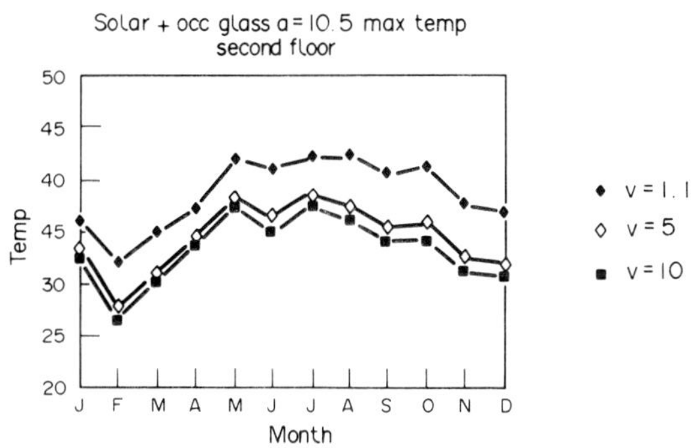

Fig. 6.3b.

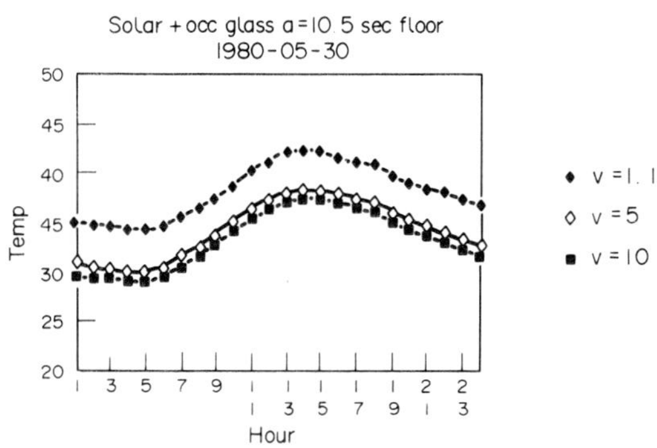

Fig. 6.3c. Influence of ventilation rate.

Indoor moisture conditions for the warm, humid and cloudy day in March are given in Table 6.2. Relative humidity in the middle of the first storey is very high. For a vapour addition of a= 0.2 kg/h, relative humidity is RH=97-98%. If vapour addition rises to a=0.6 kg/h, relative humidity increases slightly to RH=98-100%. The reason why the increase in RH is only 1-2% is that the volume of air to which vapour is added is very large (10 ach/h). However, condensation will occur on the floor.

Table 6.2. Solar + Occ Glass A=10.5 Vent v=10 1980-03-30

Time and location	Temp in first floor °C	Rel. humidity (%) in first floor with a (kg/h/storey) = a=0.2	a=0.4	a=0.8
0600				
Middle of room	21.7	98	99	100
Close to floor	20.8	(104)	(104)	(105)
1500				
Middle of room	24.1	97	98	98
Close to floor	22.0	(110)	(110)	(111)

If Table 6.2 is compared with Table 6.1b, it is obvious that a higher ventilation rate will slightly reduce relative humidity at 1500 h. On the other hand, moisture conditions at 0600 h are made slightly worse. On balance, moisture conditions are not improved by an increased ventilation rate.

6.4. Influence of interior thermal capacity

As described in Section 5, interior thermal mass is simulated by different numbers of m² of 200 mm concrete with heat transfer on both sides, such as internal walls. Figure 6.4a shows that without solar radiation or occupancy the influence of interior thermal mass is relatively small. If thermal mass is increased from 0 to 100 m² of concrete, the reduction in maximum temperature for the hot and sunny day in March is not more than 2°C. If the building is occupied and subjected to normal solar radiation, the effect of thermal mass is slightly increased to about 3°C. In both cases, a ventilation rate of v=10 ach/h is assumed.

Indoor moisture conditions for a case with 100 m² of interior concrete are set out in Table 6.3a for the cold and cloudy day in February. The case is similar to the baseline case except for the interior mass. A comparison of the baseline case (Table 6.1a) with the case with interior mass (Table 6.3a) shows that relative humidity indoors is considerably reduced on that February day. In both cases, only hygienic ventilation is used (v= 1.1 ach/h).

For a ventilation rate of v=10 ach/h, indoor moisture conditions for the warm and cloudy March day are set out in Table 6.3b.

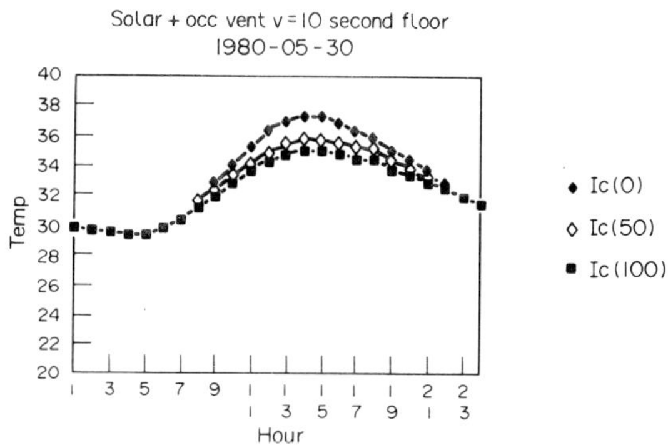

Fig. 6.4a.

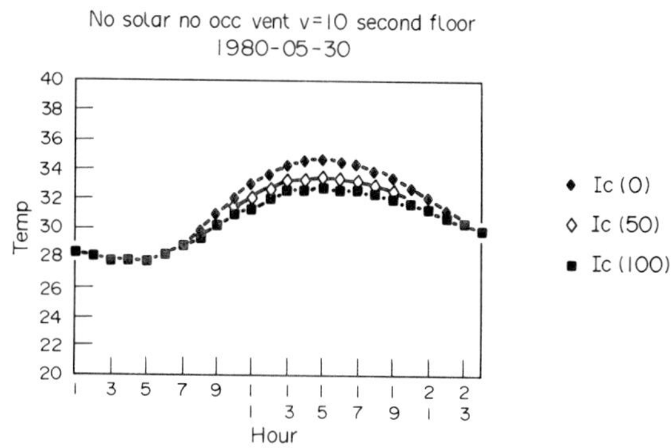

Fig. 6.4b. Influence of interior thermal capacity (m^2 of 200 mm concrete).

Table 6.3a. Solar + Occ Glass A=10.5 Vent v=1.1 1c(100) 1980-02-06

Time and location	Temp. in first floor °C	Rel. humidity (%) in first floor with a (kg/h/storey)= a=0.2	a=0.4	a=0.8
0600				
Middle of room	18.1	60	67	74
Close to floor	18.2	60	67	73
Close to windows	12.3	86	96	(105)
1500				
Middle of room	17.9	71	77	84
Close to floor	18.0	70	77	84
Close to windows	13.5	92	(101)	(110)

Table 6.3b. Solar + Occ Glass A=10.5 Vent v=10N 1c(100) White outer surface 1980-03-30

Time and location	Temp. in first floor °C	Rel. humidity (%) in first floor with a (kg/h/storey) = a=0.2	a=0.4	a=0.8
<u>0600</u>				
Middle of room	21.8	(102)	(107)	(112)
Close to floor	21.5	(103)	(109)	(114)
<u>1500</u>				
Middle of room	22.5	(111)	(116)	(122)
Close to floor	22.2	(113)	(119)	(124)

All relative humidities are RH=100% or higher. When a comparison is made with Table 6.2b, it is obvious that moisture conditions indoors are made worse by the introduction of interior thermal capacity.

6.5. Influence of night ventilation

In a hot and arid climate, night ventilation in combination with an interior thermal mass was an effective way of reducing the hig summer temperatures. Fig. 6.5a shows that the same occurs in a warm and humid climate. This figure relates to a building with a southerly glass area of A=10.5 m² and with white outer surfaces to reduce the solar heat gain through the walls and roof. Figure 6.5b for which A=6.7 m² shows the same effect. The maximum temperature during the day is slightly reduced by a smaller southerly glass area A.

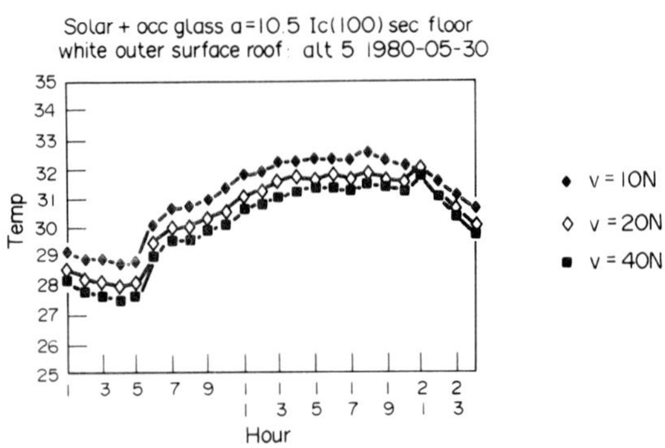

Fig. 6.5a.

Indoor moisture conditions on the warm, humid and cloudy March day are shown in Table 6.4a for a case with a southerly glass area A=10.5 m², interior mass = 100 m² concrete and night

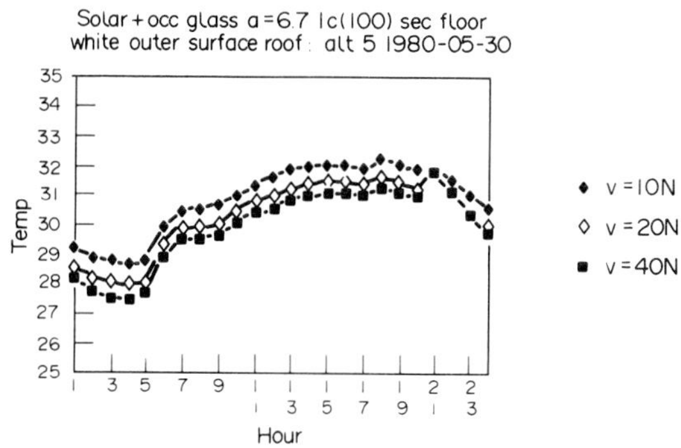

Fig. 6.5b. Influence of night ventilation.

ventilation v=10 ach/h from 2100 to 0600 h (v=1.1 ach/h from 0600 to 2100 h). All indoor relative humidities are above 100%. When a comparison is made with Table 6.3b for which a continuous ventilation of v=10 ach/h is assumed, it is eveident that on warm and cloudy spring days night ventilation makes indoor moisture conditions worse.

As seen in Table 6.4b, for a hot and sunny May day the indoor relative humidities vary between 80 and 100%.

In conclusion, it may be stated that night ventilation in combination with a large interior thermal mass is of questionable value for the transitional seasons between winter and summer.

Table 6.4a. Solar + Occ Glass A=10.5 Vent v=10N 1c(100) White outer surface 1980-03-30

Time and location	Temp. in first floor °C	Rel. humidity (%) in first floor with a (kg/h/storey)= a=0.2	a=0.4	a=0.8
0600				
Middle of room	21.8	(102)	(107)	(112)
Close to floor	21.5	(103)	(109)	(114)
1500				
Middle of room	22.5	(111)	(116)	(122)
Close to floor	22.2	(113)	(119)	(124)

Table 6.4b. Solar + Occ Glass A=6.7 Vent v=40N 1c(100) White outer surface 1980-05-30

Time and location	Temp. in first floor °C	Rel. humidity (%) in first floor with a (kg/h/storey)= a=0.2	a=0.4	a=0.8
0600				
Middle of room	28.7	94	97	(101)
Close to floor	29.0	92	96	99
1500				
Middle of room	30.1	77	80	83
Close to floor	30.1	77	80	83

6.6. Influence of windows

The influence of the southerly glass area A on the temperature on a hot and sunny day in March is shown in Fig. 6.6a. For moderate glass areas, A=6.7 and A=10.5 m², there is not a lot of difference between the temperature curves. Obviously, very large southerly glass areas will increase room temperature, but not appreciably.

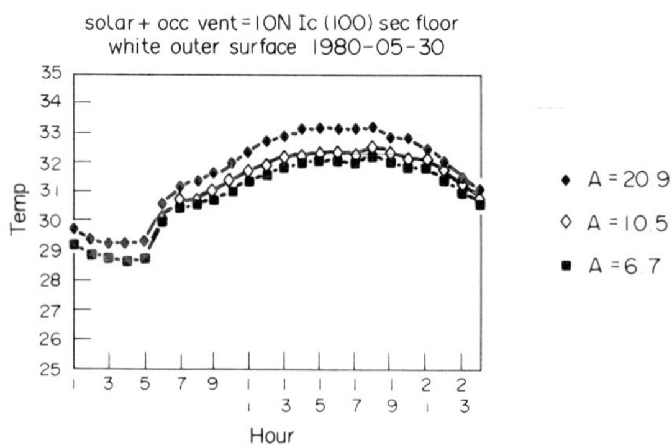

Fig. 6.6. Influence of south facing glass area A.

A study has also been made of whether double glazing will have a favourable effect on temperature and moisture conditions. In the summer, substitution of single by double glazing has no influence on maximum temperatures (less than 0.1 °C). In the winter the monthly average temperatures are increased slightly, but not more than 0.2-0.3 °C.

Indoor moisture conditions in the winter are slightly improved by double glazing. When Table 6.5b (double glazing) is compared with Table 6.5a (single glazing), it is seen that relative humidity is reduced. However, the improvement does not pay the cost of double glazing.

Table 6.5a. Solar + Occ Glass A=6.7 Single glazed Vent v=1.1 1c(100) White outer surface 1980-02-06

Time and location	Temp. in first floor °C	Rel. humidity (%) in first floor with a (kg/h/storey)= a=0.2	a=0.4	a=0.8
0600				
Middle of room	17.6	62	69	76
Close to floor	17.5	63	70	76
Close to windows	12.0	88	97	(107)
1500				
Middle of room	17.3	73	80	87
Close to floor	17.3	73	80	87
Close to windows	13.2	94	(103)	(112)

Table 6.5b. Solar + Occ Glass A=6.7 Double glazed Vent v=1.1 ic(100) White outer surface 1980-02-06

Time and location	Temp. in first floor °C	Rel. humidity (%) in first floor with a (kg/h/storey)= a=0.2	a=0.4	a=0.8
0600				
Middle of room	18.0	61	68	74
Close to floor	18.0	61	68	74
Close to windows	14.9	73	81	89
1500				
Middle of room	17.9	71	77	84
Close to floor	17.8	71	78	85
Close to windows	15.7	81	88	96

6.7. Influence of walls and roof (including solar absorption on the outside)

The influence of solar absorption in the outer surfaces of walls and roof is shown in Fig. 6.7a. In this case, southerly glass area is A=6.7 m², night ventilation is v=10 ach/h, and the interior thermal mass is equivalent to 100 m² of concrete. The walls and roof are assumed to be of light-weight structure with k=0.83 W/°Cm². Absorption is a=0.8 for a dark surface and a=0. for a white surface. The figure shows that the reduction in temperature due to the white colour is about 3°C. However, in a warm and humid climate it is very difficult to maintain a white colour. Use of a white outer surface is therefore a questionable strategy for prevention of solar gain through the walls and roof. Thermal insulation seems to be a more suitable method.

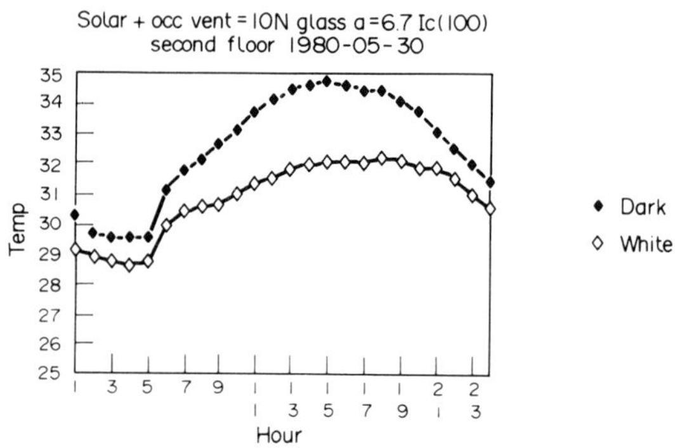

Fig. 6.7a.

For the walls and roof, the following alternatives have been studied:

Alt 7: Walls and roof of 200 mm conc ete, uninsulated, k=3.0 W/°Cm²

Alt 8: Walls of 200 mm uninsulated concrete, roof of concrete insulated on outside, k=0.63 W/°Cm²

Alt 9: Walls and roof of concrete insulated on the outside, k=0.63 W/°Cm² for the walls and k=0.35 W/°Cm² for the roof.

The building for which the walls and roof are studied is assumed to have a southerly glass area A=6.7 m², ventilation rate v=10 ach/h, interior thermal mass equivalent to 100 m² concrete, and white outer surfaces. The maximum temperature for the hot and sunny day in May is 33.7 °C for Alt 5, 33.0 °C for Alt 8 and 32.9 °C for Alt 9. See Fig. 6.7b. It is of interest to compare a case with Alt 8 in the walls and roof and with white outer surfaces, with a case with walls and roof according to Alt 9 and with dark outer surfaces. It is seen in Fig. 6.9c that the difference between the two cases is very small.

Indoor moisture conditions for Alt 8 on the cold and cloudy February day are shown in Table 6.6.

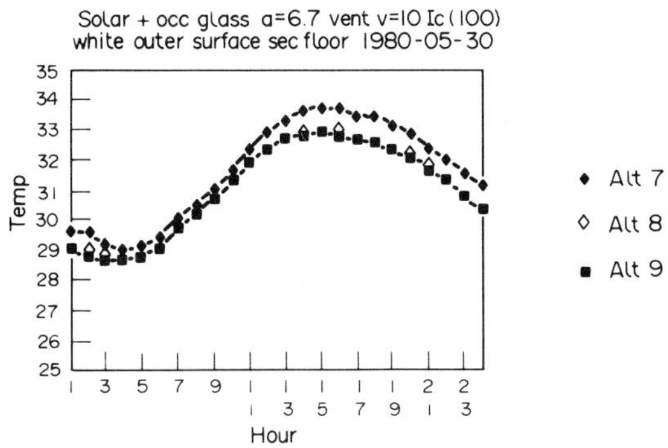

Fig. 6.7b.

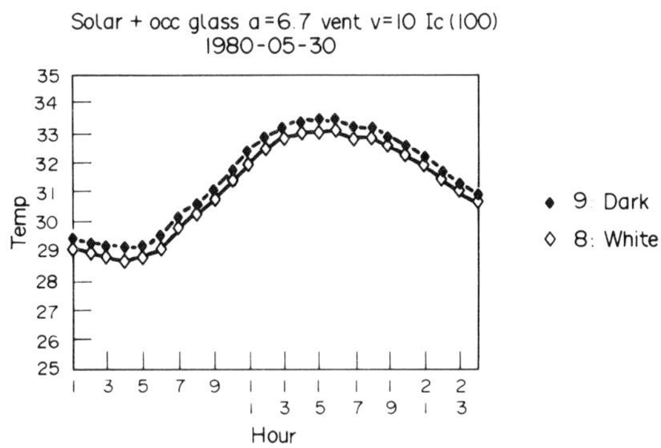

Fig. 6.7c. Influence of solar absorption and construction of walls and roof.

Table 6.6. Solar + Occ Glass A=6.7 Vent v=1.1 1c(100) Walls & roof: Alt 8 White outer surface 1980-02-06

Time and location	Temp. in first floor °C	Rel. humidity (%) in first floor with a (kg/h/storey)= a=0.2	a=0.4	a=0.8
0600				
Middle of room	16.4	67	74	82
Close to floor	16.4	67	74	82
Close to windows	11.6	90	100	(110)
1500				
Middle of room	16.2	78	86	93
Close to floor	16.2	78	86	93
Close to windows	12.8	96	(106)	(115)

6.8. The "best" passive design

The conclusion of the simulations is that a good passive design for a warm and humid climate with pronounced seasons should meet the following requirements. The building

- o should have relatively small windows so that winter conditions may be good
- o should not use night ventilation
- o should have hygienic ventilation in the "winter" (Nov-April) and 10-20 ach/h in the summer (May-Oct)
- o must have the absolute minimum of water vapour addition to the room air
- o must <u>either</u> have white outer surfaces <u>or</u> insulated walls and roof in order that solar heat gain through the walls and roof may be avoided.

Two cases, both with glass areas of 6.7 m² (southerly) and 4.5 m² (northerly), single glazed windows, walls and roof according to Alt 9, and with ventilation rates of v=1.1 ach/h from Nov to April and v=10 ach/h from May to Oct, have been chosen as the "best" design. Figure 6.8a shows a building of no interior thermal mass, and Fig. 6.8b the same building with an interior mass equivalent to 100 m² concrete.

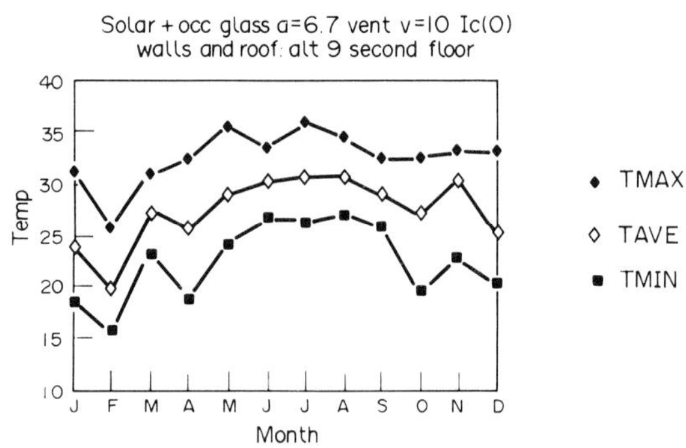

Fig. 6.8a.

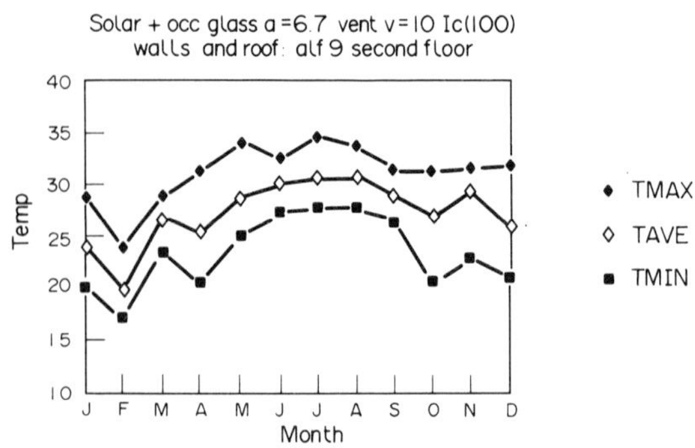

Fig. 6.8b. The "best" passive design in warm and humid climate at the chosen location.

Without interior thermal mass, the temperature T=33.8 °C is exceeded for 100 hours/year, and T=32.3 °C for 500 hours/year. Temperature drops below T=16.4 °C for 100 hours/year, and below T=19.9 °C for 500 hours/year.

With an interior mass equivalent to 100 m² of 200 mm concrete, the temperature T=33.0 °C is exceeded for 100 hours/year, and T=31.8 °C for 500 hours/year. Temperature drops below T=17.6 °C for 100 hours/year, and below T=21.5 °C for 500 hours/year. It is obvious that the interior mass has little influence on indoor thermal climate. Moisture conditions also are about the same.

Table 6.7a. Solar + Occ Glass A=6.7 Vent v=10 1c(0) 1980-03-30

Time and location	Temp. in first floor °C	Rel. humidity (%) in first floor with a (kg/h/storey)= a=0.2	a=0.4	a=0.8
0600				
Middle of room	24.5	87	92	96
Close to floor	24.2	89	93	98
1500				
Middle of room	25.1	96	100	(105)
Close to floor	24.5	99	(104)	(109)

Table 6.7b. Solar + Occ Glass A=6.7 Vent v=10 1c(100) 1980-03-30

Time and location	Temp. in first floor °C	Rel. humidity (&) in first floor with a (kg/h/storey)= a=0.2	a=0.4	a=0.8
0600				
Middle of room	22.6	97	(102)	(107)
Close to floor	24.9	85	90	94
1500				
Middle of room	25.6	93	98	(102)
Close to floor	25.2	95	100	(104)

In order to prevent addition of water vapour to the room air, the kitchen and also the washing and drying area should be placed outside living and bedrooms, for instance in a space without direct contact with the rooms. This space should be so constructed that condensation can be allowed - compare the traditional cookhouse.

6.9. Traditional design in a warm and humid climate with pronounced seasons

In warm and humid climates of the equatorial type, there is no need for heating. Diurnal temperature variation is very small, and the main strategy is to cool the human body by air movement. Relative humidity is very high both outdoors and indoors. Traditionally, the buildings are made of tree trunks and leaves. They are generally raised off the ground in order to permit air movement below the floor. This air flow below the floor provides better conditions for cross ventilation of the building. Normally, two wide doors on both opposite facades give good cross ventilation. The air movement probably also prevents mould growth indoors.

Fig. 6.9. Traditional strategies in warm and humid climate with pronounced seasons.

In warm and humid climates with pronounced seasons, winter temperatures drop to 10 °C or below. Tree trunks and leaves are still used in the villages and by minorities, but the walls of new buildings in cities are mostly made of bricks and concrete blocks. Floors are of solid concrete or concrete blocks placed between concrete beams. Mould is a problem both on the outside of the building and indoors.

The parametric study focuses on the main problem involved in designing buildings for this type of climate. The correct way seems to be passive heating in winter and a strategy for prevention of troublesome moisture conditions indoors.

7. The use of parametric studies

The intention has been to show that knowledge can be enhanced by carrying out parametric studies for the climate concerned. In combination with the good knowledge which architects and engineers have of traditional design and local conditions, these will provide a solid basis for design.

8. References

Adamson, B., 1991: Passive Climatization of Residential Buildings in Tropical Climates (University of Lund, Sweden, Lund Center of Habitat Studies) Report under publishing.

Bowen, A., Clark, E. and Labs, K., (ed), 1981: Passive Cooling (American Section of the International Solar Energy Society).

Elawa, S., 1981: Housing Design in Extreme Hot Arid Zones, with Special Reference to Thermal Performance. (University of Lund, Sweden, Department of Building Science).

Fathy, H., 1973: Architecture for the Poor. (The University of Chicago Press).

Givoni, B., 1976: Man, Climate and Architecture (Applied Science Publishers Ltd, London). Second edition.

Koenigsberger, O.H., Ingersoll, T.G., Mayhew, A. and Szokolay, S.V., 1974: Manual of Tropical Housing and Building, Part 1: Climatic Design. (Longman Group Ltd, London).

Markus, T.A. and Morris, E.N., 1980: Buildings, Climate and Energy (Pitman Publishing Ltd, London).

Mazria, E., 1979: The Passive Solar Energy Book. (Rodale Press, Emmaus, Pa, USA).

Meese, G.B., Kok, R., Lewis, M.I. and Wyon, D.P., 1984: A Laboratory Study of the Effects of Moderate Thermal Stress on the Performance of Factory Workers. Ergonomics 1984, Vol. 27, pp. 19-43.

Nyström, M., 1987: Building Climatology in Vietnam: 3. A Parametric Study of Residential Houses in the Hanoi Climate. Draft (University of Lund, Sweden, Department of Building Science) Report 1987 15 (E).

Oliver, P. (ed), 1971: Shelter in Africa (Barrie & Jenkins Ltd, London).

Revoult, J. and Maury, N., 1975: Palais et maisons du Caire I (1975), II (1975) and III (1979).

Rosenlund, H., 1989: Design of Energy Efficient Houses in a Hot and Arid Climate, including utilization of Passive Solar Energy (University of Lund, Sweden, Department of Building Science) Report 1989: 1 (E).

US Department of Energy, 1980: Passive Solar Design Handbook, Two Volumes.

CHAPTER 11

PHOTOVOLTAIC REFRIGERATION

B. McNelis

I T Power, U.K.

1. INTRODUCTION

Photovoltaic cells convert solar radiation directly into electricity. This solar generated electricity can be used to power a slightly modified, conventional refrigerator.

The first, and probably the most important, application of photovoltaic refrigeration could be in preserving medicines in the developing countries.

Over two billion people, half of the world's population, live in the rural areas and villages of developing countries. One of their major needs is for adequate health care. For example, it is estimated that throughout the developing world, six infectious diseases (diphtheria, whooping cough, tetanus, measles, tuberculosis and poliomyelitis) together kill approximately 5 million children annually.

In 1977, the World Health Organization adopted a target of providing immunization for all children by 1990 as one key element in the overall goal of "Health for All" by the year 2000. This expanded programme of immunization is achieving success, and, as a result, the number of children's lives being saved by vaccines in the developing world now approaches one million per year.

The essential component of an immunization programme is the availability of safe, effective vaccines throughout the country. To remain effective, most vaccines must be kept refrigerated between +0 °C and +8 °C up to the time of immunization. The transport and storage of the vaccines in this refrigerated state, the "Cold Chain" is a major technical and logistical undertaking.

In the rural areas of developing countries, where grid electricity is not available, vaccine storage in health centres and clinics has been achieved mainly through the use of kerosene - or bottled gas-fueled absorption refrigerators or by the use of electric compression refrigerators powered by diesel or gasoline-fueled generator sets. In addition to the problem of fuel availability and poor fuel quality, poor equipment reliability and performance coupled with spare parts distribution problems have reduced the effectiveness of the cold chain. Solar-powered refrigerators, with better performance, lower running costs, greater reliability, and longer working life, have the potential to provide a more effective and substantial cold chain.

2. PHOTOVOLTAIC ENERGY CONVERSION

The direct conversion of sunlight into electricity by photovoltaic cells is described in a number of standard texts. For a brief introduction to the theory see Backus (1) or Wilson (2), while a more comprehensive treatment of both theory and practice is provided by Rauschenbach (3).

2.1. Brief History

The photovoltaic effect was first observed by the French scientist Becquerel in 1839 when he noticed that when light was directed onto one side of a simple battery cell, the generated current could be increased. Work on the photovoltaic properties of selenium in the 1870's led to the first selenium photovoltaic cell in 1883. The photo-sensitive properties of copper and cuprous oxide structures were discovered in 1904. By 1905, it was known that the number and energy level of electrons emitted by a photosensitive substance varied with the intensity and wavelength of the light shining on it.

In the years that followed, research work continued with the objective of developing practical photovoltaic devices. Selenium and cuprous oxide photovoltaic cells were developed, leading to several applications including photographic exposure meters and other small light sensors. By 1941, selenium devices had been developed with a light-to-electricity efficiency of about 1 per cent. A new technique was later developed, known as a 'grown p-n junction', which enabled the production of single crystal cells. Using doped silicon crystals, American research workers in the mid-1950's were able to achieve solar conversion efficiencies up to 6 per cent.

Western Electric began to sell commercial licences for silicon photovoltaic technology in 1955 and there were some attempts to develop practical systems for powering specialist equipment in remote areas. It was not until the late 1950's however that crystalline silicon solar cells were developed with high enough conversion efficiencies for their use in power generators. A major impetus for the development of these cells was the space programme. The first solar-powered satellite, Vanguard I, was launched by the USA in 1958. Practically all satellites launched since then have been powered by solar arrays made up of many thousands of crystalline silicon photovoltaic cells.

Following the 1973 oil crisis, interest in photovoltaics as a terrestrial source of power increased greatly and many countries, including several developing countries, instituted photovoltaic research, development and demonstration activities as part of wider energy research programmes. Total world expenditure from all sources on photovoltaic reserach, development and demonstration activities is probably running at between $200 and $300 million per annum.

Over the last ten years, there has been more than a tenfold reduction in the real price of photovoltaic modules. This has been achieved through a combination of improved cell technologies and larger manufacturing volumes. Starting from virtually zero in 1974, sales of photovoltaic systems have grown to about 25 MWp in 1986, with a total value of at least $800 million. Worldwide there are over 20 module manufacturers of significance and there are several times this number of firms designing and marketing photovoltaic systems using bought-in components. Research efforts are continuing on a broad front to develop better photovoltaic devices and lower cost systems.

Much has already been achieved and many thousands of systems are operating reliably today, ranging in size from a few Watts to several Megawatts. From being an exotic, highly expensive technology for very specialised situations, photovoltaic generators are now an appropriate solution for a growing number of applications and in time could become a significant factor in world energy supplies.

2.2. The Photovoltaic Process

The material most commonly used to make photovoltaic cells for power applications is crystalline silicon, either in monocrystalline or, more recently, semi-crystalline form. The essential features of this type of cell are shown in Figure 1. It is made from a thin wafer of high purity silicon, doped with a minute quantity of boron. Phosphorus is diffused at a high temperature into the active surface of the wafer. The front electrical contact is made by a metallic grid and the back contact usually covers the whole surface. An anti-reflective coating (ARC) is applied to the front surface.

The phosphorus introduced into the silicon gives rise to an excess of what are known as conduction-band electrons and the boron an excess of valence-electron vacancies or 'holes', which act like positive charges. At the junction, conduction electrons from the n (negative) region diffuse into the p (positive) region and combine with holes, thus cancelling their charges. The area around the junction is thus depleted in charge by the disappearance of electrons and holes close by. Layers of charged impurity atoms (phosphorus and boron), positive in the n region and negative in the p region, are formed either side of the junction, thereby setting up a 'reverse' electric field.

When light falls on the active surface, photons with energy exceeding a certain critical level known as the bandgap (1.1 electron volts in the case of silicon) interact with the valence

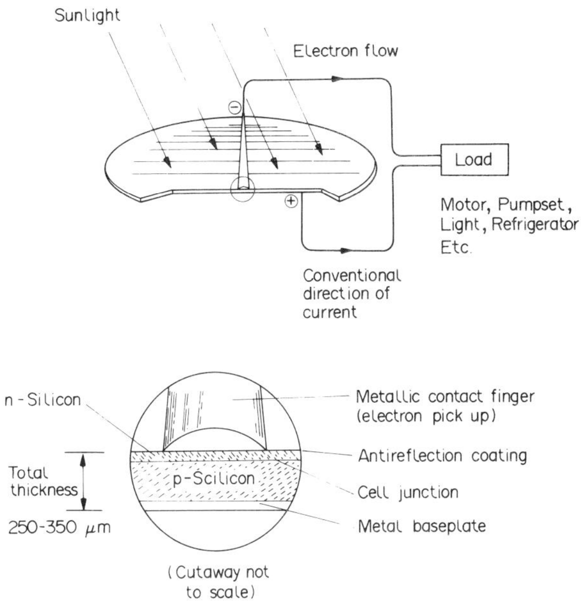

Fig. 1. Crystalline Silicon Photovoltaic Cell.

electrons and elevate them to the conduction band. This process leaves 'holes', so the photons are said to generate 'electron-hole pairs' which are generated throughout the thickness of the wafer in concentrations depending on the intensity and spectral distribution of the light. The electrons move throughout the crystal lattice and the less mobile holes also move by valence-electron substitution from atom to atom. Some recombine, neutralising their charges, and the energy is converted to heat. Others reach the junction and are separated by the reverse field, the electrons being accelerated to the negative contact and the holes towards the positive. A potential difference, or open-circuit voltage (V_{OC}), is thus established across the cell which is capable of driving current through an external load.

The current-voltage relationship (I-V) characteristics for a typical cell is dependent on irradiance and temperature, as illustrated in Figure 2 (a) and (b). For crystalline silicon cells, when illuminated by light with intensity 1000 W/m², AM 1.5 direct spectrum, at 25 °C, the open-circuit voltage (V_{OC}) is about 0.6 V and the short-circuit current (I_{SC}) about 30 mA/cm².

As the cell temperature increases, the current increases slightly as the voltage decreases significantly, in consequence, the maximum power decreases. It is therefore desirable to operate the cells at as low a temperature as possible.

The cell efficiency is the ratio of the maximum power to the

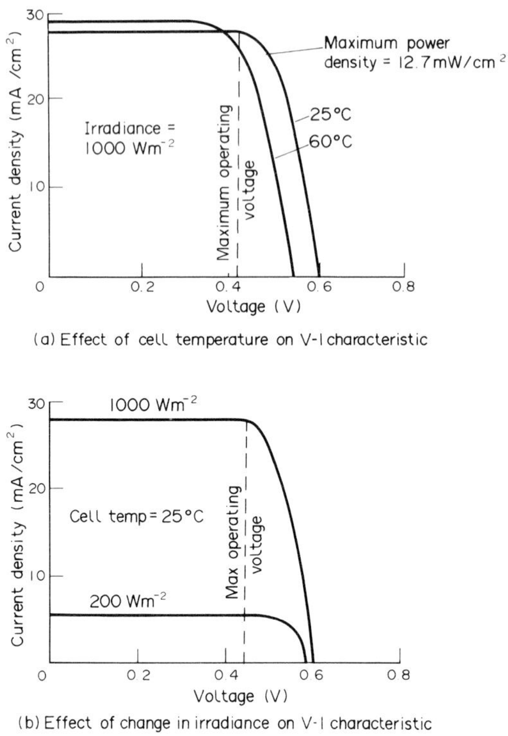

Fig. 2. Typical I-V Characteristic Curves.

product gross cell area and irradiance, usually expressed as a percentage. The photovoltaic process, like other energy conversion processes, is subject to a maximum efficiency dependent on the physical characteristics of the materials. The achievement of improved working efficiencies, closer to the practicable maximum, is therefore a major objective of research and development work. For example, the maximum practicable conversion efficiency for conventional crystalline silicon cells is about 25%, but the efficiency actually achieved for monocrystalline cells commercially manufactured is typically about 14%, although 20% has been reported for cells made in a reasearch laboratory.

2.3. Crystalline Silicon Cells

The mono-crystalline silicon-solar cell is a highly stable device and is based on well-established semi-conductor technology developed over many years for integrated circuits. Wafers about 250-350 μm thick are cut from long single crystal ingots 75 mm, 100 mm or even 150 mm in diameter. The ingots are sometimes made by the 'float zone' (Fz) process, but more usually the Czochralski (Cz) process is employed, whereby an ingot is slowly drawn out of a melt of doped silicon in an inert atmosphere. The atoms of silicon solidify into a perfect cubic lattice following the structure of a seed crystal. Commercial photovoltaic cells made from the wafers typically have efficiencies in the range 11-15 %.

Several groups have developed cast ingot processes which are less energy intensive and which are more tolerant of impurities. A melt of doped silicon is formed in a mould up to 300 mm cube and allowed to solidify under carefully controlled conditions. The resulting ingot has a semi-crystalline structure which is clearly revealed when it is sliced up to form wafers usually 100 mm square. The resulting solar cells typically have efficiencies in the range of 7.5-10 %, with some manufacturers even able to achieve 12 % using surface passivation or gettering techniques.

All ingot processes, whether for mono or semi-crystalline silicon, have the drawback that they involve sawing to form wafers. This is a time-consuming and wasteful operation, with over half the material lost.

As an alternative to the ingot processes, several research teams have been working for some years on the development of continuous sheet processes, which do not need subsequent sawing. The main problem with all these processes has been to achieve an acceptable quality of crystalline silicon sheet with a sufficiently high rate of production to render the process economic. The only commercial sheet process which has emerged to date is that developed by Mobil Solar Energy Corporation (USA), which involves drawing out a nine-sided thin-walled polygon from a silicon melt. Rectangular wafers are then cut from the walls of the polygon and made into cells. Cell efficiencies are reported to be comparable with those for ingot processes.

2.4. Amorphous Silicon Cells

Several thin film photovoltaic devices are undergoing development and commercialisation. The processes are of varying complexity and usually require highly sophisticated and expensive equipment, but because the photosensitive layers are very thin, often much less than 5 µm, very little material is used. Costs may thus be greatly reduced, given large scale production. Present research efforts are being directed to improving the cell efficiency and long-term stability of several types of thin film devices. During 1986 these devices were beginning to enter the market for developing country applications.

2.5. Modules and Arrays

Solar cells can be interconnected in series and in parallel to achieve the desired operating voltage and current. The basic building block of a flat-plate solar array is the module in which the interconnected cells are encapsulated behind a transparent window to protect the cells from the weather and mechanical damage. One or more modules are then attached to a supporting structure to form a panel and a number of panels makes up an array field which, together with the balance-of-system (BOS) components, makes up the complete system. The array field may be sub-divided electrically into a number of sub-arrays working in parallel. A selection of modules is shown in Figure 3. Flat plate arrays are normally fixed, with the modules supported by a structure such that they are oriented due South (in the Northern hemisphere) and inclined at or about the angle of latitude to maximise the amount of solar radiation received on an annual basis. The steeper angle of inclination will enhance the output in winter, at the expense of some reduced output in summer.

Fig. 3. Selection of PV Modules.

For some circumstances, it is appropriate and cost-effective to mount the modules on a support structure that tracks the sun through the day. Given clear sky conditions, the output from the array tracked in this way is more uniform and can exceed that from a fixed array by at least 20%; moreover, the extra output comes in the early morning and late afternoon, the times when demand for grid-supplied electricity is often highest. However, in view of the additional complication and the need for more skilled operation and maintenance, tracking collectors are generally not appropriate for remote sites, where fixed flat-plate arrays are preferable.

2.6. Photovoltaic Systems

Photovoltaic modules and arrays can be used to drive any device which could normally be electrically powered. For an introduction see McNelis (4) and for a review of applications see a recent report to UNESCO (5). The photovoltaic application which has received the most attention, in terms of research, development and testing is in fact water pumping. This topic is covered in detail in a handbook by Kenna and Gillett (6).

3. MEDICAL REFRIGERATORS

3.1. The Cold Chain - Potential for Solar

As noted in the introduction to this chapter, many children die because they are not immunized. In 68 developing countries the "under 5 mortality rate" (U5MR: annual deaths of children under age 5 per 1,000 live births) is more than 100 (7).

The cold chain is the system for distibuting refrigerated vaccine from the manufacturer to the point of use where the vaccination takes place. Vaccines must be refrigerated in order to remain effective.

The cold chain is illustrated in Figure 4. There are some important factors to consider:

1. Maintaining the cold chain is more critical at higher distribution levels (closest to point of manufacture) because the quantities of vaccine stored are highest.

2. The highest distribution levels (central, district, and regional stores) are easiest to supervise and hence are more reliable.

3. At the health centre level, supervision and logistical problems are greatest, creating a compelling need for reliable equipment.

4. The cold chain includes both the equipment and the people responsible for it, hence training aspects related to equipment use are crucial.

In the past, kerosene cooled absorption refrigerators have been used in the cold chain. If solar powered refrigerators replace kerosene refrigerators, there is potential for the following benefits:

* Reduced running costs as a result of:
  - elimination of kerosene fuel costs
  - elimination of kerosene transportation costs
  - reduced vaccine losses and hence vaccine replacement costs
  - lower refrigerator maintenance costs
  - reduced need for backup refrigerators where there are fuel supply or repair problems;

* Improved vaccine availability (reduced refrigerator down time) as a result of:
  - elimination of fuel supply problems

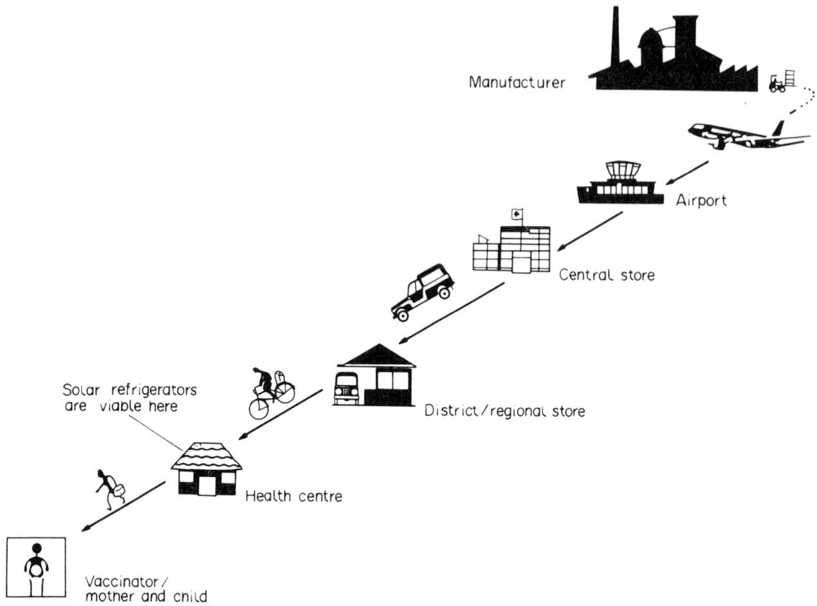

Fig. 4. The Cold Chain.

- elimination of fuel quality problems
- greater refrigerator reliability
- reduced down time due to spare parts non-availability
- longer refrigerator life
- better refrigerator performance (greater temperature control);

* Staffing benefits from:
- reduced workload for medical staff (user maintenance)
- reduced workload for technical staff (less repairs required); and

* Cold chain management benefits due to:
- longer equipment life and hence less re-ordering
- reduced logistical problems arising from fuel shortages
- reduced logistical problems arising from non-availability of working refrigerators
- reduced logistical problems arising from lower vaccine losses.

The above operational advantages of introducing solar refrigerators into the cold chain indicate that solar refrigerators can provide a more effective and sustainable cold chain leading to reduced incidence of disease.

It should be noted that the benefit from increased availability can be greater than immediately apparent. For example, if a mother brings her children to a clinic for vaccination only to find that the vaccines are not available, she may be reluctant to return at another time.

In addition, some immunizations require revaccination or boosters after specified periods. If these boosters are not available at the correct time, the effort and cost of the initial vaccination is largely wasted.

3.2. Choice of Solar Technology

Solar cooling methods are reviewed by the World Health Organisation (WHO) as part of its Expanded Programme on Immunization in 1980 (8). Five alternative approaches were identified:

- photovoltaic/vapour compression
- photovoltaic/thermoelectric
- solar thermodynamic - solid absorption (zeolite/water)
- solar thermodynamic - solid absorption (calcium chloride/ ammonia)
- solar thermodynamic - liquid absorption (water/ammonia)

and a total of 14 projects studied.

It was concluded that photovoltaic/vapour compression refrigerators would be available for field testing within one year and prices in 1984 were projected as being $1,000 to £2,000.

Several advantages including potentially lower costs and possibilities for local manufacture were identified for the solar thermodynamic systems. It was hoped that solar thermodynamic refrigerators could be further developed but these are still not yet considered ready for use in the field.

The photovoltaic/thermoelectric system which uses a solid state Peltier effect element for cooling is of interest because no moving parts are involved although a fan is usually required to assist heat transfer from the hot side. Small thermoelectric refrigerators are available commercially for use in cars, and three examples were tested by WHO in 1979. These could not maintain sufficiently low internal temperatures or make ice.

In 1981 WHO issued an outline specification for photovoltaic refrigerators (9). WHO asked manufacturers to supply systems meeting the specification for laboratory testing. The results of WHO's work and plans were presented at a meeting of the UK section of the International Solar Energy Society, in London during 1982 (10).

3.3. The Photovoltaic Refrigerator

Refrigerator technology is not discussed here, but a review of refrigeration is available in several publications including Brinkworth (11).

Photovoltaic refrigerators operate on the same principle as normal alternating current (AC) compression refrigerators used in domestic households and AC vaccine refrigerators used in health centres and hospitals where grid electricity is available. The difference with a solar refrigerator is that the electricity supply is obtained from solar cell modules - the solar array. These devices produce direct current (DC) electricity, hence a special DC-powered compressor is required. In addition, the electricity produced by the solar array will vary as the solar input varies, necessitating the use of regulators and employing batteries to store electricity.

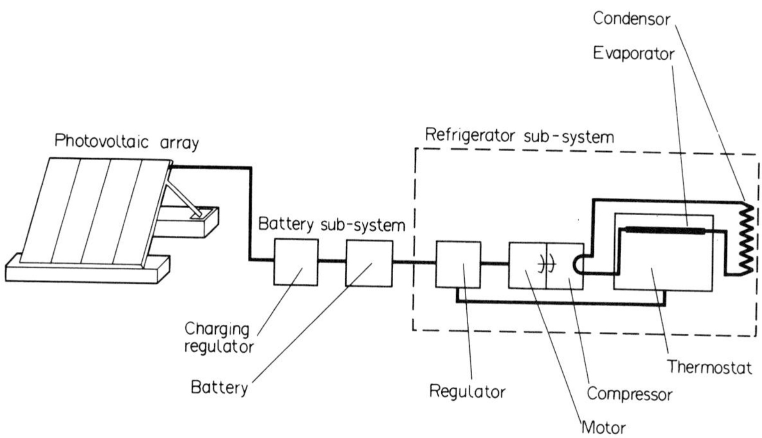

Fig. 5. Schematic of a solar powered refrigerator.

A schematic of a photovoltaic refrigerator is provided in Figure 5. A photovoltaic array charges a battery via a voltage regulator which is employed to prevent the battery from being overcharged (Figure 6). The battery powers a DC motor which is coupled directly to the compressor. The motor/compressor is usually manufactured as a hermetically sealed unit. The motor is of the electronically commutated, brushless type. A second regulator is employed to ensure that the motor/compressor is run only within its rated power range, and to prevent overdischarge of the battery. Freon refrigerant is used in the cooling cycle in the normal way, i.e., the cooling effect is achieved by the heat absorbed by the refrigerant as it evaporate

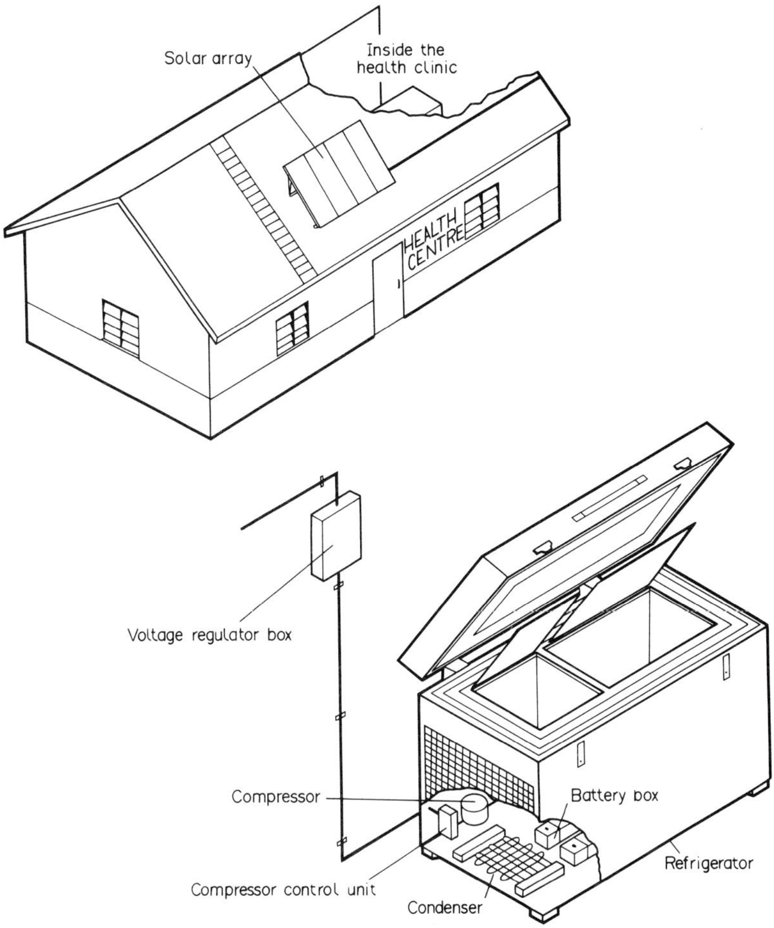

Fig. 6. Components of the power supply.

in the evaporator. A thermostat is used to switch the compressor on and off as required. Some models have two compressors and thermostats, one each for the refrigerator and the freezer. Refrigerator components are indicated in Figure 7.

The insulation around the refrigerator is normally of the expanded polyurethane type and of high thickness, from 100 mm to 200 mm, to reduce heat gain, decrease the energy consumption, and increase the time the refrigerator can maintain temperatures in the event of power cut-off or no sunshine.

The power (i.e., size) of the photovoltaic array is determined by the daily energy demand of the refrigerator and the insolation

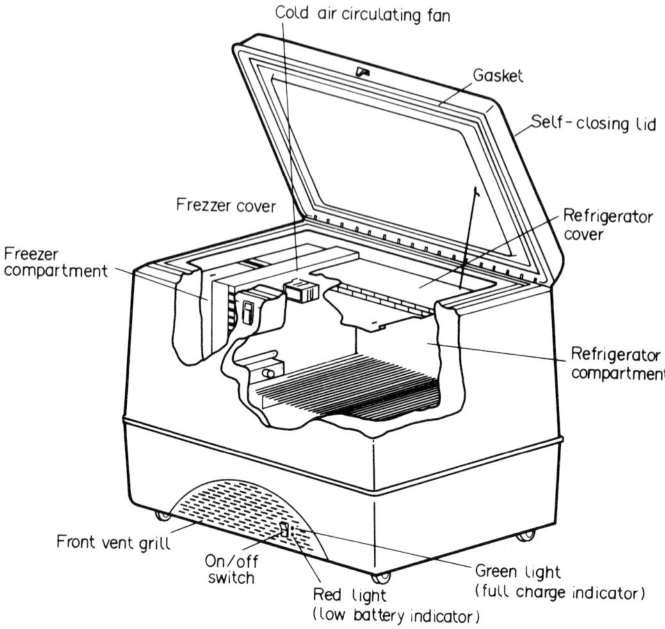

Fig. 7. Components of the refrigerator.

at the site at which the system is to be operated.

With current technology, the battery is an essential component because a higher current is required to start the compressor than to keep it running. The battery is therefore required as a buffer, and to provide energy storage to operate the refrigerator on cloudy days. It would be technically possible to produce a system with electronic power conditioning which would eliminate the need for a battery when the compressor switches in. However thermal storage would then be required for cloudy days.

Photovoltaic refrigerators are commercially available from several suppliers not only for vaccine storage but also for leisure and recreational uses. There are many 12 volt, DC-powered refrigerators available for use is recreational vehicles, campers and boats, but these have been designed for low initial capital cost without consideration of energy consumption or internal temperature variation. Such systems, although easily adapted for photovolatic power, are not suitable for vaccine storage.

The World Health Organization Expanded Programme for Immunization (EPI) in association with UNICEF publishes Cold Chain Product Information Sheets (12). Inclusion of a product in effect means that, based on the information and experience available to WHO-EPI, the product is considered suitable for the Cold Chain.

Table 1. Commercial Photovoltaic Powered Vaccine Refrigerators

MANUFACTURER	MODEL	COUNTRY OF ORIGIN	REFRIGERATOR VOLUME		FREEZER VOLUME		ICE MAKING CAPACITY at 32°C (Kg/24h)	ENERGY CONSUMPTION at + 32°C + Ice Pack freezer (kWh/24h)
			GROSS (liters)	NET USABLE (liters)	GROSS (liters)	NET USABLE (liters)		
B.P. Solar	VF 50	U.K.	45	38	5	5	2.6[1]	1.0
Electrolux	RCW24DC	Luxembourg	27	14	Note 3		1.9	0.45
FNMA	FNMA 75	Zaire	50	27	13	10	2.7	0.25
Leroy Somer	R50+IF50	France	46	16	46	16	2.1[2]	0.82
Marvel	RTD 4	USA	93	80	18	10	2.1	0.53
Polar Products	RR-2	USA	100	80	27	20	2.1	0.62
S.E.T.		Germany	104	56	55		2.4	1.1

Note 1: in 10 hours

Note 2: in 9 hours

Note 3: moveable divider volume varies

Institutional purchasers generally select equipment from this publication.

The 1986/87 edition includes 14 solar refrigerators for vaccine storage. Other system suppliers have been approved since the publication of the product information sheets, but overall the systems offered are based on the refrigerator units presented in Table 1.

3.4. System Sizing

The term "sizing" means estimating the required size or capacity of the components of a photovoltaic system so that it will perform satisfactorily with the intended load. The sizing methodology described in this section is recommended to achieve a target level of photovoltaic system reliability that exceeds conventional alternatives, which in this case are kerosene refrigerators. The steps for a sizing methodology are summarized below:

(i) Calculate the Load. The solar refrigerator average daily energy load is calculated for each month of the year. The energy load will be dependent on the ambient air temperature and the intended usage of the system. In particular the amount of ice to be made and removed, plus any refrigeration of material of significant thermal capacity, should be calculated. The energy consumption of the PV refrigerator may then be determined from test data for the particular model, making allowances for ambient temperature and usage. If the load totals vary, an average daily load over each month of the year will be required.

Measured energy consumption of eight photovoltaic refrigerators is shown in Figure 8.

(ii) Determine the Local Solar Irradiation (Insolation). The solar energy incident on the photovolatic system at the application site may be obtained from local monthly insolation data for various array tilt angles.

(iii) Calculate "Worst Month" Insolation and Load Value. The sizing approach requires identification of the load and insolation values expected during the "critical month" of the year. This is done by constructing a table of average insolation and refrigeration/freezer load values for each month of the year and then determining the month with the lowest ratio of insolation to load. The insolation and load values for the selected month are used in subsequent steps to calculate required array size and batter-storage capacity for the solar refrigerator/freezer.

(iv) Calculate Array Power. Calculating array size means calculating its required peak power output. The array sizing calculations incorporate the critical-month ratio of insolation to load, array sizing factor based on the critical-month insolation, efficiencies of all system

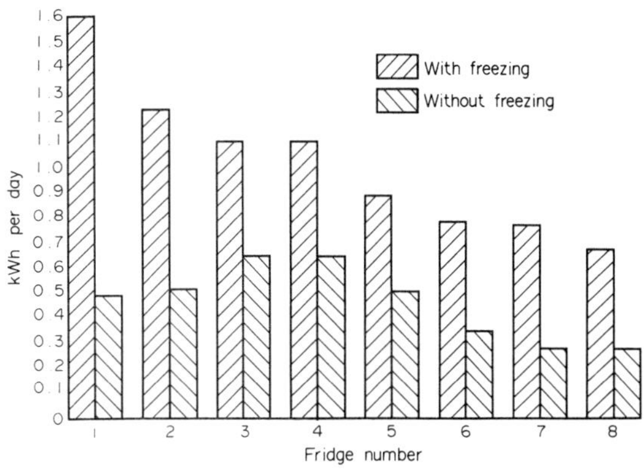

Fig. 8. Energy consumption of 8 solar refrigerators at 32 degrees C. Ambient. (Source: WHO tests).

elements, (batteries, regulators, etc.) and a term to account for long-term array degradation and sizing uncertainty. It is not within the scope of this chapter to give a detailed description of array sizing, a topic which is covered in other texts such as Rauschenbach (3).

(v) Determine Boundary Storage Requirements. For solar refrigerators it is recommended that a usable battery capacity of at least 5 times the calculated daily load is used. In some applications it may be necessary to increase this factor, for example highland locations subject to more than 5 days cloud cover. In these cases the battery should be sized to accommodate the maximum number of sunless days statistically predicted. All rated capacities are adjusted to ensure battery operation is within the acceptable depth of discharge limits.

(vi) Calculate the Voltage Regulator Size. The battery charging voltage regulator is sized to handle the maximum amount of array output that is not being used to supply the load. For conservatism, this will be considered to be the full rated peak array output power to account for situations in which the load has been disconnected.

4. EXPERIENCE WITH PHOTOVOLTAIC REFRIGERATORS

The potential benefits of using solar refrigerators in vaccine cold chains are so great that several major field trial programmes were initiated at the beginning of the decade. The programmes include some 35 systems evaluated in 25 countries by NASA Lewis Research Centre, the U.S. Agency for International Development (AID), and the Centres for Disease Control and several units evaluated by the World Health Organization. The overall results of the field trails were analyzed by WHO.

Even though field trials followed exhaustive laboratory testing to reduce the possibility of field testing unsuitable equipment, the decision to undertake field trial programmes rather than widespread introduction of solar refrigeration appears well justified. Important technical and implementation factors were derived from the field trials vital for the successful use of solar refrigerators in the cold chain (13).

The world's largest programme to use solar refrigerators (100 systems) and lighting systems (750 installations) is currently underway in Zaire. Systems are being installed in clinics and dispensaries throughout the country under the auspices of the Department de la Sante Publique and financed by the European Community (European Development Fund). The systems utilise a refrigerator manufactured in Zaire. The first installations of the refrigerators have been very successful and have received good user response. More than 300 health centres and hospitals have been equipped with solar refrigerators and/or lighting systems with good experience reported. The results of this work were presented by Durand (14).

In the Gambia, a World Bank pre-investment study on photovoltaic applications in the health sector (15) reported that, for existi installations "the experience with solar refrigerators can be considered successful". In Ecuador, users have reported being very impressed by the reliability of the solar refrigerators.

In Mali, the importance of preventative maintenance has been identified and, with over 6 years of trials in the country, the Laboratoire de l'Energie Solaire has concluded that, with period checks by a trained technician, rates of equipment failure are negligible (16).

In French Polynesia, there are more than 300 solar refrigerators in use, and in the South Pacific as a whole there are now severa thousand successful photovoltaic installations (mainly lighting systems).

5. ECONOMIC CONSIDERATIONS

A recent study for the US Department of Energy/Sandia National Laboratories investigated the economic and financial viability of photovoltaic refrigeration (17).

The main motivation for the procurement of PV refrigerators is generally to overcome the widespread problems of fuel shortages and distribution difficulties associated with maintaining keroser fuelled vaccine refrigerators in remote and sometimes inaccessab locations. Hence a photovoltaic refrigerator competes directly with kerosene refrigerators in the market, and at present there are no other realistic options for clinics remote from a grid (available diesel generators are grossly oversized, wind generators powered refrigerators are technically feasible but not commercially available at present).

It is important for potential users to ensure that investment in solar photovoltaic vaccine refrigerators represents a sensible use of development funds when compared with the costs of the existing kerosene system. Therefore it is essential to be able to present a rational analysis of the economics of switching to solar. The WHO/EPI programme is not an "economic activity", and it is not possible to carry out a cost-benefit analysis. The only meaningful quantifyable results relate to the relative cost of the options and also their likely influence on the achievement of immunization programme goals. This latter point is particularly important since the fixed overheads for any immunization programme are generally large compared with direct vaccine refrigeration costs.

The financial benefits to be expected from the use of solar vaccine refrigerators may be summarized as follows:

(i) more doses of vaccine successfully delivered per unit of money invested.
(ii) savings from kerosene substitution
(iii) savings from eliminating the need for paying for kerosene transportation
(iv) savings from reduced maintenance and wages
(v) savings by reducing the requirement for back up transporting of vaccines due to unserviceable refrigerators.

Improved refrigerator reliability will increase the effectiveness of immunization both with initial and follow up "booster" shots.

The methodology used for financial/economic analysis consists of calculating lifecycle costs for each option by taking the summed Present Vales of their respective cash flows. This can then be annualized and either divided by the number of vaccinations per year to give the cost per dose or by the available refrigerator capacity to give the cost per litre-month (a 100 litre refrigerator operational for 8 months per year would have 800 litre-month capacity in a year).

Kerosene fuelled refrigerators used in the vaccine cold chain have an initial capital cost of only $300 to $800, considerably less than a PV refrigerator. With transportation and installation this may rise to $1,500 installed compared with typically $6,000 for an installed PV refrigerator. The operation and maintenance costs are however very large and their reliability very low. As shown in Figure 5, the World Health Organisation have estimated that 71% of the life cycle cost of a Kerosene refrigerator is for recurrent costs (fuel and maintenance).

The availability of Kerosene refrigerators in field use, i.e. the percentage time the vaccines are maintained at correct temperatures is on average only 50% because of fuel and spare part supply problems. This poor reliability of Kerosene refrigerators compared with PV refrigerators affects the comparative costs significantly. An exhaustive search for data on the performance and operating costs of Kerosene refrigerators was therefore undertaken and "Cold Chain" managers responsible for vaccine refrigerators in Kenya, Mali, Pakistan, Tanzania and Zaire were contacted. The conclusions on availability and O & M costs are summarized in Table 2 along with similar data on PV refrigerators concluded from the field trial programmes described above.

A general cost comparison has been undertaken based on the data given in Table 2 to compare the annualised life cycle costs of Kerosene and Photovoltaic Refrigerators. This is summarised in Figure 9 as unit costs per litre-month of refrigerated capacity.

Table 2. Comparative Cost Data for Photovoltaic and Kerosene Refrigerators.

REFRIGERATOR TYPE	PHOTOVOLTAIC	KEROSENE		
PARAMETER		LOW CASE	TYPICAL	HIGH CASE
NET VACCINE CAPACITY (LITRES)	100	100	100	100
INITIAL CAPITAL COST ($)	4500	300	500	1000
CIF AND INSTALLATION ($)	1500	500	800	1000
FUEL COSTS ($/DAY)	-	0.25	0.70	3.00
MAINTENANCE COSTS ($/YEAR)	150	50	100	150
LIFE TIME (YEARS)	15	10	5	2
AVAILABILITY (% TIME IN SERVICE)	95	80	50	20

The factors that most affect the life cycle costs are the reliability (and hence availability) of the Kerosene refrigerators along with their operating and maintenance costs. Data collected from Kenya, Mali, Pakistan, Tanzania and Zaire suggests that photovoltaics would be most economic in many locations within these countries.

It is important to note that the recurrent costs (fuels, maintenance and spares) are mostly borne by the host country not donors. Hence in terms of providing a sustainable vaccine cold chain, solar powered refrigerators are most beneficial.

An analysis relating to an actual immunization programme in the Republic of the Gambia was undertaken as part of a World Bank pre-investment study (15). The study concluded that the introduction of photovoltaic refrigerators into the Gambian Cold Chain would improve the productivity of the immunization programme by almost 20%, i.e. the annualised life cycle unit cost per dose of vaccine used would drop by almost 20%.

A detailed guide to photovoltaic refrigeration for the vaccine

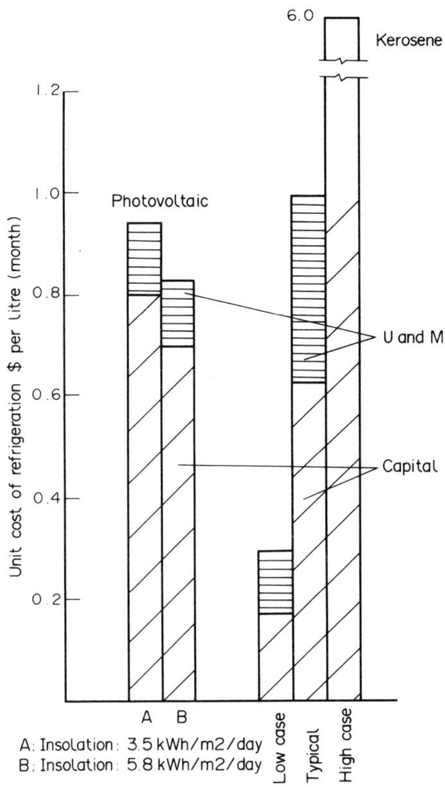

Fig. 9. Cost comparison of Kerosene and Photovoltaic Refrigerators.

cold chain has been issued by the World Health Organization (18) and I T Power have prepared 'A Handbook for Decision Makers' (19).

6. SUMMARY

The body of experience, both of the technical performance and economic viability of photovoltaic refrigerators, has led major international agencies such as UNICEF to implement major programmes to install these systems.

As the cold chain reaches further into remote areas of developing countries, the problems associated with kerosene quality and costs and logistical problems will result in an increasing number of decisionmakers considering the use of solar power in the cold chain. Major projects are presently being initiated in Africa and the South Pacific.

Solar powered refrigeration has the potential to make a huge impact on the health of the world.

7. ACKNOWLEDGEMENTS

Most of the work desribed in this paper was undertaken by our colleagues in I T Power. Thanks are due to Anthony Derrick, Peter Fraenkel, Jeff Kenna and Sarah Lancashire.

8. REFERENCES

1. Backus, C.E., Photovoltaic Conversion. Chapter 12. Sayigh A.A.M. (Ed) Solar Energy Engineering, Academic Press, 1977.
2. Wilson, J.I.B., Photovoltaics, Wykeham Press, 1979.
3. Rauschenbach, H.S., Solar Cell Array Design Handbook, Van Nostrand Reinhold Co., 1980.
4. McNelis, B., Photovoltaics for Autonomous Power Supplies - A Review. Proc. UK-ISES Conference Applications of Photovoltaics. Newcastle Photovoltaics Applications Centre, UK, 1985.
5. McNelis, B., Derrick, A., Shaw, M.R., An Evaluation of Photovoltaic Applications for Developing Countries, Contract Report to UNESCO, (1986). To be published by I T Publications Ltd., London, 1987.
6. Kenna, J.P., Gillett, W.B., 'Solar Water Pumping: A Handbook', I T Publications, Element Books, Longmead, Dorset, UK, 1985.
7. UNICEF The State of the World's Children - 1986, New York, 1986.
8. World Health Organization. Solar Refrigerators for Vaccine Storage and Icemaking. EPI/CCIS/81.5 (1981).
9. World Health Organization. Specification for Photovoltaic Refrigerator. EPI/9.82/CC (June 1981).
10. McNelis, B., Lloyd, J.S., 'Evaluation of Solar Refrigerators for use in the Vaccine Cold Chain', Proc. UK-ISES Conference London, January 1982.
11. Brinkworth, B.J., Refrigerator & Air Conditioning, Chapter 16, Sayigh, A.A.M., Solar Energy Engineering, Academic Press, 1977.
12. World Health Organization. The Cold Chain Product Information Sheets 1986/87. EPI Technical Series No. 1, WHO/UNIPAC, 1987.
13. Ratajazak, A.F., Photovoltaic-Powered Vaccine Refrigerator-Freezer Systems - Field Test Results. NASA LeRC TM - 86972, April 1985.
14. Durand, J.M., "Solar-Powered Refrigerators - The Zaire Experience", Photovoltaics: Investing in Development Conference, New Orleans, May 1987.
15. Fraenkel, P.L., Pre-Investment Report on Photovoltaic Applications in the Health and Telecommunucations Sectors (of Gambia), World Bank, 1986.
16. Lloyd, J.S., Solar Solutions for the Cold Chain, Africa Health Journal, September 1986.
17. Derrick, A., McNelis, B., I T Power Report: Evaluation of International Photovoltaic Projects: Photovoltaic Refrigerators. Prepared for Meridian Corp/Sandia National

Laboratories for USAID and USDOU (1986).
18. WHO/UNICEF/UNDP/UNDTDC, Guide in Implementation of Solar Energy for the Vaccine Cold Chain DRAFT, 1986.
19. Derrick, A., Lancashire, S.L., Solar Refrigerators for Vaccine Storage - A Handbook. Draft for Comment. Sandia National Laboratories, New Mexico, USA, 1987.